ECOHYDROLOGY

VEGETATION FUNCTION, WATER AND RESOURCE MANAGEMENT

ECOHYDROLOGY

VEGETATION FUNCTION, WATER AND RESOURCE MANAGEMENT

Professor Derek Eamus
Professor of Environmental Sciences
University of Technology, Sydney

Dr Tom Hatton
Director
CSIRO Water for a Healthy Country Flagship Program

Peter Cook
Principal Research Scientist
CSIRO Land and Water

Christine Colvin
Groundwater Science Research Group Leader
CSIR, South Africa

CSIRO
PUBLISHING

National Library of Australia Cataloguing-in-Publication entry

Ecohydrology: vegetation function, water and resource management

Bibliography.
Includes index.
ISBN 0 643 06834 1.

1. Ecohydrology – Australia. 2. Woody plants – Effect of water levels on – Australia.
3. Natural resources – Management. I. Eamus, Derek.

333.91

Available from
CSIRO PUBLISHING
150 Oxford Street (PO Box 1139)
Collingwood VIC 3066
Australia

Telephone:	+61 3 9662 7666
Local call:	1300 788 000 (Australia only)
Fax:	+61 3 9662 7555
Email:	publishing.sales@csiro.au
Website:	www.publish.csiro.au

Front cover: Murrumbidgee River scene near Darlington Point, NSW.
Photograph by Willem van Aken, © CSIRO Land and Water.

Back cover: Cattle Creek wetland scene, near Ingham, Queensland.
Photograph by Willem van Aken, © CSIRO Land and Water.

Set in Adobe Minion and ITC Stone Sans
Edited by Adrienne de Kretser, Righting Writing
Cover and text design by James Kelly
Typeset by Desktop Concepts Pty Ltd, Melbourne
Printed in Australia by Ligare

Contents

Preface

What is ecohydrology?

This book is not a textbook on Australian ecology. That has been admirably dealt with in three recent texts – *Plants in Action* (Atwell et al. 1999), *Australian Plant Communities* (Specht & Specht 1999) and *Ecology: An Australian Perspective* (Attiwell & Wilson 2003). Nor is it a textbook on hydrology and groundwater or groundwater quality; there have been many books dedicated to these topics (e.g. Brooks et al. 1997; Cech 2005; Nielsen 2005). The purpose of this text is to reveal and discuss the links between vegetation function and water in landscapes – that is, to discuss ecohydrology. However, we must, through demands of space, confine ourselves to a subset of the totality of ecohydrology. Thus we focus primarily (but not exclusively) on the interactions among the woody components of vegetation, rainfall and changes in groundwater availability. Woody vegetation is the focus because of the centrality of changes in woody vegetation cover to the ecohydrology of Australia over the past 100 to 200 years. A similar focus was taken by Eagleson (2002) in his intensely mathematical description of relationships among trees, forests, climate and soils. Furthermore, we focus on vegetation function (ecophysiology) rather than structure, because it is the functioning of vegetation that influences hydrology in the first instance.

Although this book uses Australian examples, the principles, philosophy and methodological approach are applicable worldwide.

At its broadest, ecohydrology can be described as the study of how the movement and storage of water in the environment and the structure and function of vegetation are linked in a reciprocal exchange. However, the practice of ecohydrology requires integration across the traditional disciplines of meteorology, plant ecophysiology and hydrology. Traditionally, meteorology has dealt with the where, when and how of precipitation (usually, but not exclusively, rain), ecophysiology has dealt with the relationships among plant function and the environment, and hydrology has dealt with the storage and movement of water to and from surface and groundwater stores. Clearly, no single text can hope to comprehensively address all three disciplines. What this book attempts is to provide a grounding in the language, techniques and underlying knowledge base that will allow readers to move forward from their single discipline base into a broader understanding of landscape function. Ecohydrology transcends the boundaries between disciplines and leads to an improved, more holistic and explanatory, rather than descriptive, understanding of landscapes. Ecohydrology is encapsulated in Figure P1.

Water and vegetation resources

Water resources are, of course, critically important for human consumption. However, problems arise due to the competing demands for domestic, commercial and industrial uses. Also, the importance of water to the maintenance of environmental health is becoming increasingly accepted. Managing the tensions between competing users, for example, those dependent on water for broad-acre irrigation, those who need river water for drinking purposes,

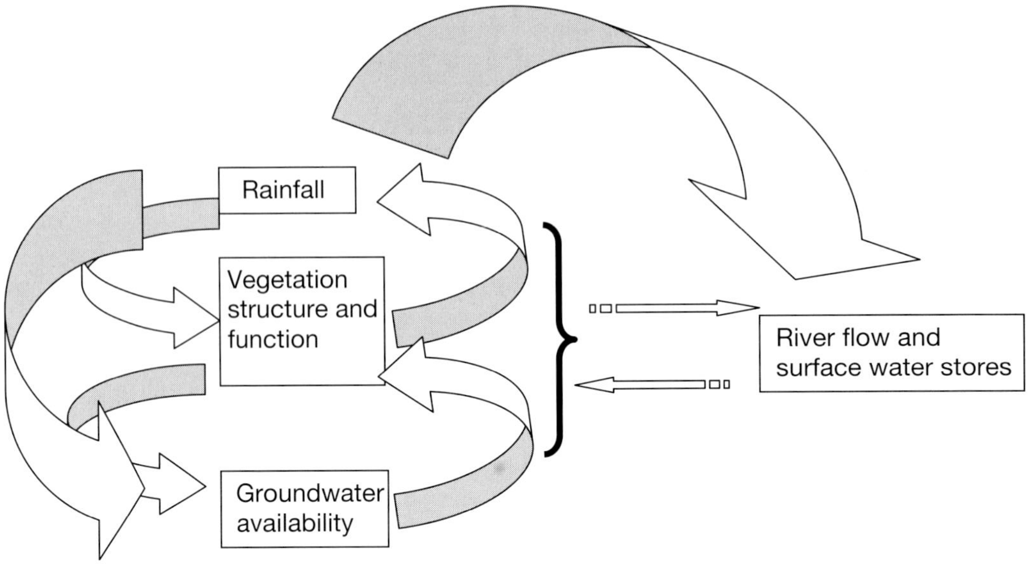

Figure P1 Ecohydrology is the integrated study of water and vegetation in landscapes. It requires information from ecophysiology, hydrology, soil science and micro-meteorology.

and the need to maintain environmental health, is increasingly problematic. Furthermore, most large water resources (aquifers, rivers, lakes) cross state or national boundaries. This leads to conflicting demands and rights of ownership, which exacerbate the management problems. As a result, water resources in the 21st century are set to become the equivalent of oil reserves in the 20th century – a globally limiting resource with the potential to cause regional and international conflict. Indeed, conflict over the management and distribution of water is already apparent in the Arab world, Africa, South and North America and Australia (especially in relation to the Murray-Darling). We must be mindful of past tragedies – ocean fisheries provide a striking reminder of the way jointly shared resources can become devastated, mismanaged and fought over (the 'tragedy of the commons').

The safe management of vegetation resources is equally important. Vegetation resources include the food we eat, the fibres we dress in and the wood we use in construction and the paper industries, and they are important for a huge range of commercial activities. Vegetation resources, however, are much more than the direct economic value of products. Vegetation plays a central role in most of the services provided by ecosystems, such as reducing soil erosion, fixing carbon from the atmosphere and preventing the development of dryland salinity (Eamus et al. 2005). Ecosystem services, a recent concept discussed in this book, are those functions that an ecosystem provides by its very existence and without which the economic, aesthetic, social and cultural values of society would fail (and are failing, in some places). The prevention of dryland salinity is one topic of particular importance in Australia and it is given its own chapter because of the large spatial and economic scale of its impact. It also highlights the central importance of synthesising and integrating ecological and hydrological perspectives for the management of healthy landscapes.

The scale of study of ecohydrology

This book is, by definition, concerned with landscapes, primarily focusing on the role of trees within landscapes in influencing the movement and use of soil and groundwater within landscapes. Trees are the principal focus because of their central role in the changed hydrological balance that characterises Australia. However, reference is made to wetlands and mound springs because of their close links with groundwater and their importance to biodiversity in Australia. Vegetation responses tend to be the focus of many studies of ecohydrology because it is vegetation that most often (but not always) responds first to the utilisation of groundwater. The fauna associated with the flora can be said to be more distally, or secondarily, dependent on groundwater through their dependence on the vegetation. An obvious exception to this are fish that rely on groundwater input to keep rivers flowing in the dry season of monsoonal Australia.

Small-scale measurements of individual plant parameters, for example the pre-dawn water potential of leaves, are only of value (in the context of this book) in understanding how canopies, stands, vegetation assemblages and the landscape are behaving. Similarly, point measures of hydrological value (streamflow, groundwater discharge or rainfall) are only useful in how they contribute to describing landscape-scale processes. Thus, an understanding of small-scale plant ecophysiological measurements and processes, coupled to an understanding of small-scale hydrological measurements and processes, are used as a base from which to build to a landscape-scale understanding of how large-scale changes in vegetation cover, for example, influence the hydrological balance of landscapes. The reverse is also true. Large-scale changes in hydrological features (e.g. groundwater depth) can have a profound effect on vegetation behaviour across large areas.

Science has traditionally used a reductionist approach, coupled, in experimental disciplines, to controlled experiments, for discovering new knowledge. This worked admirably for centuries in many disciplines, including plant physiology and hydrology. However, management of landscapes and natural resources such as water requires a different approach – one that is more holistic and integrated; a synthetic mindset is required. This book certainly does not attempt to integrate and synthesise all the knowledge required in ecohydrology, but it does attempt to show that someone who grasps the link between water and trees is better able to understand resources and landscapes than someone who does not.

In summary, this book deals mostly with the linkages between the functioning of trees (using plant physiological and ecophysiological knowledge), and the movement, availability and location of water in the landscape of Australia. We expect that readers of this book will comprise students and practitioners of plant physiology, ecophysiology, ecology, hydrology, and landscape managers.

This book aims to answer such questions as:

- What is the water status of the soil, vegetation and atmosphere at a site and how do these measures change daily and seasonally?
- What is the rate of water movement through vegetation? What is the transpiration rate of individual plants and canopies of plants?
- How do we know from where vegetation is extracting water?
- What is overland flow and how might we measure it?
- If vegetation cover across a landscape changes, what are likely impacts on overland flow, streamflow and groundwater recharge?
- What is the rate of movement of water through soils and aquifers and how does this influence stream flow? How do we measure stream flow?
- Why is Australia suffering from extensive dryland salinity?
- What are groundwater dependent ecosystems, where do they occur and why should we worry about them?
- What policies and principles can be applied to manage water and vegetation resources?

This book will provide plant scientists (predominantly ecophysiologists and ecologists) with a broad understanding of the concepts, language and techniques of hydrology and how hydrological information can be used to further our understanding of plant function and landscape management. Equally, it aims to provide hydrologists with a broad understanding of the concepts, language and techniques of plant ecophysiology and how these can be used to assist in sustainably managing water resources for the benefit of the environment. Finally, it is hoped that land managers, resource managers and policy makers will gain a deeper understanding of landscape functioning, especially but not exclusively in Australia, and that this will influence management and policy decisions.

Structure of the book

This book can be divided into three sections. The first section encompasses Chapters 1 to 5. These chapters provide an overview of the water, vegetation and climate of Australia (Chapter 1), the basic concepts, tools and language of plant–water relations (Chapter 2), and basic hydrology (Chapter 3) and the techniques and concepts used in plant ecophysiology and hydrology (Chapter 4). Chapter 5 integrates Chapters 2 and 3 by presenting models of vegetation–hydrology interactions. In particular, this chapter builds from simple concepts of soil moisture and plant water use, to present some models being used in the ecohydrological domain. Plant–water relations is the field of study concerned with how water moves in the soil–plant–atmosphere continuum and how water use and water status of plants varies in a daily and seasonal pattern. Ecophysiology is the study of plant function in the natural environment and how plants respond to changes in the environment (climate, soil water content). Hydrology is the study of the movement and storage of water in landscapes, including surface waters (rivers and lakes) and groundwater. A clear understanding of the subject matter of these chapters is required before readers move on to the second part of the book.

The second section encompasses Chapters 6, 7 and 8. Chapter 6 is a specialised chapter on groundwater dependent ecosystems, an increasingly important subject of study in ecohydrology. This chapter examines the ecophysiology and hydrology of groundwater dependent ecosystems by presenting four case studies from around Australia. Chapter 7 deals with five case studies of practical applications of ecohydrology, excluding groundwater dependent ecosystems. Topics covered include the impact of fire on the structure and function of vegetation and hydrology of Australian landscapes; a study of Mountain Ash forests as understood from an ecological and hydrological perspective; the influence of mining on local mound springs; links between hydrology and functioning of wetlands and floodplains; and, finally, Lake Toolibin of Western Australia – an exercise in restoration ecohydrology. Chapter 8 is devoted to the issue of salinity and the links between land use, forest cover and landscape–water balance.

The third section encompasses Chapters 9 and 10. Chapter 9 provides a review of the policies and guidelines governing the allocation of water and management of groundwater dependent ecosystems in Australia. Chapter 10 offers a case study of South African ecosystem and water management. It provides a synthesis and big-picture overview of the management of water and vegetation in South Africa and draws upon the language, concepts and practicalities discussed in Chapters 2 to 9. It is designed to show how ecohydrology and the sustainable management of water and vegetation are important not only in Australia but in all arid and semi-arid countries of the world, including much of Africa and the Middle East.

Enjoy!

Chapter 1

Setting the scene: water and vegetation resources in Australia

This chapter provides a broad overview of the water resources and vegetation of Australia. The chapter starts with a review of surface and groundwater stores and their uses. We present a continental-scale water balance and highlight the fact that Australia is precariously balanced in its water use. Vegetation resources and the factors that influence vegetation structure and function are then described. In particular, the influence of climate (especially rainfall) and the Southern Oscillation on variability of rainfall and the adaptations that vegetation has evolved in order to cope with the Australian climate, are discussed. Finally, we show that simple vegetation classification systems can be applied to almost all Australian landscapes and explain how such systems have direct relevance to ecohydrology and allow effective communication between hydrologists and ecologists. Changes in Australian landscape hydrology are primarily a function of changes in landscape use, especially changes in land cover, and therefore a brief comparison of pre-and post-European colonisation vegetation cover is made.

After reading this chapter, readers should be familiar with the following topics:

- the amount and fate of rain that falls on the Australian continent;
- the major uses of water in Australia;
- the principal climate zones of Australia;
- the Southern Oscillation index;
- sclerophylly and other adaptations to the Australian climate and soils exhibited by Australian plants;
- distinguishing features of Australian soils;
- foliage projected cover: life form, stratum, leaf area index, structural and functional attributes of vegetation;
- a classification system for vegetation types in Australia;
- vegetation change over the past 200 years.

Introduction

A broad knowledge of the water and vegetation resources of Australia is a necessary prelude to a text on the ecohydrology of Australia. In particular, it is important that a continental-scale context is given to water resources (both surface and groundwater) and vegetation. Because climate and soils also influence both water resources and vegetation, and because vegetation influences how water moves through the landscape, it is important to include an overview of these broad areas to act as a backdrop for later chapters. The following section provides a brief overview of the water resources of Australia. Subsequent sections in this chapter deal with climate and vegetation.

Water resources in Australia

Water resources in Australia consist of surface and groundwater stores. The supply of water to these stores comes principally in the form of rainfall (direct input) and run-off (the movement of surface water from one place to another along a slope, i.e. redistribution). Understanding these flows at a continental scale provides the background required to consider water resources at a local scale.

Rainfall and run-off

Australia is well known as the driest of all permanently inhabited continents. Mean annual rainfall is about 350–450 mm, considerably lower than the mean annual rainfall for Africa (750 mm), North America (800 mm), South America (1800 mm), Europe (820 mm) and Asia (650 mm). The total volume of water received by Australia as precipitation has been estimated to be about 3 320 000–3 390 000 GL per year (Foran & Poldy 2002). Because of the high **evaporative demand** of the climate (high levels of solar radiation lead to warm-to-hot temperatures and relatively dry air for much of the year) and the flatness of the continent, little of this water reaches the sea as river flow. About 350 000 GL (c. 10%) of this precipitation is lost to sea as river flow. While this seems a large number, it is the smallest fraction of total rainfall for any permanently inhabited continent (see Figure 1.1, Colour Plate 1). Much, but not all, of the water flowing in rivers is derived from **run-off**, that is, overland or shallow subsurface lateral flows of water. However, in many rivers, a component of river flow is derived from the discharge of groundwater (termed **base flow**). This is discussed more extensively in Chapter 3.

Twelve drainage divisions cover all of Australia and Table 1.1 shows the volume and percentage run-off from each of these. In six drainage divisions (half of the total), less than 2.5% of rainfall is lost as river flow. The remainder is lost as evaporation, transpiration and seepage into groundwater or other storage (lakes). The low amount of rainfall that is 'lost' from the

Table 1.1 Run-off from 12 drainage divisions in Australia

Drainage division	Mean annual run-off (GL)	Annual run-off (%)	Mean annual outflow (GL)
North-east coast	73 411	19	69 580
South-east coast	42 390	10.9	40 366
Tasmania	45 582	11.8	45 336
Murray-Darling	23 850	6.2	5750
South Australian Gulf	952	0.2	787
South-west coast	6785	1.8	5925
Indian Ocean	4609	1.2	3481
Timor Sea	83320	21.5	81 461
Gulf of Carpentaria	96615	24.7	96 066
Lake Eyre	8638	2.2	n/a
Bulloo-Bancannia	546	0.1	n/a
Western Plateau	1486	0.4	n/a
Total	**387 184**	**100**	**348 752**

Source: Data from the Australian Water Resources Assessment (2000)

continent as river flow sets an upper limit to how much water can be extracted from rivers for human use.

Australia's continental water budget

Of the total volume of water received by Australia as **precipitation** (3 320 000–3 390 000 GL per year; Foran & Poldy 2002), about 10% is lost to sea as river flow (see above) and about 3 100 000 GL is lost as **evapotranspiration**, that is, evaporation from wet soil, lakes and other wet surfaces, plus transpiration from vegetation. This means that there is very little spare capacity in the water budget for new activities or for using more water in existing activities; in other words, the Australian continental water budget is precariously balanced. It also means that when rainfall is significantly below average, as is often the case in Australia, river flow and evapotranspiration decline too. Declines in river flows and evapotranspiration have three serious knock-on effects. First, river health declines when low river flows are maintained for too long. Second, a reduction in evapotranspiration results in increased fire frequency and reduced productivity (growth and yield) of crops and native vegetation. Finally, when rainfall is low there is an increased reliance on groundwater (water stored underground) for irrigation and human consumption. Too many aquifers are already being **over-extracted**, with extraction at rates greater than natural rates of groundwater **recharge** (see Table 1.3). The argument that additional water can be taken from rivers is difficult to sustain when it is accepted that the maintenance of river health is important and that doing so requires adequate river flows. There is minimal scope for taking more water from rivers of the southern half of the continent (although more could be taken from rivers in the northern half). As discussed in Chapters 6 and 8, regulation of river flows and extraction from rivers has generated significant problems for the hydrology, ecology and sustainability of many activities in Australia. There are many lessons here, regarding managing extractions and flows, that are applicable to other arid and semi-arid zones in Africa, the Middle East and Southern America.

Surface and groundwater stores

Surface water is water in rivers, lakes, dams and other locations on the surface of the earth.

Australia possesses about 450 large dams with a combined storage of about 80 000 GL. These are used principally for urban water use, irrigation and, in Tasmania, hydroelectric power generation. New South Wales has the largest surface water storage capacity, followed by Tasmania. On-farm dams are large in number (the estimate is more than 2 million) but their total volume is small – less than 10% of the total surface water stored. In addition to 'static' stores of surface water, the volumes of water passing out to seas in rivers are also viewed as surface water stores. However, river flows represent only a small fraction of rainfall and are highly variable between years.

There is another store of water upon which Australia is heavily reliant – **groundwater**. Groundwater is water located below the surface of the earth in porous soil and rock (an **aquifer**). Examples of the major **artesian basins**, including details of their depth and salt content, are given in Table 1.2. Salt contents vary widely across each of these systems; tabular values are indicative only. Some 72% of Australia's sustainable supply of groundwater has less than 1.5 g L^{-1} of salt, with an additional 16% of groundwater in the range of 1.5–5 g L^{-1}. Groundwater can reach the surface in some locations if recharge of the aquifer is very high and the aquifer is close to the surface, or if the aquifer intersects a hill slope or cliff face.

Australia has extensive extractable groundwater reserves. The Australian Water Resources Assessment (2000) cites a sustainable yield of almost 26 000 GL nationally, and estimates that Australia currently uses about 10% of this. Note that this is an estimate of the **sustainable yield** of all the aquifers in Australia (see below). It is not an estimate of the total volume of

Table 1. 2 Major artesian basins of Australia, their areal extent, depth and salt content

Basin	Area ('000 km^2)	Thickness of water layer (m)	Salt content (g L^{-1})
Great Artesian Basin	1751	0–2100	c. 6
Desert	388	30–550	0.3
Murray	282	30–400	1.5–1.8
Eucla	191	90–610	6–37
North-western	77.5	60–1220	4–5
Coastal plain of Perth	54	60–760	
Ord-Victoria	31	60–330	

Source: Shiklomanov & Rodda (2003)

groundwater in Australia, which is much larger. The Great Artesian Basin alone is estimated to store a total of 8 700 000 GL of water.

Safe yield has generally been defined as the volume of water that can be extracted from an aquifer without depleting the reserves and is therefore determined by the rate of recharge (see Chapter 3). However, this is a flawed concept because in the long term, under conditions of equilibrium, natural recharge is balanced by discharge to the surface (through vegetation, as transpiration, or as liquid water flow) from an aquifer (Sophocleous 2000). Therefore, if groundwater extraction is equal to recharge, discharge will decline to zero and groundwater dependent ecosystems (rivers and mound springs; see Chapter 5) will be deprived of water and suffer. Safe yield is now generally replaced by **sustainable yield**, which is lower than safe yield as it allows for the provision of water for the maintenance of ecosystem health (Sophocleous 2000).

Surface and groundwater use

In total, approximately 22 000–24 000 GL of water per year are used to support human activities in Australia. Almost three-quarters of this is taken from rivers (about 20 000 GL), or, combining all sources (rivers, lakes and reservoirs) about 80% is derived from surface waters. Water extracted from rivers is actually a mixture of rainwater and groundwater because most rivers in Australia receive some groundwater input (termed **base flow**). About 20% of the water

Table 1.3 Average annual surface and groundwater use for the eight states and territories of Australia

	Total annual surface water use (GL)	Total annual groundwater use (GL)	Ratio of surface to groundwater use
NSW	9000	1008	9.0
Vic	5166	622	8.3
Qld	2969	1622	1.8
WA	658	1138	0.6
SA	746	419	1.8
Tas	451	20	22.6
NT	51	128	0.4
ACT	68	5	13.6
Total	**19 109**	**4962**	**3.9**

Source: Australian Water Resources Assessment (2000)

Table 1.4 Groundwater use in Australia, listed by state or territory – n/a not available

	Total use 1983/84 groundwater (GL)	Total use 1996/97 groundwater (GL)	% change in groundwater use 1983/84–1996/97
NSW	318	1008	217
Vic	206	622	202
Qld	1121	1622	45
WA	373	1138	205
SA	542	419	-22
Tas	9	20	122
NT	65	128	97
ACT	n/a	5	–
Total	**2634**	**4962**	**88**

Source: Reproduced from the Australian Water Resources Assessment (2000), National Land and Water Resources Audit

used in Australia comes directly from groundwater via bores sunk into aquifers. In the Northern Territory and Western Australia, about half of all the water used is groundwater; in all other states and territories, use of surface water is between 2 and 22 times larger than groundwater use (Table 1.3).

About 70–75% of water used in Australia is used in agriculture. Only about 20% is used for urban and industrial use. It is the rural sector's use of water and landscapes that forms a backdrop to much of this text.

Groundwater use across Australia almost doubled between 1983 and 1996, from about 2600 GL to about 5000 GL (Table 1.4). In some states (e.g. New South Wales and Western Australia) the increase was even larger; a 200% increase occurred in that period. Most of this increase was for irrigation. In most states groundwater extraction exceeds licensed allocations (Australian Water Resources Assessment 2000). Furthermore, in many aquifers, the rate of abstraction exceeds the rate of recharge (Table 1.5). **Recharge** is the rate at which water moves from the surface into an aquifer. In some locations, we are mining (extracting without putting back) water that is several thousands or millions of years old. The fact that this water is extremely old reveals that this groundwater is not being replenished (recharged) very quickly and that it should be treated as a finite, not infinite, resource.

The commercial return on the water used in many irrigated industries is often small. For example, the beef cattle industry uses about 800 L of water to generate $1 of product value, and the seed cotton industry uses about 1600 L of water to generate $1 of output. Rice in the husk uses almost 7500 L to produce $1 of output (Foran & Poldy 2002). Similarly, the livestock, pasture and grains industries return approximately $300 000 per GL of water used, while the rice industry returns $200 000 per GL used (Australian Water Resources Assessment 2000). These are compared to the returns on other agricultural industries in Table 1.6. It is questionable whether the rates of return of many irrigation industries in Australia are sustainable given

Table 1.5 Rates of over-extraction from aquifers

System	Main use	Abstraction (GL y^{-1})	Natural recharge (GL y^{-1})
Burdekin Delta, Qld	Irrigation	263	200
Namoi, NSW	Irrigation	160	110
Condamine Valley, Qld	Irrigation	87	13

Table 1.6 Gross value, water use, irrigated area and value accrued per GL of water used for a range of irrigated industries in Australia

	Gross value ($m)	Net water use (GL)	Irrigated area (ha)	Value/ha ($/ha)	Value/GL ($m/GL)
Livestock, pasture, grains	2540	8795	1 174 687	2162	0.3
Vegetables	1119	635	88 782	12 604	1.8
Sugar	517	1236	173 224	2985	0.4
Fruit	1027	704	82 316	12 476	1.5
Grapes	613	649	70 248	8726	0.9
Cotton	1128	1841	314 957	3581	0.6
Rice	310	1643	152 367	2035	0.2
Total	**7254**	**15 503**	**2 056 581**		

Source: Australian Water Resources Assessment (2000)

the amount of water they use and the environmental damage that excessive irrigation of treeless landscapes can produce (irrigation-induced salinity). The topic of dryland salinity is discussed extensively in Chapter 6.

Climate, soil and fire in Australia

The previous sections provided an overview of the stores and uses of surface and groundwater resources. This section briefly discusses Australia's climates and soils as a prelude to a description of Australia's vegetation, because climate and soil have a significant impact on vegetation. The importance of fire is also discussed because fire influences both vegetation structure and the local water balance. When fires devastate vast areas of vegetation in the landscape the local water balance is affected for months, years or many decades. The impacts of fire on vegetation and hydrology are discussed more extensively in Chapter 6; this section introduces these topics at a continental scale.

Climate

Australia has several **climatic zones**. The key features of climate, in relation to ecology, are the seasonal mean maximum and minimum temperatures (and hence the temperature range) and the amount and timing of rainfall. Broadly speaking, six or seven climate zones are recognised in Australia. If we use temperature and humidity as the dominant indicators of climate in Australia (which to a large extent mirror rainfall patterns across the continent) then six zones are recognised, as shown in Figure 1.2 (Colour Plate 2).

The key differences among different climate zones are the amount and timing of rainfall and the average summer and winter temperatures. For example Darwin, in the Top End of the Northern Territory of Australia, has uniformly high daytime temperatures (28–32°C in summer and winter; Table 1.7) and receives an average of 1666 mm of rainfall in the wet season (summer), which extends from November to April inclusive. In the dry season (winter), June to September, essentially no rainfall occurs. In contrast, in Perth, Western Australia, rainfall (annual rainfall is 869 mm) occurs predominantly in the winter. Summers are hot and dry (mean temperature is 31.5°C) and winters are cool (mean temperature is 16.8°C).

Significant rainfall is essentially confined to the north and east coasts of Australia, Tasmania and the south-west coast (Fig. 1.3a, Colour Plate 3). Most of the southern coast, most of the

Table 1.7 Mean maximum and minimum winter and summer temperatures for the major cities of Australia, with mean annual rainfall

	Mean temperatures (°C)				Annual rain (mm)
	January		July		
Location	Max	Min	Max	Min	
Adelaide	28.5	16.6	14.9	7.5	553
Alice Springs	36.1	21.2	19.5	4.0	274
Brisbane	29.2	21.0	20.6	9.5	1189
Canberra	27.8	12.9	11.1	–0.2	631
Darwin	31.8	24.8	30.4	19.3	1666
Hobart	21.5	11.7	11.5	4.5	624
Melbourne	25.7	14.0	13.3	5.8	661
Perth	31.5	16.8	17.7	8.1	869
Sydney	26.3	18.5	16.9	6.7	1220

Source: Bureau of Meteorology

central western coast and all of the interior of Australia can be considered arid and semi-arid, receiving less than 400 mm of rainfall per year.

Mean daily temperatures are, on average, warm to hot. Only a small fraction of the continent receives any snow and all snowfall occurs at altitude, mostly in Tasmania, the Kosciuszko plateau in New South Wales and the Bogong High Plains of Victoria. Average maximum temperatures increase with decreasing latitude (i.e. moving towards the equator) and with increasing distance from the coast. Average minimum temperatures, by contrast, increase only with decreasing latitude – they decline with increasing distance inland (deserts have cold nights).

Discussion of average rainfall and temperatures is useful only to a limited extent in Australia and provides only very broad-brush descriptions of average climate zones. In reality, Australia's climate variability is the largest of any continent. Rainfall at a site can vary by orders of magnitude between years and average annual rainfall for Australia ranged from 310 mm to almost 800 mm during the 20th century (Fig. 1.3b, Colour Plate 3). This huge variation is principally because of the impact of the **El Niño/La Niña Southern Oscillation**. The Southern Oscillation Index (SOI) is a measure of the strength of the influence of the El Niño/La Niña weather patterns. A strongly negative value for the index is associated with an El Niño event, which is characterised by drought, especially in eastern Australia, increased fire frequency, warmer than average temperatures and greatly reduced crop yields in much of Australia. El Niño events are typically said to occur every two to seven years, but this frequency is extremely variable. Both the frequency and intensity of the occurrence of negative values of the SOI have varied considerably over the past 100 years (Fig. 1.5). Consequently the intensity and duration and areal extent of the increased temperatures, increased fire risk and decreased rainfall associated with these negative values are highly variable. Most of the period between 1927 and 1939 was characterised by a small positive SOI while the 1990s were mostly characterised by periods of large negative SOI. This variability in the SOI explains much of the variability in climate of eastern Australia.

Because Australia's rainfall is highly variable between years (more variable than any other continent) perennial vegetation such as trees have to be able to cope with periods of ample rain followed by extended periods of drought. Furthermore, some regions always receive low levels

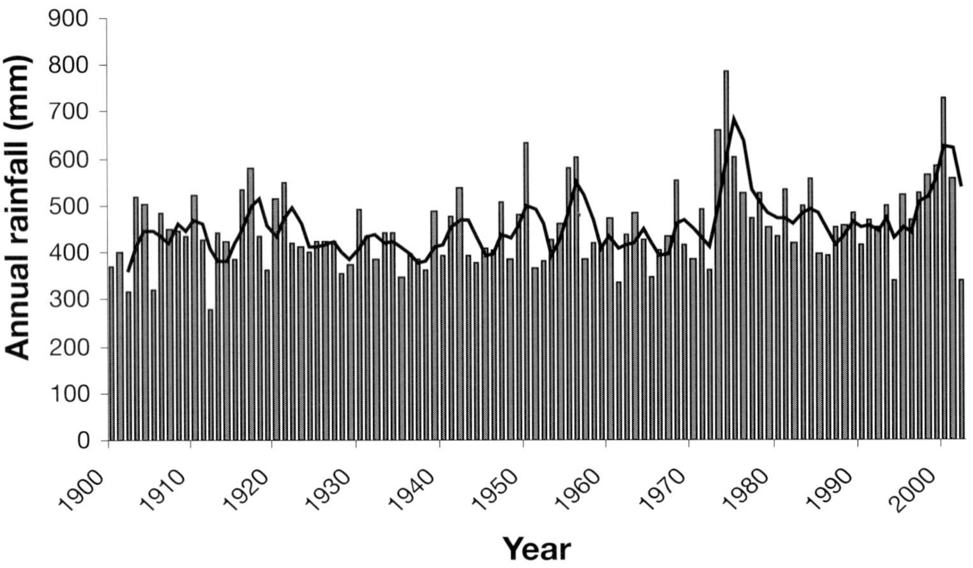

Figure 1.4 Average annual rainfall of Australia has varied between a low of 310 mm and a peak of almost 800 mm. There are more years receiving significantly less rainfall than the 5-year average than there are above the 5-year average. Below average rainfall is the norm for Australia

Source: © Bureau of Meteorology

of rainfall (less than 250 mm per year) while others receive copious quantities (1500–3000 mm per year). On average the height, plant density and complexity of vegetation increases as annual rainfall increases. Adaptations to low rainfall include the following plant attributes:

- tough sclerophyllous leaves (defined below) which can withstand periods of drought without dying;
- sunken stomata to reduce rates of water use by leaves;
- hairy or pendulous leaves which reduce the amount of solar radiation absorbed by leaves, thereby reducing the heating and drying effects of solar radiation;
- storage of water in stems and root stores (lignotubers). Lignotubers are storage organs located underground which store water, carbohydrates and nutrients;
- ability to be drought deciduous, whereby the entire canopy of leaves is dropped during drought, followed by recovery when rain occurs;
- deep roots which access groundwater stores. The topic of groundwater dependent ecosystems is discussed extensively in Chapter 6.

Soil type and plant adaptations

Australian soils have three common features. The first is that they are very old, and hence extremely weathered and therefore low in nutrient content. The second feature is that in many places they tend to be thin. In some locations where deep 'soil' occurs, such as Western Australia, it is almost entirely composed of sand, which has low water-holding capacity and low organic nutrient content. The third feature is that most Australian soils contain significant amounts of salt. Often this is located at depth in the soil profile, but the amount of salt can be up to 1 tonne per square metre. Consequently, when the groundwater level rises towards the ground surface, it often brings salt up and results in salinity within the landscape (see Chapter 8).

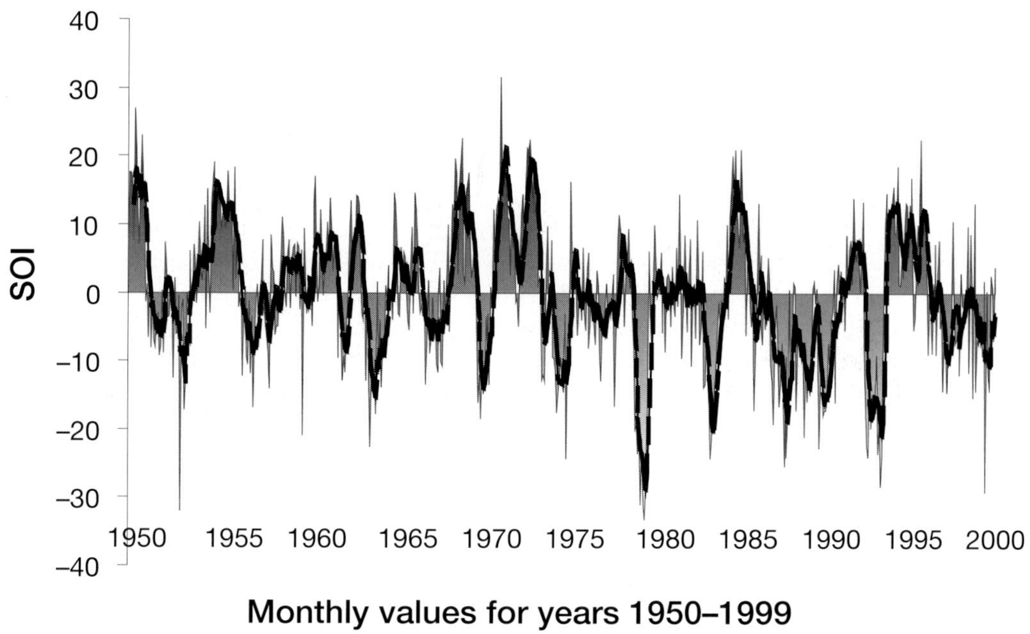

Figure 1.5 The frequency, duration and intensity of variation in the monthly SOI has varied enormously over the past century. Here, data for the period 1950–2000 is presented. Periods of largely negative and largely positive SOIs can be seen, with large inter-annual variation often observed between consecutive years

Source: © Bureau of Meteorology

The old, depauperate (**nutrient-poor**), shallow soils of Australia also often have poor water storage capabilities. This is because of their shallowness and low organic content. The Australian flora has developed several adaptations in response to these attributes.

- **Cluster roots**. These are clusters of finely divided and highly branched roots. They greatly increase the surface area for nutrient uptake, but more importantly cause a set of chemical modifications of the soil which increases the bioavailability of nutrients, especially phosphorus. Cluster roots, also known as **proteoid** roots in the family *Proteaceae* (e.g. *Banksia*), have evolved in many species, especially in Western Australia.
- **Mycorrizas.** These are fungal associations with roots. Most plant species in Australia have mycorrizal fungi growing on and/or through their roots. The fungi gain carbon as an exudate from the root. The plant gains additional access to water and nutrients from the fungi because the fungal hyphae extend out into the surrounding soil further than the root and there is some transfer of water and nutrients to the root from the fungus.
- **Sclerophylly.** A sclerophyllous leaf is thick, stiff and leathery with low water content per unit dry mass of the leaf. Sclerophylly is an attribute of the leaves and phyllodes of most species of trees in Australia. Sclerophyllous leaves are resistant to drought and herbivorous attack and might also be an adaptation to the low nutrient status of Australian soils, especially low P availability. Phyllodes look superficially like leaves, and function the same way as leaves. They fix carbon dioxide in photosynthesis and transpire water through stomata, but developmentally they are different from leaves. *Acacias* have phyllodes, eucalypts have leaves. *Acacia* phyllodes are very sclerophyllous, a lettuce leaf is not at all sclerophyllous.

- **Tubers** and **lignotubers.** Lignotubers and tubers are structures that allow perennial plants to survive extended droughts (and fires) by lying dormant below ground until the fire or drought has passed.

Fire

Fire is a highly significant feature of the Australian landscape (Fig. 1.6, Colour Plate 4), with large areas of Australian bushland burning every year or two in some (northern) savanna regions. Massive conflagrations of smaller areas occur less frequently in southern regions. Fire appears to have increased in frequency across the Australian landscape since *Homo sapiens* arrived about 60 000 years ago and has shaped the evolution, composition and distribution of much of the vegetation. Increased fire frequency favours the dominance of eucalypts and other fire resistant genera (e.g. *Banksia*, *Casuarina*) and pushes fire sensitive ecosystems (rainforests, *Allosyncarpia* forest) into fire protected refugia. Plant adaptations to and hydrological impacts of fire are discussed in detail in Chapter 7.

Classification of Australian vegetation

There are many ways to classify vegetation, based on various measures of structure and composition. However, to appreciate changes in vegetation structure and composition at the landscape scale, and to have relevance to the interpretation of ecohydrological studies, a broad system based on **functional** and **structural attributes** and the dominant vegetation type is optimal. In the following sections, such a classification system is explained.

Functional and structural attributes

Functional attributes refer to the habit of the plant. Trees, grasses, herbs and forbs, lianas and vines, epiphytes and shrubs are the major functional groups in Australian landscapes. Within a broad group such as 'trees' it is possible to divide further, for example into deciduous and evergreen species, or nitrogen fixing and non-nitrogen fixing groups. However, within this book, such splitting is not warranted and we refer only to the broadest functional types (trees, grasses, etc).

Structural attributes are those attributes of individual plants or assemblages of plants (e.g. a forest) that relate to the physical form of the plant or assemblage. One example is the number of canopy layers. Within a pure grassland community, one layer is assumed to be present: that of the grass canopy. Within a savanna of the Northern Territory, we recognise three layers – the tree overstorey, the mid-layer or mid-stratum of the subdominant shrubs and short trees and finally the grass understorey. Within a rainforest of far north Queensland we might recognise six layers, including epiphytes, vines, upper, mid and lower storey trees and understorey shrubs.

The number of canopy layers is one structural feature of vegetation. Others include:

- height of the various layers;
- plant density (the number of plant stems per unit area, for example the number of trees per hectare);
- **leaf area index** (the amount of leaf area (in m^2) directly above a square metre of ground);
- the **foliage projected cover** (FPC) of the canopy, which is a measure of the degree to which the upper canopy is open or closed. FPC is a measure of the fraction of ground overlain by leaf and it varies between zero (desert) and 1 (rainforest).

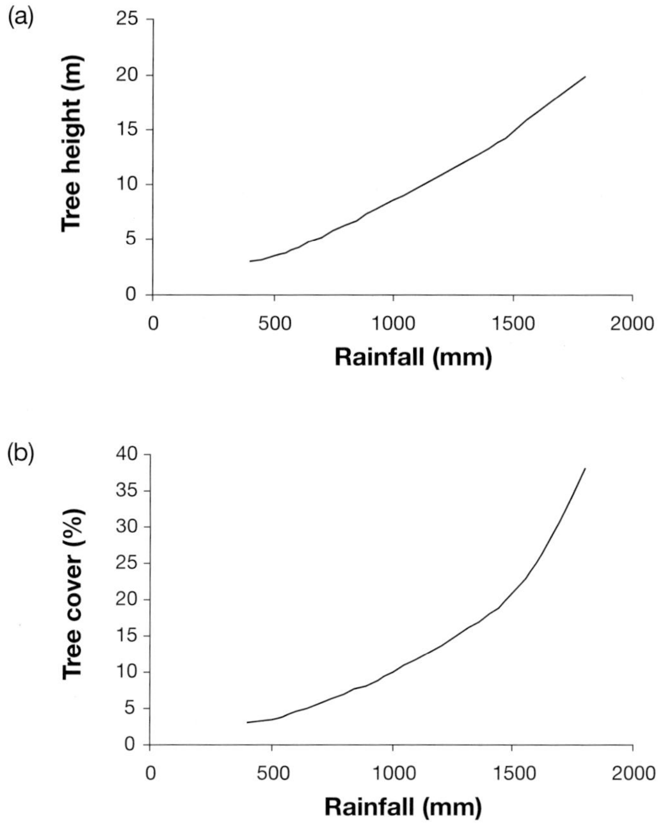

Figure 1.7 As rainfall increases (a) the average height of the upper tree canopy increases and (b) the percentage tree cover also increases. Data are for tropical northern Australia

Source: Redrawn from Williams et al. (1996)

Plant density, degree of canopy closure (expressed as FPC) and leaf area index are of particular relevance to ecohydrology because these attributes have a significant impact on the water balance of a site (since leaves transpire water) and because they are very much determined by climate, especially rainfall. Thus average canopy height (which is usually proportional to leaf area index) and tree cover are positively correlated with rainfall (Fig. 1.7a, 1.7b). For example an open woodland, found in a rainfall zone of 1200 mm per year, does not have overlapping canopies of adjacent trees. A closed forest in a rainfall zone of 2400 mm per year, in contrast, has an overlapping set of tree canopies. Consequently the leaf area index (LAI) of an open woodland and closed forest are very different. The LAI of the open woodland is likely to be in the range 0.6–1.2 m^2 leaf per m^2 ground, while that of the closed forest is likely to be 2–4 m^2 leaf per m^2 ground.

A structural attribute of an ecosystem similar to LAI is the foliage projected cover (FPC) of the canopy. This is a measure of the openness of the canopy. Imagine 100 vertical lines at 5 m intervals, projecting upwards from the ground to a point above the canopy. If 25 lines intersect a leaf, 15 lines come into contact with twigs, branches or flowers and 60 lines do not touch any part of the canopy, the FPC is 25% (25/100). If 100 vertical lines projected upwards from the ground to the sky have 65 lines intersecting a leaf, 25 lines intersecting a twig or branch or flower and 10 not intersecting any part of the canopy, then the FPC is 65%.

As FPC increases, the degree of closure of the canopy increases – we see less of the sky when we look up through the canopy.

Plant density, LAI and FPC are important structural attributes of vegetation because they all correlate well with the amount of water used by the vegetation. When all else is constant, the amount of water transpired or evapotranspired (evaporated from wet surface after rain, plus the water transpired during the day) increases as the LAI, FPC or plant density increase.

Native, unmanaged equilibrium vegetation (e.g. remnant forest or remnant coastal heath) (see Figs 1.8, 1.9, Colour Plate 4) in Australia has come into equilibrium with the water balance of the site on which it grows. Thus, the average height and density of trees and the FPC of a site increases from the arid interior of Australia towards the coast, where rainfall is higher (Fig. 1.8). Generally, as the availability of water at a site increases, the LAI and rate of water use by vegetation at the site increases, and the vegetation will use almost all of the water that arrives as rainfall. This is an important point to keep in mind throughout this book.

Classification systems based on dominant vegetation type

For there to be effective communication between ecologists, hydrologists and scientists of other disciplines, it is important that an accepted and widely applicable classification system of vegetation types be adopted. While there are many ways to classify vegetation types, the needs for simplicity and wide applicability recommend the two related schemes proposed by Specht and Carnahan. These two schemes use a very small combination of functional and structural attributes plus, in the Carnahan system, reference to the dominant genus in the vegetation (represented by a floristic code) which can be memorised and then applied very quickly. Furthermore, the attributes used in both schemes fortuitously have direct relevance to ecohydrology, as discussed above.

The simple yet effective classification system of Specht (Specht & Specht 1999) is based on just three attributes: height, dominant functional type (e.g. tree, grass or forb; also called plant habit, see above; also called structural or life form) and foliage projected cover (FPC). A simple table is available (Table 1.8) which can effectively classify all vegetation types in Australia.

The Carnahan system incorporates a floristic code in addition to the structural attributes used in the Specht system. Classifying a vegetation system using the Carnahan system requires answers to three simple questions.

1 What is the dominant life form (or growth form)? For example, tree, shrub or tussock grass. Tall shrubs are differentiated from low trees by the slightly arbitrary definition that shrubs are multi-stemmed at or close to the ground and trees are not. The following growth form codes, utilising upper case letters, are used:
   - T = tall trees
   - M = medium trees
   - L = low trees
   - S = tall shrubs
   - Z = low shrubs
   - H = hummock grasses
   - G = tussock/tufted grasses
   - F = other herbaceous plants.

2 What is the FPC of the upper stratum? Four foliage classes are used:
   - 1 = <10%
   - 2 = 10–30%
   - 3 = 30–70%
   - 4 = 70–100%

Table 1.8 Specht classification system for Australian vegetation

Plant functional type and height of the tallest layer	Code	Foliage projected cover			
		100–70 (4)	**69–30 (3)**	**29–10 (2)**	**<10 (1)**
Trees >30 m	T	Tall closed forest	Tall open forest	Tall woodland	Tall open woodland
Trees 10–30 m	M	Closed forest	Open forest	Woodland	Open woodland
Trees 5–10 m	L	Low closed forest	Low open forest	Low woodland	Low open woodland
Trees <5 m	VL	V low closed forest	V low open forest	V low woodland	V low open woodland
Shrubs >2 m	S	Closed scrub	Open scrub	Tall shrubland	Tall open shrubland
Shrubs 0.25–2 m Sclerophyllous Non- clerophllous	 Z C	 Closed heathland Low closed scrub	 Heathland Low open scrub	Open heathland Low shrubland	Sparse heathland Low open shrubland
Shrubs <0.25 m Sclerophyllous Non-sclerophyllous	 D W	 – –	 – –	Dwarf open heathland Dwarf open shrubland	Dwarf sparse heathland Dwarf sparse shrubland
Hummock grasses	H	–	Dense hummock grassland	Hummock grassland	Open hummock grassland
Herbaceous layer graminoid & grasses	G	Closed tussock grassland	Tussock grassland	Open tussock grassland	Sparse tussock grassland
Sedges	Y	Closed sedgeland	Sedgeland	Open sedgeland	Sparse sedgeland
Herbs	X	Closed herbland	Herbland	Open herbland	Sparse herbland
Ferns	F	Closed fernland	Fernland	–	–
Reeds and rushes	R	Closed reedland	Reedland	–	–

Source: Specht & Specht (1999)

3 What is the dominant genus present in the upper stratum? If the upper stratum has less than 10% cover and the foliage cover of the lower stratum is much higher than this, the lower stratum is usually included too. The following floristic codes (using lower case letters) are commonly used, though others exist:
   - e = eucalypts
   - w = *Acacia* (wattle)
   - b = *Banksia*
   - c = *Casuarina*
   - f = Fabaceae (including clover)
   - g = graminoids
   - h = *Hakea*
   - k = Chenopod shrubs
   - m = *Melaleuca*

An ecosystem could therefore be given the code:

wL1kZ

This means it is an acacia (w = wattle) dominated low (L) woodland with a foliage cover of less than 10% (class 1) and the understory is a chenopod low shrub (Z). Similarly:

eM3L

indicates a eucalypt (e), medium (M) height, open (3) forest with a low tree understorey (L) (see Fig. 1.9, Colour Plate 6).

Climate, water availability and vegetation function

A cursory comparison of maps of rainfall distribution and vegetation distribution across Australia shows that at a continental and subcontinental scale, many structural and functional attributes of vegetation change in a predictable manner in parallel with changes in rainfall. Thus, the height and FPC (Fig. 1.7) and tree density (Fig. 1.10a) decline with decreasing rainfall. **Net primary productivity** (NPP; the net annual accumulation of biomass by vegetation) also increases at continental (Table 1.9; Fig 1.10c) and species (Fig. 1.10d) scales, while tree water use similarly increases with rainfall (Fig. 1.10b). It is principally the balance between rainfall (water input), evaporative demand (measured as potential evaporation) and soil storage capacity that determine the amount of vegetation (measured in terms of tree

Table 1.9 Relationship between increasing rainfall and NPP

	NPP (GtC y^{-1}) of natural vegetation	Land area (km^2)	NPP/land area (tC y^{-1} km^{-2})	Mean annual rainfall (mm)	NPP/rainfall (tC y^{-1} km^2 mm)
Africa	12	30 300 000	396	720	0.55
North America	6	9 529 000	630	800	0.79
South America	14	17 800 000	787	1800	0.44
Asia	12	44 800 000	268	620	0.43
Australia	2	7 686 850	260	480	0.55
Europe	2.5	9 900 000	253	820	0.31

Source: Eamus (2003)

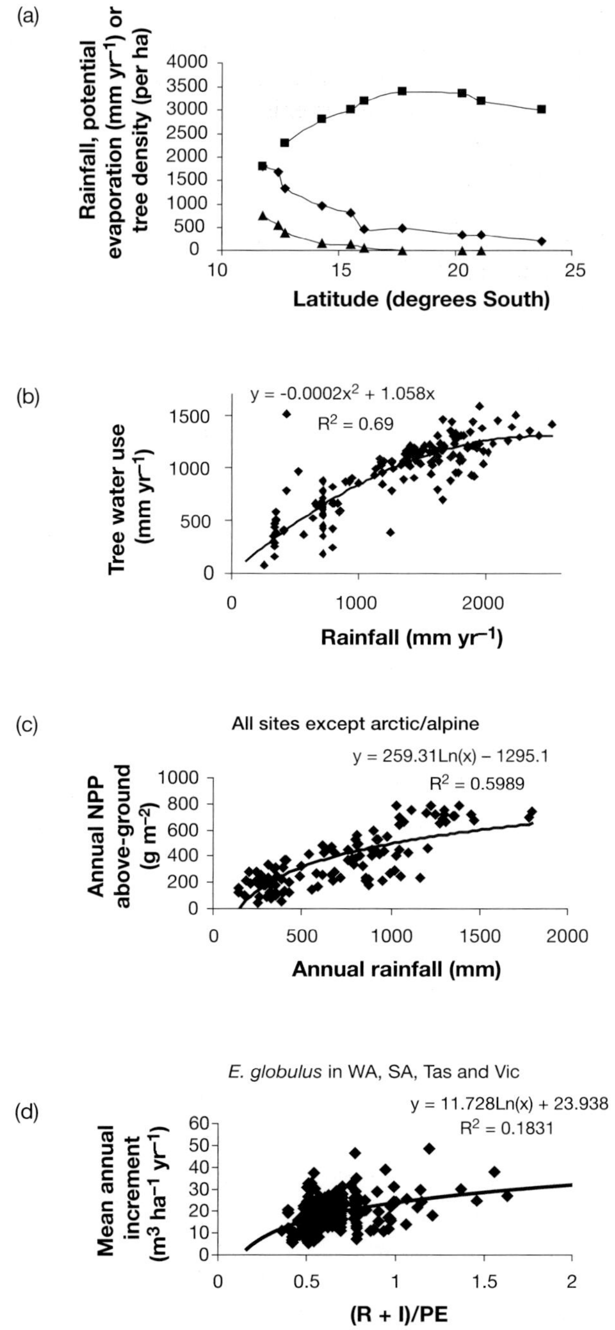

Figure 1.10 (a) Along a north–south transect in north Australia, potential evaporation (squares) increases as distance from the northern coast increases, rainfall declines (diamonds) and tree density (triangles) declines (Schulze et al. 1998); (b) Tree water use increases with rainfall for a range of Australian species growing in a range of sites across Australia; (c) Within the US, as rainfall increases, above-ground productivity increase (data from the Long-term Ecological Research webpage – http://intranet.lternet.edu/cgi-bin/anpp.pl – excluding alpine and arctic sites); (d) Within a species, growing on a number of sites, a similar relationship is observed (data supplied by D Barrett)

Source: Eamus (2003)

density, standing biomass, leaf area index, or productivity) that can occur at a site. Local variations in topography (slope) and soil water storage capacity (determined by soil depth and soil type) give rise to local variation in these attributes within a catchment. Thus, access to a shallow water table, or low points in the landscape that receive significant surface run-on of water, will sustain more vegetation than would be predicted from knowledge of rainfall and potential evaporation alone (Eamus 2003).

In addition to changes in vegetation structure with rainfall, and changes in vegetation function (primary productivity, water use) with rainfall, there are significant changes in life form. For example, the proportion of deciduous species or species with storage roots declines with declining rainfall across the Northern Territory of Australia (Egan & Williams 1996). Clearly, vegetation structure and function, water use and water availability are strongly linked. In the rest of this book, we examine various aspects of these linkages.

Generalities about Australian vegetation, climate and catchment water balance

The relative hydrological sensitivity of the Australian landscape to changes in vegetation cover has led to the development of fairly robust generalities about the relationships among vegetation cover, climate and the residual of rainfall and evapotranspiration (run-off or groundwater recharge).

The foremost generality is that fully stocked stands of trees (natural or planted) use more water than any other kind of vegetation at a given location. The rates of evaporation from forests or tree plantations at sites where water is not limiting tend to approach **equilibrium evaporation rates**; this is equivalent to 1000–1400 mm per year in southern Australia. Where annual rainfall is less than this, only minimal amounts of water tend to escape the root zone of trees other than as local evapotranspiration over the long term.

The second useful generality about tree hydrology is a close relationship between rainfall and the LAI of stands of woody perennial vegetation. In essence, the LAI of a site comes into a predictable, dynamic equilibrium with the amount of water available. This relationship seems to be independent of vegetation type, at least among Australian sclerophyllous vegetation types (Fig. 1.11). Thus, if we know the equilibrium tree LAI of a site in an area with less than 900 mm of annual rainfall, we can assume that virtually no recharge reaches local groundwater systems at this value. The degree to which a catchment is cleared of trees (as expressed in the remnant LAI) is a good approximation of the enhancement of groundwater recharge resulting from clearing.

An important corollary to the LAI–rainfall relationship is that where landscapes generate lateral water flow either as surface or subsurface discharge (as a result of clearing upslope or natural discharge processes), stands downslope may develop a (higher) LAI which reflects the total supply of water from local rainfall and the additional water supply from upslope. Local examples of relatively lush vegetation at the break of slope of hills abound. More profound and extreme examples occur where tree plantations (e.g. pines) occur over shallow, local fresh aquifer systems. In such cases, trees have been shown to use more than local rainfall and actually draw upon groundwater. In doing so, the trees can create a local groundwater depression (i.e. equivalent to a **cone of depression**; see Chapter 3) that induces lateral groundwater flow towards the plantation. If the aquifer is sufficiently transmissive, large amounts of water can be used by the trees (potentially the difference between local rainfall and potential equilibrium evaporation rates). This can translate to a net local groundwater discharge of hundreds of millimetres per year. Such cases have been identified in association with pine plantations in South Australia (growing over fresh karst hydrogeological systems) and in Western Australia (pines growing on highly transmissive and fresh sand aquifers).

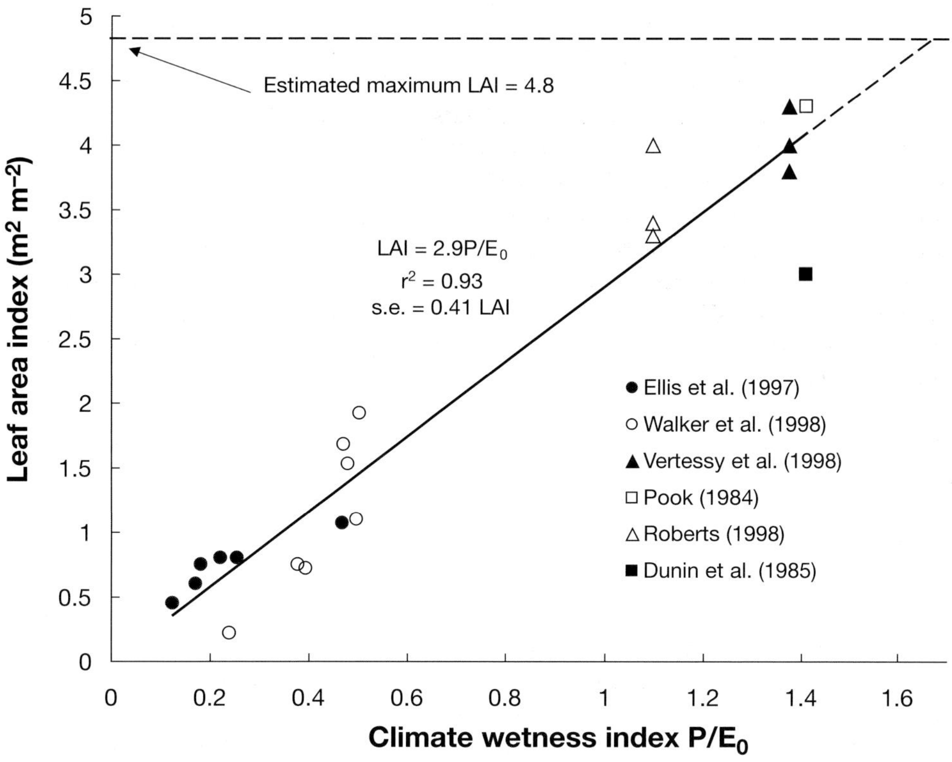

Figure 1.11 LAI of natural eucalypt forest in southern Australia versus a climate wetness index; average annual rainfall (P) divided by average annual pan evaporation (E_0)

Source: Ellis et al. (1999)

Zhang et al. (1999) examined a large number of datasets from around the world relating annual evapotranspiration to annual rainfall, and found clear differentiation between forests and grasslands (Fig. 1.12).

In those parts of Australia where widespread clearance of perennials was followed by replacement with annual plant-based agriculture (typically the zone with 250–750 mm annual rainfall), concern is focused largely on groundwater recharge. The relationship of Zhang et al. (1999) for 'excess water' may provide a useful indication of the upper limit to the long-term average recharge measurements, but does not differentiate run-off from recharge.

George et al. (1999) considered recharge data from 60 sites in the wheatbelt of Western Australia, and developed distinctive relationships among annual rainfall and recharge for agricultural landscapes and native perennial vegetation. Recharge was shown to be effectively nil over a large range in annual rainfall (up to 900 mm) under native vegetation, while annual recharge under annual plant-based agriculture ranged between 10 and 100 mm over a range of 300–900 mm of annual rainfall.

Petheram et al. (2002) investigated the potential for developing generic relationships from measurements of recharge made in previous studies that would allow the assessment of the impact of land use change on recharge (Fig. 1.13). The following relationships for annual vegetation were derived.

Ln(recharge) = -19.03 + 3.63 ln(rainfall) [for sandy soils]; $F(1, 96) = 149.03$; $R^2 = 0.60$
Ln(recharge) = -12.65 + 2.41 ln(rainfall) [for non-sandy soils]; $F(1,151) = 46.87$; $R^2 = 0.23$

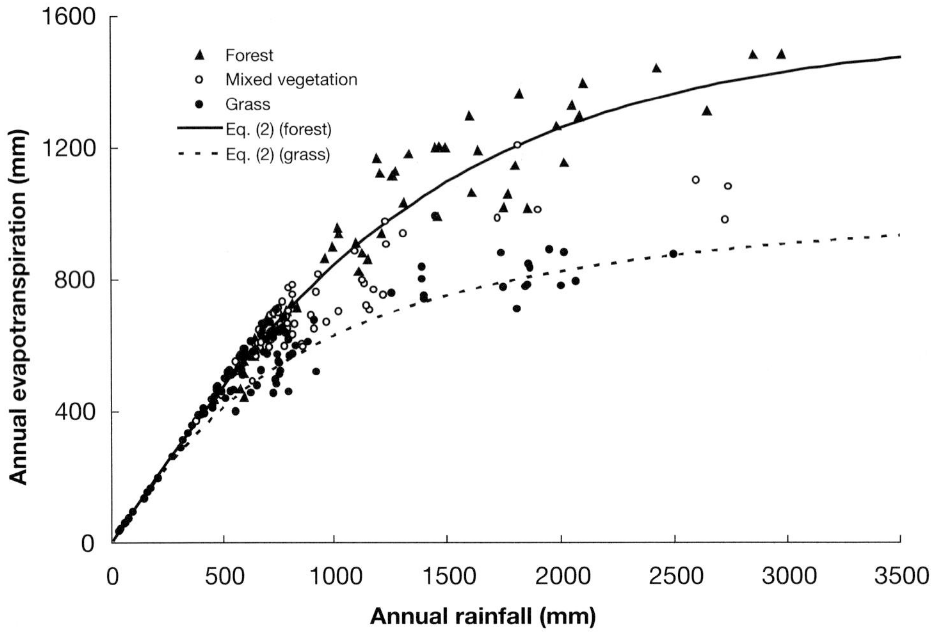

Figure 1.12 Relationship between annual evapotranspiration and rainfall for different vegetation types

Source: After Zhang et al. (1999)

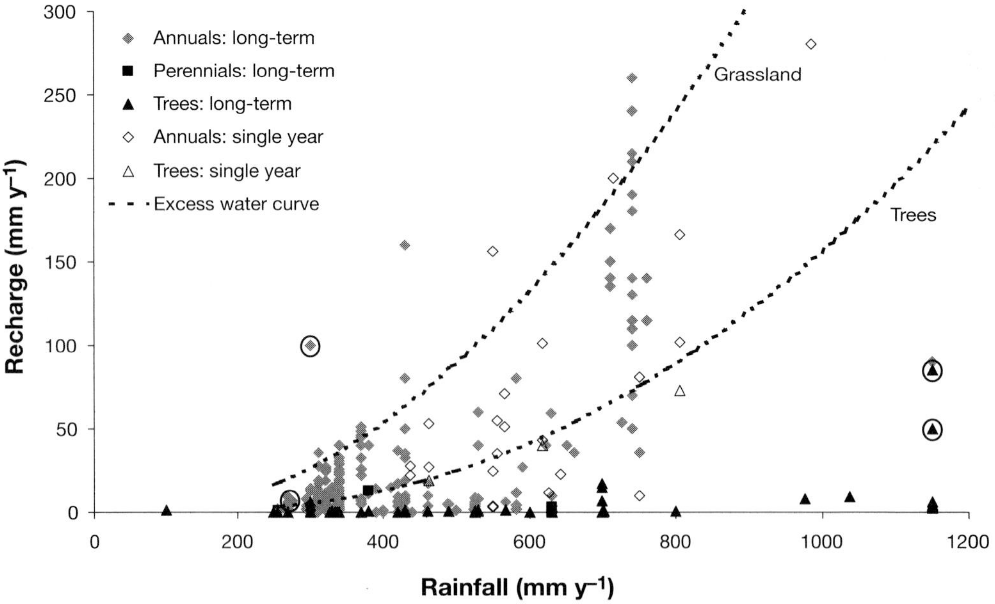

Figure 1.13 Annual recharge vs annual rainfall for different vegetation types and time-scales of measurement

Source: Petheram et al. (2002)

The low degree of explanation of rainfall for the annual non-sandy data suggests that it is likely that soil structure becomes more important for higher clay content soils. Recharge under trees was negligible compared with that under annuals.

Collectively, the results showed that:

- rainfall explains a significant proportion of the observed recharge variation;
- there is a significant difference between mean recharge under trees and annual vegetation;
- there is a significant difference between mean recharge under annual vegetation on sandy soils and non-sandy soils;
- across a broad range of locations, recharge is higher under shallow-rooted annual vegetation than deep-rooted vegetation.

Changes in Australia's vegetation pre- and post-European settlement

Much of ecohydrology is concerned with studies of changes in ecology and hydrology following changes in land use (e.g. converting open woodland into agricultural land). Two major changes in vegetation have occurred as the result of human activity. The first was when humans first arrived about 60 000 years ago, and fire was used as a landscape management tool. The influence of fire on ecohydrology is discussed in Chapter 7. The second occurred when Europeans arrived in the 18th century. This section briefly reviews how European settlement influenced vegetation.

Australia is experiencing widespread changes to its landscape hydrology. This has arisen because of significant changes in land use over the past 200 years. Principally, changes in vegetation cover (e.g. replacing woods and forests with crops and pasture) are responsible for significant alterations in the hydrologic (water) cycle (Chapter 3). While vegetation composition and distribution vary over geological time as continents move towards or away from the equator and as climates become drier, wetter, colder or warmer, we are not concerned with such long time-frames. Instead, we consider only the distribution and amount of vegetation immediately prior to European settlement in comparison to that observed at the end of the 20th century.

Australian vegetation is **highly diverse** at the species level and at the level of vegetation assemblage (ecosystem). There are about 20 000 species of higher plants in Australia and about 3000 vegetation types can be recognised. In addition, Australian vegetation has very **high levels of endemism**. Endemism is a measure of how many species are found only within Australia. About 85% of Australian plant species are endemic (found only in Australia). Australian vegetation is more diverse with a higher level of endemism than the flora of Europe and North America, for example.

Immediately prior to European settlement, Australian vegetation was dominated by seven vegetation types (Fig. 1.14, Colour Plate 5; Table 1.10). These are (in rank order based on area):

1 hummock grasslands;
2 eucalypt woodlands;
3 acacia shrublands;
4 acacia forests and woodlands;
5 tussock grasslands;
6 chenopod shrublands and forblands;
7 eucalypt open woodlands.

Table 1.10 Area (km^2) of major vegetation groups pre-European settlement of Australia

Major vegetation group	Area (km^2)	Major vegetation group	Area (km^2)
Rainforest and vine thickets	43 493	Mallee woodlands and shrublands	383 399
Eucalypt tall open forests	44 817	Low closed forests and closed shrublands	15 864
Eucalypt open forests	340 968	Acacia shrublands	670 737
Eucalypt low open forests	15 066	Other shrublands	115 824
Eucalypt woodlands	1 012 047	Heath	47 158
Acacia forests and woodlands	657 582	Tussock grasslands	589 212
Callitris forests and woodlands	30 963	Hummock grasslands	1 756 962
Casuarina forests and woodlands	73 356	Other grasslands, herblands, sedgelands and rushlands	100 504
Melaleuca forests and woodlands	93 501	Chenopod shrubs, samphire shrubs and forblands	563 389
Other forests and woodlands	125 328	Mangroves, tidal mudflats, samphires and bare areas, claypan, sand, rock, lakes, lagoons, lakes	
Eucalypt open woodlands	513 943		112 063
Tropical eucalypt woodlands/ grasslands	256 434		
Acacia open woodlands	117 993		

Together these vegetation types occupied about 75% of Australia's area. Aboriginal fire practices, the generally hot and arid climate of Australia and the low nutrient content of Australian soils favour sclerophyllous, sparse and generally low (<20 m) vegetation. Grasslands, woodlands and shrublands dominate, rather than rainforests or tall closed forests, which show a very restricted distribution. However, forests and woodlands (tree dominated ecosystems) of all types covered about 45% of the land area immediately prior to European settlement (Australian Native Vegetation Assessment 2001).

After more than 200 years of European settlement, the top seven vegetation types occupied about 83% of Australia's land. Note that the second largest single system is cleared or modified vegetation (Table 1.11; Australian Native Vegetation Assessment 2001):

1 hummock grasslands;
2 cleared/modified vegetation;
3 eucalypt woodlands;
4 acacia shrublands;
5 acacia forests and woodlands;
6 chenopod shrublands and forblands;
7 tussock grasslands.

The largest single vegetation type (hummock grassland) covers about 23% of Australia and is little changed by European settlement. This is because land supporting hummock grassland has low economic value. In contrast, cleared or modified vegetation occupies about 1 million ha, or 13% of the land. Cleared and modified vegetation is most significant in the south-west of Western Australia, the south and east of New South Wales, almost all of Victoria, the eastern

Table 1.11 Area of major vegetation groups of Australia in the late 20th century

Major vegetation group	Area (km^2)	Major vegetation group	Area (km^2)	Major vegetation group	Area (km^2)
Cleared/modified native vegetation	982 051	Casuarina forests and woodlands	60 848	Acacia shrublands	654 279
Rainforests and vine thickets	30 231	Melaleuca forests and woodlands	90 513	Other shrublands	98 947
Eucalypt tall open forests	30 129	Other forests and woodlands	119 384	Heath	25 861
Eucalypt open forests	240 484	Eucalypt open woodlands	384 310	Tussock grasslands	528 998
Eucalypt low open forests	12 922	Tropical eucalypt woodlands/grasslands	254 228	Hummock grasslands	1 756 104
Eucalypt woodlands	693 449	Acacia open woodlands	114 755	Other grasslands, herblands, sedgelands and rushlands	98 523
Acacia forests and woodlands	560 649	Mallee woodlands and shrublands	250 420	Chenopod shrubs, samphire shrubs and forblands	552 394
Callitris forests and woodlands	27 724	Low closed forests and closed shrublands	8749	Mangroves, tidal mudflats, samphires and bare areas, claypan, sand, rock, salt lakes, lagoons, lakes	106 999

Source: Data from Australian Native Vegetation Assessment (2001)

half of southern and central Queensland and eastern Tasmania (Fig. 1.14b, Colour Plate 5). These are regions where rainfall and irrigation can support extensive agriculture.

Forests and woodlands now occupy about 35% of the land area. This decline is more significant than the 10% difference implies because the decline in area occupied by forests and woodlands expressed as a percentage of the land area that previously supported forests and woodlands is actually a much larger percentage decline – about 20%. This loss of trees from Australia's landscape is a major cause of the change in the hydrological balance of Australian landscapes. Chapter 8 deals with the salinity problems resulting from this change.

Conclusion

This chapter has shown that water resources in Australia consist of surface and groundwater stores and that the continental water balance is precariously balanced. There is little scope for sustainably using much more water in the southern half of the continent but significant opportunity may exist for using more of the run-off that occurs in the northern half of the country. Whether agricultural development in the north can ever equal that observed in the south remains to be seen; certainly water resources are currently more available in the north than in the south. However, developing agriculture without extensive detrimental environmental impacts has yet to be shown to be feasible in Australia.

Vegetation in Australia shows a suite of adaptations that allow the flora to cope with a harsh environment. Australia's rainfall is low or very low across much of the continent and extremely variable between years; its soils have poor water storage capacity and are low in nutrients. Fire is a recurrent threat throughout much of Australia. Because of this relatively unique combination of features and the long geographical isolation of Australia, the flora of Australia show high degrees of endemism and are worthy of protection and conservation.

A simple yet effective means of classifying Australia's vegetation is based upon two or three structural attributes (height of the upper canopy, foliage projected cover and life form) plus a reference to the dominant genus. Such a simple system that uses attributes with direct relevance to ecohydrology facilitates effective communication between diverse scientific disciplines and highlights structural attributes that provide information about likely rates of water use of vegetation.

Australia is undergoing a profound change in landscape hydrology, principally through changes in land cover. A comparison of pre-and post-European land cover showed that after more than 200 years of European settlement the second largest single system is cleared or modified vegetation. This fact is explored in detail in many of the following pages.

Chapter 2 provides a detailed description of the theory and practice of plant water relations. Understanding the movement of water through vegetation and the influence of changes in water availability upon vegetation is central to the sustainable management of landscapes. This is the subject matter of Chapter 2.

Further reading

Attiwell P & Wilson B (2003). *Ecology: an Australian perspective.* Oxford University Press, Melbourne.

Atwell B, Kriedemann P, Turnbull C, Eamus D & Bieleski RL (1999). *Plants in action: adaption in nature, performance in cultivation.* Macmillan Education Australia, Melbourne.

Dyer R, Jacklyn P, Partridge IR-SJ & Williams D (2001). *Savanna burning: understanding and using fire in northern Australia.* Tropical Savannas CRC, Darwin.

Eagleson PS (2002). *Ecohydrology: Darwinian expression of vegetation form and function.* Cambridge University Press.
Eamus D, Hutley LB & O'Grady AP (2001). Daily and seasonal patterns of carbon and water fluxes above a north Australian savanna. *Tree Physiology* **21**: 977–988.
Morgan G (2001). *Landscape health in Australia. A rapid assessment of the relative condition of Australia's bioregions and subregions.* Environment Australia.
State of the Environment Committee Report (2001). CSIRO, Melbourne.

Chapter 2

Water relations of plants

Plant water relations encompass the study of the water content of soil, plant and atmosphere and the movement of water from soil to atmosphere via a plant. An understanding of such topics is central to the study of ecohydrology. In this chapter, the concepts of plant water relations are explained. For those with prior experience in plant physiology and ecophysiology, this chapter is merely a means of refreshing past knowledge. To those who are ecologists or hydrologists, much of this material will be new. The chapter provides the foundations of the theory and practice of plant water relations, thereby allowing readers to study the relationships among hydrology, vegetation structure and plant behaviour.

The water status of vegetation is closely linked to the water content of the soil, the water content of the atmosphere, the climate (especially solar radiation input) and the structure (hydraulic architecture) of the vegetation. Whether the objective of a given management regime is the management of the water resources in a catchment or the management of the vegetation for wood production (forestry), it is essential that the links between vegetation water status and the other factors listed above are understood. An understanding of the water relations of plants provides a pivotal link between hydrology and ecology.

After reading this chapter, readers should be able to examine a data set about the water relations of plants in the field and provide an interpretation of those data. In addition, they should be familiar with the following concepts:

- water movement in the soil–plant–atmosphere continuum (SPAC);
- how we can measure and compare the water status of plants, soil and atmosphere. The role of solar radiation, soil water content and atmospheric water content in determining the flux of water through the SPAC;
- mechanisms used by plants to regulate water movement in the SPAC;
- sources of water used by plants and how to determine what sources are being used;
- hydraulic lift and the advantages conferred to plants that exhibit the phenomenon;
- hydraulic architecture of plants and how this matters when thinking about water movement through vegetation;
- typical patterns of water movement (transpiration) and plant water status in Australian plants in the field. What it tells us if we know how these patterns change with time or in different ecosystems;
- finally, whether we can use measurements of individual trees to provide reliable information about the behaviour of whole canopies and whole catchments of trees. What do we need to know to reliably scale from individual trees to whole stands of trees?

Introduction

Water is fundamental to life on earth. From subcellular processes such as photosynthesis and enzymic reactions to continental-scale processes such as sedimentation and erosion, water plays a key role. Indeed, without the very special properties of water, life would be very different. The hydrogen bonding that occurs between water molecules explains many of its important properties, such as the high latent heat of vaporization which allows it to act as an effective coolant for plants and animals through evaporation and also allows water to act as a thermal buffer to prevent large variations in cell temperature. The high boiling point of water is a reflection of the high latent heat of vaporization and extensive hydrogen bonding while the high surface tension of water allows some aquatic animals to skate upon it. The high surface tension also contributes to the ability of water to ascend up the tallest trees through capillary action (see below). Water possesses a small negative and positive charge on the oxygen and hydrogen atoms respectively; this polarity, plus the hydrogen bonding ability of water, makes it a potent solvent for ionic and polar compounds. Ice (solid water) floats on liquid water, an important and unusual attribute. Few solid forms of a compound float on the liquid form of the same compound. This contributes to the continuity of life in oceans and rivers when they freeze over. When ice forms and floats in oceans and deep rivers, the ice acts as a thermal insulator from the cold air above that causes the freezing. This ensures that some liquid water remains below the ice in which fish and other aquatic and marine life can survive. Water also has several distinctive roles in life. It acts:

- as a **solvent**, within which almost all chemical reactions of life occur;
- as a **matrix** within which membranes and proteins can fold;
- to allow **molecules to diffuse** and interact within cells;
- as a **transport medium** in plant and animal bodies;
- as a **coolant** (through evaporative cooling);
- as a means of **movement** (e.g. hydraulically controlled limbs in insects and the movement of stomata in leaves);
- as a means of **maintaining shape** of cells (through hydrostatic pressure, which drives plant growth);
- and takes part in many biochemical reactions, including photosynthesis.

Water is also involved in a myriad of hydrolysis reactions, whereby hydrolytic enzymes (hydrolases) insert the constituents of water (H_2O) into proteins, carbohydrates and nucleic acids to split these macromolecules into smaller units. Furthermore, water is also the most ubiquitous compound in living organisms, making up 70–95% of the mass of most organisms (the main exclusions are seeds and wood, which have lower water contents). Its abundance and special properties are central to the evolution of life on earth.

Water is therefore important to life at a subcellular scale, but also at a whole plant (Chapter 2), ecosystem (Chapters 4, 5, 6, 7) and landscape (Chapters 6, 7, 8) scale.

In order to think about water in the landscape, it is important to have a means of quantifying the amount, availability and movement of water through the landscape. One of the principal pathways for the movement of water in the hydrological cycle (Chapter 3) is through the **soil–plant–atmosphere continuum** (SPAC). The SPAC represents the physical continuity of water as it moves from soil, through roots and up the plant to leaves, and out into the atmosphere. This chapter describes the SPAC and explains the concepts required to understand and discuss the movement of water through this pathway. The discharge of rainfall and groundwater through vegetation can amount to thousands of millimetres per year. Thus, almost all of the rainfall on a catchment can be transpired through the vegetation of that catchment.

Movement through the SPAC is an important process for the discharge of water from catchments and has an impact on catchment water balance. Indeed, it is the widespread loss of trees from the Australian landscape that has changed the hydrological balance of vast areas of rural land and given rise to salinity in the landscape (see Chapter 8).

Plant water potential and its component parts

The study of ecohydrology requires the study of the availability and movement of water in the environment. Water is present in soils, plants and the atmosphere in both liquid and gaseous forms. One of the most important developments in plant physiology and ecophysiology in the 20th century was the application of thermodynamics to describe and explain the energy status of water in biotic (living) and abiotic (non-living systems). By providing a unifying framework for talking about water and by providing a unified nomenclature for describing the gradients in the energy status of water in soil, plants and the atmosphere, it became possible to compare the energy status of water anywhere within the soil–plant–atmosphere continuum. The SPAC represents the physical continuity between water in the soil, water in the plant and water in the atmosphere and represents a central pathway by which soil water and groundwater is returned to the atmosphere. Indeed, trees may be considered as wicks or hydraulic pumps, moving water up into the atmosphere. To understand the movement of water through the SPAC we must understand the meaning of water potential and all of its component parts. This section provides that understanding.

A ball rolls downhill because gravity acts upon it: rolling down the hill does not require an input of energy (unlike rolling uphill, which does require an input of energy). In fact, rolling downhill releases energy and this energy can be used to do work, just as water falling downhill can be used to turn a turbine in a hydroelectric power plant to generate electricity. Rolling downhill is termed an 'energetically downhill' process and rolling downhill occurs because of the gradient in gravity between the top of the hill and the bottom of the hill.

Similarly, water moves within a plant, or from wet soil into the roots of a plant, because the movement is energetically downhill. In the place of gravity in the ball-and-hill example is a **gradient of water potential** that must exist between two adjacent cells or between soil and roots to cause water to move. It is a difference in water potential that causes water to move from a region of high water potential (e.g. wet soil) to a region of lower water potential (e.g. root xylem). Xylem is the network of pipe-like cells that connect roots to leaves via the stem. The movement of water through the xylem of a plant is termed **sapflow.**

The concept of water potential is based on the concept of free energy. Free energy is defined as the energy available to do work (e.g. lifting water up a tree constitutes work). Work is done when something moves against gravity or against some other force that opposes that movement. The chemical potential of water is defined as the free energy per mole of water and water potential is the chemical potential of water divided by the partial molar volume of water (i.e. the volume occupied by 1 mole of liquid water at standard temperature and pressure). Thus water potential is a measure of the energy available for doing work, expressed per mole of water. This rather esoteric definition is less important than the need to understand two things: first, water moves in the SPAC from regions of high (close to zero) to regions of low (more negative) water potential; and second, that the water potential of a plant cell has two principal components, namely solute potential and turgor potential. Water potential can be zero or negative and the reason for this is explained below. The meanings of solute and turgor potential are discussed shortly.

Water potential is given the symbol Ψ_w (pronounced Psi W), turgor potential (a measure of the hydrostatic or water pressure in a cell) has the symbol Ψ_p and solute potential (a measure of the concentration of solutes in cell) has the symbol Ψ_s. At this stage we can note that:

$$\Psi_W = \Psi_p + \Psi_s \quad (1)$$

The water potential of water in a root or in a leaf, or of any system, is measured relative to an internationally agreed reference value. The reference value has been arbitrarily defined as the water potential of pure water at sea level, sitting in a beaker at standard temperature and pressure (a temperature of 25°C and 0.1 MPa atmospheric pressure) and has been arbitrarily given a water potential equal to zero. Within a plant (and for soil and atmosphere too) all other water potentials are either lower than this value (i.e. negative) or equal to this (i.e. zero).

Solute potential of plant cells

A single root or leaf cell can be thought of as a small container containing water and solutes. Thus, consider a second beaker of pure water, sitting next to the reference beaker described above. The addition of a solute such as NaCl to this second beaker reduces the water potential to a value lower than that of the reference water potential of zero, i.e. to a negative value. It is the **presence of solutes** that has reduced the water potential of the water in the salty beaker, relative to the water in the reference beaker. The water potential of the water in the salty beaker is reduced because the Na^+ and Cl^- ions electrostatically attract the negative and positive ends of the polar water molecule and restrict the movement of these water molecules (Fig. 2.1). It is this restriction of the free movement of the water molecules that reduces the water potential. The reduction in free movement of a proportion of water molecules is said to reduce the ability of the water to do work, that is, there is less energy available to do work, and therefore water potential is reduced.

As the concentration of solutes in the water increases, the water potential of the solution decreases further. To a good approximation in dilute solutions, the reduction in water potential can be calculated from:

$$\Psi_W = -R \times T \times C \quad (2)$$

where R is the gas constant, T is the temperature (in Kelvin) and C is the **osmolal** concentration of the solution. Molality is similar to molarity, but molality is a measure of the amount of solutes present per kilogram of solvent, while molarity is a measure of the solute content per litre of solution.

Because the presence of salts (solutes) has reduced the free energy of the water in the salty beaker, we can say that the solute potential of the water has decreased. Solute potential (Ψ_s) is also called osmotic potential because changes in osmotic potential of cells can influence the osmotic movement of water between cells. When solutes (e.g. salts) are present in water, the concentration of water molecules, and hence the free energy per volume of water, is reduced and therefore so is water potential. Furthermore, the electrostatic interaction between salt ions and water molecules restricts the free movement of the water molecules and so their free energy is reduced. A definition of solute potential is therefore:

> **Solute potential is a measure of the change in water potential of a system due to the presence of solutes. The more solute molecules are present, the lower (and more negative) is Ψ_s.**

Solute potential is always zero (no solutes present) or negative (solute present).

The reduction in water potential (Ψ_w) of the salty beaker due to the presence of solutes is due to a reduction in the solute potential (Ψ_s) of the solution. Thus, water potential and solute potential are equal in the salty beaker above. Figure 2.1 schematically outlines the situation in the two beakers. In the beaker of salty water, water potential and solute potential are equal.

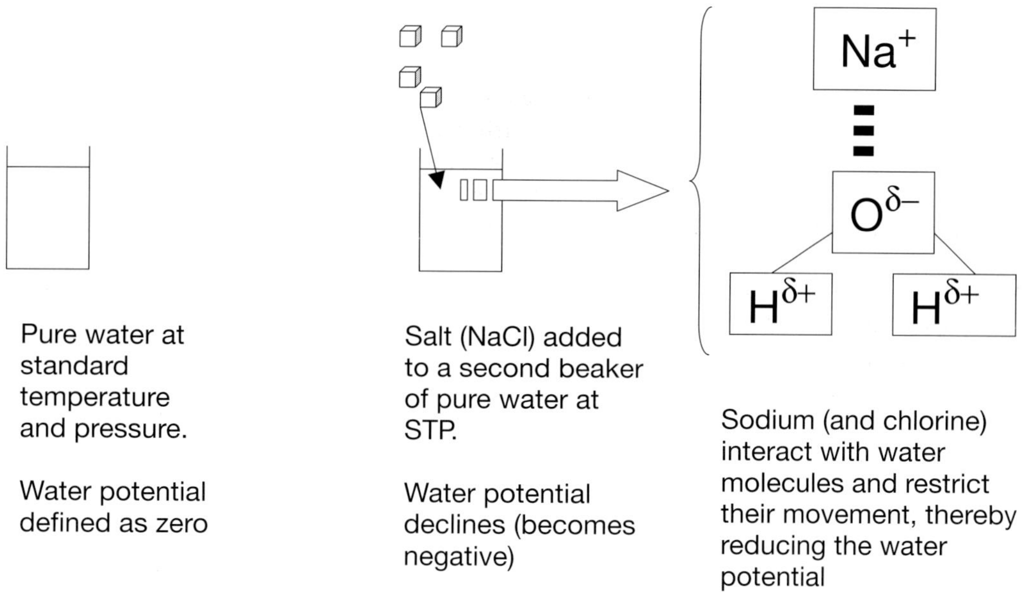

Figure 2.1 A schematic outline of the changes in solute (and hence water) potential of a beaker of pure water following addition of salt crystals to the water. The horizontal bars between the $O^{\delta-}$ and Na^+ on the right side of the diagram represent the electrostatic interaction between an Na^+ ion and a water molecule. Only one water molecule is shown but in reality a large number of water molecules are held in a cage-like structure surrounding each ion

The solute potential of a plant cell is always lower (more concentrated) than the solute potential of the soil around the roots of the plant. This is because the plant continuously absorbs salts (K^+, Cl^-, Mg^{2+} and many others) and deposits them in plant cells. In addition, plant cells actively produce solutes, such as sugars and amino acids, which further reduce the solute potential of cells to a value much lower than the solute potential of soils (see below).

Pressure and matric potential of plant cells

The pressure (of the atmosphere or that of a cell wall) exerted on water also influences water potential. The reference beaker discussed above is at standard temperature and pressure at sea level. If water is placed in a sealed container and pressurized, the water molecules are compressed together and a repulsive force is created between pairs of negatively charged oxygen atoms and pairs of positively charged hydrogen atoms on adjacent molecules. This force results in an increase in the available energy within the water and so water potential increases. Pushing two water molecules together is conceptually equivalent to compressing a spring – there is energy stored in the compressed spring which is available to do work. Within a plant cell, water is generally under considerable pressure. This pressure can exist because of the presence of a strong cell wall comprised of cellulose microfibrils pushing inwards. The air pressure in an inflated car tyre is typically in the region of 0.2–0.3 MPa (2–3 bar or 28–42 psi air pressure). A plant cell can generate hydrostatic pressures of up to 2–3 MPa. This hydrostatic pressure is called **turgor potential** (Ψ_p) or **pressure potential,** when we are talking about water in plant cells. Turgor potential is usually zero (e.g. in a flaccid cell in a wilted leaf) or positive (e.g. in a turgid cell in a turgid leaf).

Thus, water potential of plant tissues has two main components – solute potential and turgor potential:

$$\Psi_W = \Psi_s + \Psi_p \tag{3}$$

These are the two main components of importance when considering water in plants. Later, we shall see that an additional component, matric potential (Ψ_m), is added in relation to soils.

In addition to hydrostatic pressure and the presence of solutes, the water potential of a leaf is also influenced by temperature and height above sea level (and hence the magnitude of gravity). However, water potential is expressed relative to a standard temperature (25°C) and pressure (0.1 MPa) at sea level and the gravity potential and corrections for temperature are often omitted in practical assessments of water potential in the field.

The **matric** potential resulting from the interaction of water and surfaces (such as cell walls, membranes and proteins) also contributes to water potential of plant cells but only becomes significant when the water content is so low that cell death is imminent. It is usually ignored in practical measurements. Matric potentials are sometimes considered to be accounted for in measurements of solute potential. Matric potentials for plant tissues are rarely cited in plant physiology or ecophysiology texts.

It is possible to measure the water potential of a volume of soil, a root cell, an excised section of root, the xylem of a stem, a leaf cell, a whole leaf or the atmosphere using a range of techniques (see Chapter 4). Water potential decreases (becomes more negative) from the moist soil, through the roots, stems and leaf, to the atmosphere. Some indicative values of water potential of the components of the SPAC are shown in Table 2.1.

Water potential is zero or negative, solute potential is zero or negative and turgor potential is zero or positive, in living cells. The unit for all potentials is Pascals (but usually in multiples of millions, hence MPa; note that 1 MPa = 10 bar = 140 psi). When water potential of a cell is zero the solute potential is equal, but of opposite sign, to the turgor potential of the cell. Thus in a turgid cell the solute potential could be -1.2 MPa, the turgor potential could be 1.2 MPa and the water potential is therefore zero (equation 3). As a cell loses water, its turgor potential declines towards zero (the cell deflates) and its solute potential gets more negative because the solute concentration of the cell increases. Consequently water potential of the cell declines. The relationships among water content, water, solute and turgor potentials are discussed in depth in Chapter 4.

Table 2.1 Representative values of water potential for different components of the SPAC. Values of water potential decline from soil to atmosphere

Component of the SPAC	Indicative ranges or values of water potential (MPa)
Soil	
Wet	–0.001
Moist	–0.1
Dry	–2.0
Plant	
Root cells in moist soil	–0.1 to –0.2
Root xylem in moist soil	-0.2 to –0.4
Leaf xylem of turgid plant	–0.5 to –1.0
Turgid leaf	–0.1 to –2.0
Wilted leaf	–2.0 to –4.0
Atmosphere (at 25°C)	
99% relative humidity	–1.36
90% relative humidity	–14.2
50% relative humidity	–93.6

The importance of differences in water potential between two points in the SPAC and the movement of water from soils to the atmosphere through the SPAC are now discussed.

Gradients of water potential, water movement in the SPAC and energy balances

It is easiest to think about water flow in the SPAC by starting at the upper leaves of the canopy and working down the plant to the soil. In the following treatment a moist soil is assumed. After discussing how water moves through plants, this section proceeds to a consideration of energy balances of leaves.

As the sun rises, leaf and air temperatures increase, and stomata in leaves open. Stomata open in response to the increase in light level and in response to the decline in CO_2 concentration within the leaf which is caused by the onset of photosynthesis, itself driven by light.

The walls of the cells within the leaf are saturated with water – they act as cellulose wicks just as blotting paper does. The water potential of the water in the cell walls is very close to zero immediately before sunrise as the leaf has spent all night rehydrating (assuming soil water is readily available). Immediately below the stoma (singular of stomata) in the leaf is the **sub-stomatal cavity**, which is saturated with water vapour (relative humidity (RH) at least 99.8%) and hence its water potential is close to zero (-0.27 MPa for RH of 99.8% at 20°C). The sub-stomatal cavity is saturated with water vapour because it is in equilibrium with the water potential of the walls of the cells lining the sub-stomatal cavity.

The concentration of water vapour inside the sub-stomatal cavity is much larger than that in the atmosphere. This gradient in concentration means that the rate of diffusion of water from the leaf exceeds that into the leaf. This diffusion of water out of the leaf is called **transpiration** and given the symbol E.

Because of water loss from the sub-stomatal cavity, a gradient in water vapour pressure is generated between the water at the surface of the walls of cells surrounding the sub-stomatal cavity and the cavity itself. Water therefore evaporates from the cell walls to replace the water lost as transpiration. This causes the water potential of the cell wall to decline, which causes water to move from regions of cell walls further away from the sub-stomatal cavity and also from **cytoplasm** of the cells adjacent to the sub-stomatal cavity. As water moves from regions of cell walls and cytoplasm further away from the sub-stomatal cavity, a gradient of water potential is created expanding from each stoma through the leaf. Eventually the water potential of the region immediately adjacent to the xylem of the leaf becomes lower than the water potential of the xylem and water moves out of the xylem into the cell walls and cytoplasm of leaf cells, and eventually out of the leaf as transpiration.

As water is lost from the xylem of leaf, adjacent water molecules are 'dragged' up the xylem. Each water molecule brings another along behind it due to the hydrogen bonds between adjacent water molecules. Thus, water shows **cohesive properties** – water is said to be sticky – and tensions are generated within the xylem as water is pulled up the xylem in the stem against gravity (i.e. against the weight of the water further down the stem) and against the resistance to flow that exists because of the rough walls of the xylem. Walls of pipes exert a drag on the movement of fluid through them. At very low flow rates, water can also ascend up the xylem due to capillary action of polar water molecules up the hydrophilic cell walls of the xylem. This explanation is generally known as the cohesion-tension theory.

The **cohesion-tension theory** of water flow up xylem is generally accepted as a valid description of the flow of water up the xylem. The tension of water in the xylem of a transpiring plant is conceptually similar to pulling on two ends of an elastic band: one end represents evaporation of water from the leaf, the other represents the weight of the water column down to the roots. The columns of water in xylem conduits occasionally snap and this snapping can be detected as acoustic emissions using sensitive microphones. Snapping of the water column results in the formation of a bubble within the xylem. This bubble is called an **embolism.** A xylem embolism is bad news for a plant as it blocks further water movement up that file of xylem cells (discussed later in this chapter).

The reduction in water potential that was first generated in the leaf cells through evaporation of water is now expressed across the root, from the xylem to the outer epidermal cell walls and out further into the soil immediately around the root. Thus, the evaporation of water from leaves causes water to enter the roots because of the gradient of water potential created through the plant and into the soil and we see a decline in the values of water potential between soil and atmosphere, as exemplified in Table 2.1.

When a living cell has a water potential lower than an adjacent living cell, water will move from the cell with the highest (closest to zero) water potential to the cell with the lowest. For this to happen a living membrane must separate the two cells, otherwise the cells cannot maintain a difference in water potential between them for long periods because they come into equilibrium through free diffusion of solutes and water. Similarly, for the soil and xylem of a root to have a difference in water potential between them, a living membrane must separate the two compartments (soil and xylem). A dead root cannot maintain a water potential gradient between the soil and root xylem, thus water does not move from soil into a dead root. The membrane in both examples is the plasmalemma or plasma membrane which lies immediately adjacent to the cell wall and forms the outer boundary of the cytoplasm of all living plant cells. A living membrane separates the water in soil from the water in the xylem of roots. Similarly, a living membrane separates the water in the cytoplasm of mesophyll cells in leaves from the water in walls of the cells around the sub-stomatal cavity from where water evaporates in the process of transpiration. The process of evaporation and transpiration is now discussed in detail as it is central to the study of the water relations of vegetation and vegetation water use.

Solar radiation balance, energy balance and evaporation

Evaporation of water occurs from wet surfaces such as cell walls of leaves or the surface of a lake. It requires a significant input of energy for the phase transition from liquid water in the cell walls to gaseous water in the sub-stomatal cavity. **Solar radiation** is shortwave radiation (with wavelengths up to 1.0 µm) and is the principal supply of energy to leaves and the driver of evaporation.

A leaf surface can receive two types of solar radiation:

1 direct solar radiation: (W_s) radiation that has not materially interacted with another mass;
2 diffuse solar radiation: (w_s) radiation that has interacted with another mass, such as being reflected from a cloud or another leaf.

Thus the total incoming shortwave radiation is $W_s + w_s$. Some of this solar radiation is **absorbed** by the leaf, some is **reflected** and some is **transmitted** through it. The **albedo** (α) of a surface is the fraction of total solar radiation reflected from a surface.

In addition to shortwave radiation, long-wave radiation (wavelengths longer than 1 µm) is also present as it is emitted from all surrounding objects (soil, vegetation and the atmosphere). All objects above a temperature of absolute zero (–273°C) emit long-wave radiation. This is 'black body radiation' and leaves lose long-wave radiation because of their low (compared to the sun) temperature. A leaf or vegetation canopy can both receive and emit long-wave radiation, but can only receive (not emit) shortwave radiation (because leaves are too cold to emit shortwave radiation). The **radiation balance** of a surface such as a leaf is the net balance between incoming shortwave radiation plus incoming long-wave radiation (I_a) and emitted long-wave radiation (I_g). Thus:

$$R_n = (W_s + w_s)(1 - \alpha) + I_a + I_g \tag{4}$$

where R_n is net radiation of the surface.

Net radiation can be positive or negative. When positive, a leaf loses energy through several processes, but especially the evaporation of water. When negative, a leaf is gaining (absorbing) energy and is getting warmer. When leaf temperature exceeds air temperature, the leaf will lose energy through three processes. First, it loses long-wave radiation. The amount of radiation emitted (B) increases with leaf temperature and is described by the equation:

$$B = e\sigma T^4 \tag{5}$$

where e is **emissivity** of the object (close to 1 for leaves), σ is the Stefan-Boltzmann constant (the proportionality constant between the energy radiated from a surface per unit time per unit area and the fourth power of the temperature (T, expressed in Kelvin) of the surface). Small increases in the temperature of an object results in large increases in the amount of energy radiated from that object. Cold objects (anything on the earth's surface is cold compared to the sun) emit long-wave radiation. Very hot objects (e.g. the sun) emit short-wave radiation.

Second, leaves can also lose energy through transfer of heat from the leaf to the air. This loss of energy is also known as **sensible heat flux** (H). Energy loss through sensible heat flux is increased by wind and is larger from small needle-shaped leaves than from large flat leaves.

Finally, leaves can lose energy through the evaporation of water since the conversion of 1 g of liquid water into 1 g of water vapour requires about 2.45 kJ of energy (supplied principally by solar radiation). This loss of energy is given the symbol λE, where λ is the heat required to evaporate water (latent heat of vaporization 2.45 kJ g^{-1}) and E is the rate of evaporation. No consideration is given to the use of absorbed energy in photosynthesis as it is such a small fraction of the total incoming radiation. Similarly, over the short term or long term (but not intermediate) the storage of energy (as an increase in temperature of vegetation) is small enough to be ignored.

Thus, the **energy balance** of a leaf is given by:

$$R_n = B + H + \lambda E \tag{6}$$

Rearranging equation (6), we get:

$$\lambda E = R_n - B - H \tag{7}$$

This is an important equation from an ecohydrological perspective as it describes how evaporation from a leaf (transpiration) is determined by the net radiation receipt of the leaf, the loss of energy from the leaf through black body radiation (determined itself by leaf temperature which is determined by radiation input) and sensible heat flux (which is influenced by leaf temperature).

The relative contributions of transpiration and sensible heat flux to the energy balance of a leaf vary according to the availability of soil water. When water is readily available to a canopy, most (up to 90%) of the available energy is used to evaporate water from leaves. Transpiration is therefore very much determined by the amount of radiation received by the leaf. By having a maximum opening of stomata (and hence unregulated transpiration), carbon gain through photosynthesis is maximised. However, when water availability is limited, partial and eventually complete stomatal closure occurs and the amount of water transpired, and hence the amount of energy lost from the leaf through transpiration, is much reduced. This results in heating of the canopy and surrounding air.

Canopies can receive significant inputs of energy as sensible heat from surrounding areas. The most obvious case is an oasis, where the moist and green but very isolated canopy is surrounded by a hot dry desert. This input of energy (as sensible heat) is in addition to the solar radiation received by the canopy, and can be a large proportion (10–30%) of the total energy budget of the canopy.

Radiation (energy) input to a leaf or canopy is thus the major determinant of transpiration. The maximum amount of radiation received by a canopy is very much determined by its location on the earth's surface, with tropical regions receiving far more radiation than temperate zones, which in turn receive far more radiation than polar regions. These differences in solar radiation input determine regional climate (along with differences in rainfall).

Evapotranspiration

Evapotranspiration is the sum of evaporation from wet surfaces such as soil, a wet leaf and bark surfaces, *plus* transpiration through **stomatal** pores of leaves (stomatal transpiration) *plus* transpiration through the cuticle of leaves (**cuticular** transpiration). Evapotranspiration is influenced by several factors. While the input of solar radiation is the largest and principal factor determining the rate of evapotranspiration, it is not an absolute requirement. Water will evaporate from a wet surface in the dark as long as the bulk air above the wet surface is not saturated with water vapour and the temperature of the surface is above absolute zero (–273°C). Wet leaves, soil and bark can lose water through evaporation because of the energy derived from the warmth of the objects and surrounding air.

In the dark, stomata of trees and the majority of all other plants (i.e. all plants except those possessing Crassulacean Acid Metabolism such as pineapple) are closed. Consequently transpiration through stomata is generally zero at night. Cuticular transpiration occurs all day and night but is a very small fraction of total evapotranspiration (0.1–5%) and usually assumed to be zero. When measuring total transpiration from leaves, which is the sum of stomatal and cuticular transpiration, it should be recognised that some water loss is occurring across the cuticle. Micro-meteorological treatments of evapotranspiration tend to explicitly consider only two of the three pathways for evaporation of water – evaporation through stomata, and evaporation from wet surfaces such as wet leaves, wet soil and wet bark – and ignore cuticular water loss.

It is worth noting that evaporation of water from a wet surface (such as a wet soil or the wet cell walls of leaves inside a leaf) into the bulk atmosphere is a two-step process. The first step is an input of energy to convert liquid water into water vapour. The amount of energy required is called the latent heat of vaporization and for every kilogram of water converted from liquid to vapour, 2.45 MJ of energy is required. At absolute zero (–273.13°C), vaporization of water ceases because there is no energy available. As stated above, solar radiation is the largest single source of energy driving this process but the warmth of leaves, soil and atmosphere also contribute. The second step in this process is the diffusion of molecules of water vapour from the wet surface into the bulk air. The driving force for diffusion of water vapour molecules is a gradient of water vapour concentration (alternatively, measured as a gradient in water vapour pressure) between the evaporating surface and the bulk air above the surface. In the absence of a removal of water vapour molecules, the air adjacent to the wet surface becomes saturated and evaporation rate declines. Thus, a large gradient of water vapour pressure arising from the bulk air being relatively dry, and the presence of wind (which removes the saturated air) both increase evaporation rates from a wet surface (or leaf). This is discussed below in more detail.

There are many factors that influence the rate of evaporation from a wet surface. These include:

- wind speed;
- turbulence;

- vapour pressure difference;
- temperature.

Wind speed influences evaporation and transpiration by determining the thickness of the boundary layer (the layer of stationary air adjacent to all solid surfaces). As wind speed increases, the thickness of the boundary layer decreases. The boundary layer imposes a resistance to diffusion of water molecules because the boundary layer tends to become saturated with water molecules and these impede the free diffusion of water molecules from the wet surface into the bulk air. A thinner boundary layer imposes a smaller resistance to diffusion. Since resistance and conductance are the inverse of each other, this means that as wind speed increases, boundary layer conductance increases and so evapotranspiration increases. This is why we like to feel a cooling breeze on a hot sticky day. We lose heat through converting liquid sweat into water vapour and the efficiency of this is enhanced when the boundary layer is small on a breezy day.

Wind speed, leaf temperature and the rate of evaporation interact so that a change in one results in alteration of the other two. As wind speed over a leaf declines, leaf temperature increases because boundary layer conductance declines and hence the evaporative cooling effect of transpiration is reduced. This increase in leaf temperature increases the energy available to cause evaporation of water from the leaf and the gradient in water vapour concentration between leaf and air is increased, so transpiration will tend to increase with increased leaf temperature. Thus changes in wind speed, temperature and rate of evaporation interact.

Turbulence can also influence evaporation from wet surfaces and leaves. At a given wind speed the amount of turbulence experienced by the leaf surface can vary according to factors such as the hairiness and size of the leaf. Hairs on the surface of leaves increase the boundary layer thickness of leaves by trapping a thick layer of stationary air around the leaf. Thus, turbulence over the leaf surface is reduced and evaporation rate is reduced. Increased turbulence increases evaporation by bringing pockets of relatively dry air from the bulk atmosphere above the canopy into close contact with the leaf, thereby increasing the gradient of vapour pressure between inside the leaf and outside the leaf. Leaf size influences turbulence over leaf surfaces – small diameter (narrow) long needles experience more turbulence than large flat leaves, have a larger boundary layer conductance and are generally cooled more effectively than large flat leaves growing under the same conditions.

Turbulent transfer of heat (energy) and momentum (momentum is the physicist's way of saying 'exchange of mass in the form of water vapour and CO_2') is the mechanism underlying the measurement of canopy gas exchange using eddy covariance. This is discussed in detail below.

The **vapour pressure** of the air adjacent to leaves influences the rate of evapotranspiration by influencing the rate at which molecules of water vapour diffuse away from the evaporating surface. Vapour pressure is often intuitively thought of in terms of the relative humidity of air. A low humidity corresponds to dry air and a low water vapour pressure; high humidity air at the same temperature has a larger water vapour pressure and a lower evaporative demand. Evaporative demand is a measure of the ability of the atmosphere to evaporate water from a surface. Dry air exerts a larger evaporative demand than moist air.

The **temperature** of a leaf is a function of the amount of solar radiation received by the leaf, the absorbance of the leaf, the size of the leaf, the amount of evaporative cooling occurring through transpiration and wind speed.

The following **basic principles** tend to apply when thinking about interactions among wind speed, solar radiation input, leaf temperature, long-wave emissions by leaves, leaf water status, vapour pressure gradient between leaf and air, leaf size and rates of evapotranspiration from leaves.

- As solar radiation input to a leaf increases, leaf temperature tends to increase until an equilibrium is reached whereby the incoming radiation receipt is balanced by heat loss through advection, conduction, re-radiation of long-wave radiation and transpiration. These four processes cool the leaf.
- As leaf water status declines, stomatal closure increases and hence transpiration declines. This decline in transpiration with reduced stomatal opening is partially offset by the increase in leaf temperature that occurs when transpiration is reduced and evaporative cooling is reduced. As leaf temperature increases, the gradient in water vapour pressure between the inside and outside of the leaf increases, which acts to increase diffusion of water through stomata.
- As wind speed increases, boundary layer thickness decreases and evaporation from wet surfaces and transpiration from leaves increase.
- Leaf temperature increases with leaf size (when all else is constant) because boundary layer thickness increases with leaf size. Conifers, which have needles with very small surface area compared to a broad-leaved tree, have needle temperatures close to air temperature. This is because they are highly coupled to air temperature which means they have a thin boundary layer, a small thermal mass and a larger surface area-to-volume ratio and therefore are better able to 'dump' (transfer) heat to the atmosphere. Broad leaf temperature can often be higher than air temperature because broad leaves are less coupled to the atmosphere.
- As water vapour pressure of the air declines, the evaporative demand of the air increases and transpiration tends to increase because of increased rates of diffusion of water vapour molecules. However, stomata are sensitive to the rate of transpiration and therefore as the air becomes drier, stomatal regulation of transpiration increases and stomata close when atmospheric water vapour pressure is too small.

Water and energy fluxes can be inter-converted

Rates of transpiration can be expressed in many ways. Plant physiologists express **transpiration** as mmol of water transpired per square metre of leaf per second (mmol m^{-2} s^{-1}). One mole of water is 18 g and hence 1 mmol of water is 18 mg of water.

Hydrologists express **water use** by vegetation as mm water per day (mm d^{-1}). A rate of water use of 1 mm d^{-1} for 1 ha of forest corresponds to 10 m^3 ha^{-1} d^{-1} (10 000 m^2 multiplied by 0.001 m = 10 m^3). Since 1 ha is 10 000 m^2 and 1 m^3 is 1000 L, this is equivalent to 10 000 L ha^{-1} d^{-1} which is equivalent to 1 L m^{-2} d^{-1}. One litre (= 1000 g) is 55.5 moles and therefore an evaporation rate of 1 mm d^{-1} per hectare is equivalent to 55.5 moles m^{-2} d^{-1}. Since it takes 2.45 MJ of energy to evaporate 1 L (or 1 kg) of water (at 20°C and 0.1 MPa atmospheric pressure), then 1 mm d^{-1} evaporating from 1 ha of forest requires 2.45 MJ energy input per m^2 d^{-1}, or 24 500 MJ ha^{-1}. Therefore we can express a 1 mm d^{-1} rate of water use by vegetation as equivalent to an energy flux of 2.45 MJ m^{-2} d^{-1}. Daily solar radiation input to most parts of the world for much of the year exceeds this amount. In Australia a typical sunny day receives in the order of 25–35 MJ m^{-2} d^{-1} of solar radiation.

Daily and seasonal patterns of leaf water potential

Knowledge of inputs (rainfall and solar radiation) and evaporative demand allows a meaningful interpretation of plant water relations. This section provides a description and explanation of daily and seasonal patterns of leaf water potential to illustrate this point. Data are taken from studies in the wet and dry seasons of northern Australia as these provide useful and large contrasts for comparison.

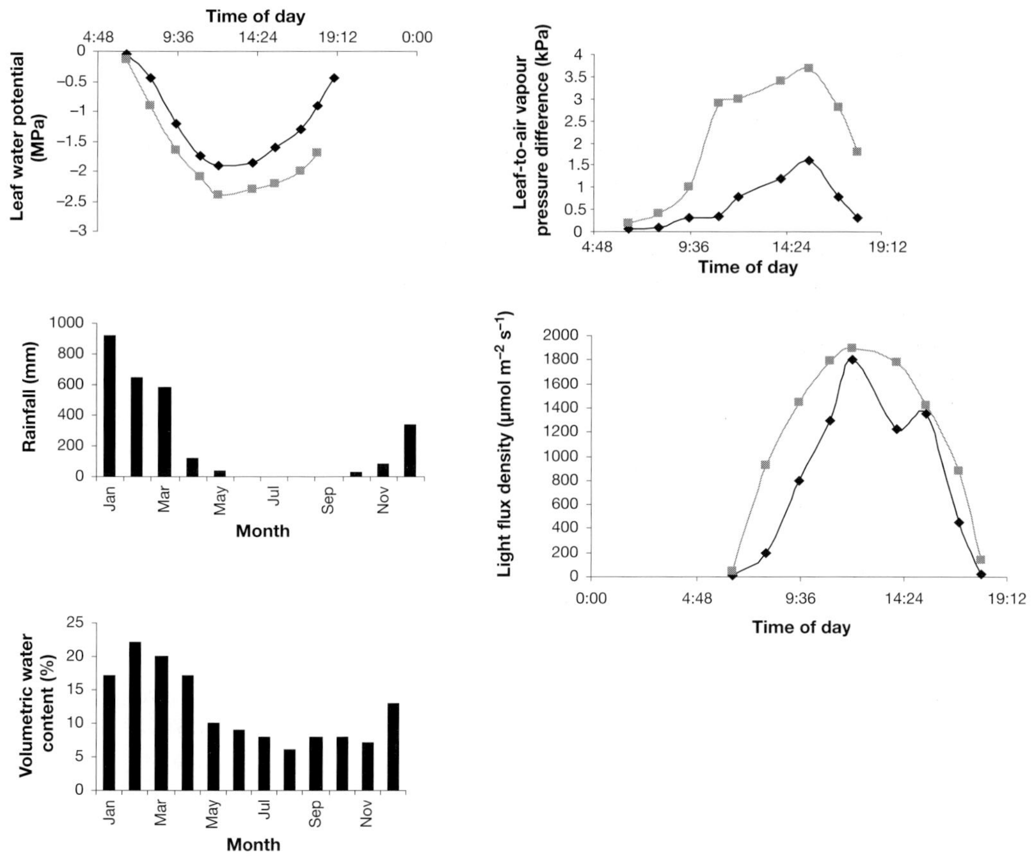

Figure 2.2 Seasonal patterns in rainfall, light flux density, soil moisture and leaf-to-air vapour pressure difference are observed throughout monsoonal Australia. These data pertain to a site 20 km south-east of Darwin. Squares represent winter data and diamonds represent summer data. The reduction in light levels in summer is because of cloud cover

Source: Data reproduced from Duff et al. (1997)

The amount of solar radiation incident on the top of a canopy varies from zero at night to a maximum at solar noon. As solar radiation increases in the morning then decreases in the afternoon we would expect transpiration to follow the same pattern (discussed below). However, because of the resistance to water flow that exists between the soil and leaf, there is a time lag between increasing rates of transpiration and increasing rates of water uptake up by roots. A resistance to water flow exists within the soil (because of the tortuous path of movement around soil particles) and within the roots (because of the presence of membranes and cell walls). There is also a resistance within the xylem (because of the frictional drag of xylem cell walls on water molecules) and within the leaves (because of the presence of membranes and cell walls). Because of these resistances, leaf water potential declines during the first part of the day and can increase during the later part of the day and early part of the night.

A comparison of patterns of leaf water potential in highly seasonal climates can serve to illustrate the interplay between environmental factors (especially solar radiation, soil water content and atmospheric water content) and plant water status. For example, rainfall in northern Australia (Fig. 2.2) or south-western Australia is highly seasonal (Fig. 2.3). Ninety percent of annual rainfall in Darwin occurs in the summer months of November to March inclusive, with no effective rainfall from June to September (effective refers to having a significant impact on

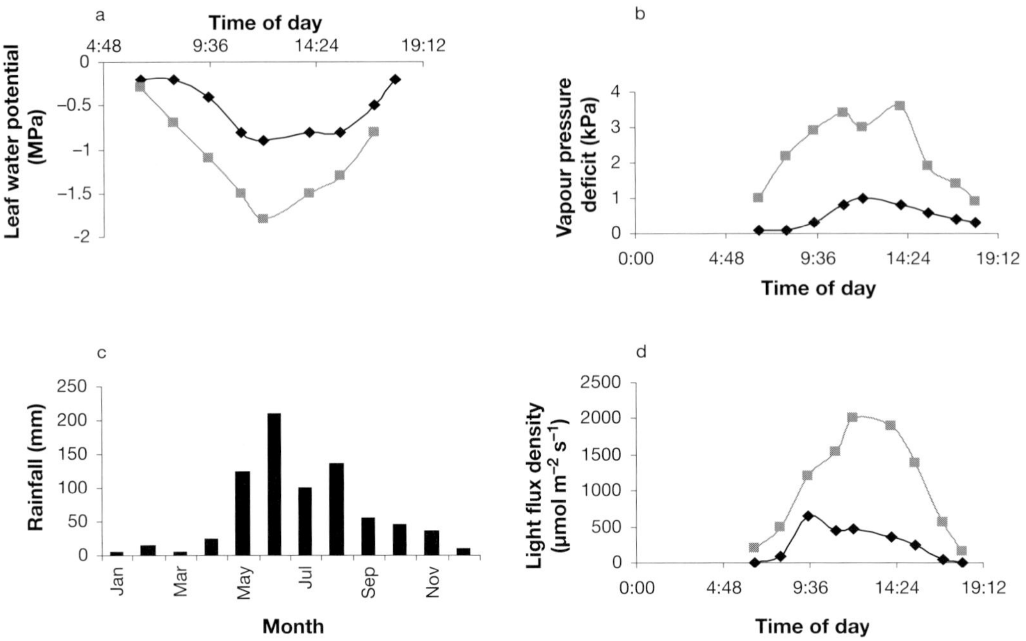

Figure 2.3 Changes in (a) leaf water potential, (b) vapour pressure, (c) rainfall for Perth in Western Australia and (d) light flux density. Squares represent summer data and diamonds represent winter data. Western Australia observes the same pattern as the Northern Territory (Fig. 2.2) except that rainfall is predominantly confined to winter. These graphs shows typical patterns for water potential, rainfall, light flux density and vapour pressure deficit

Source: Dodd & Bell (1993)

soil water stores). The pattern of seasonal rainfall is reflected in seasonal changes in soil water content (Fig. 2.2), with large reductions in soil water content occurring during the dry season. This decline in soil moisture is the result of water use by vegetation (especially trees), lateral flow to low points in the landscape (such as paperbark swamps) and deep percolation to groundwater.

Rainfall is confined to the wet season and light levels in the wet season fluctuate due to cloud cover (Fig. 2.2). Evaporative demand is much larger in the dry season than the wet season. Evaporative demand, when used in relation to vegetation, is best expressed by the leaf-to-air vapour pressure difference; that is, the gradient in water vapour pressure between the inside of the leaf and the ambient air. The larger evaporative demand in the dry season occurs because evaporation from soil and vegetation is less then so the air is drier (see below for discussion of tree water use) and because winds come from the hot dry interior of Australia. In contrast, wet season winds come from the north and are moist oceanic winds. In the wet season, leaf-to-air vapour pressure difference (LAVPD) rarely exceeds 2 kPa while in the dry it can reach almost 4 kPa. The fact that rainfall is confined to the wet season, plus the smaller LAVPD, means that the minimum leaf water potential is higher (closer to zero) in the wet season than in the dry season (Fig. 2.2). Conversely, the daily decline in leaf water potential in the dry season is larger than that in the wet season because the evaporative demand is larger and soil water content is smaller in the dry season than the wet season. As soils dry out progressively during the dry season, a larger gradient in water potential (and hence a lower leaf water potential) is required in the dry season to extract water from a drier soil.

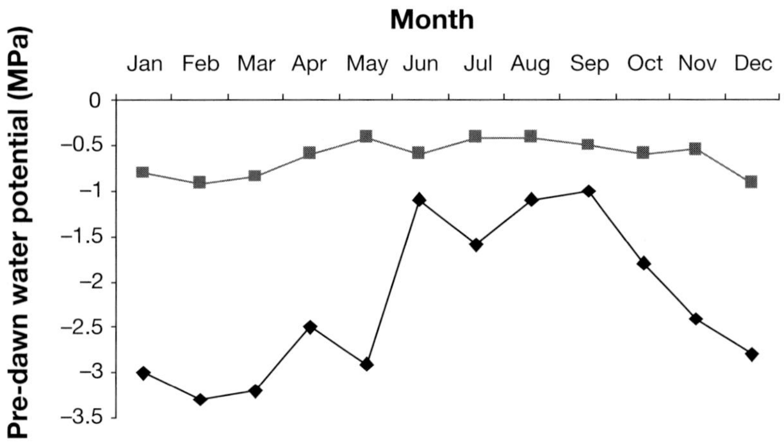

Figure 2.4 Pre-dawn water potentials of two co-occurring species – a shallow rooted short shrub (diamonds) and a deep rooted tall tree (squares) – differ despite receiving the same solar radiation, rainfall and air temperature

Source: Data from Western Australia (Froend pers. comm.)

Similar relationships among rainfall, vapour pressure deficit and leaf water potential can be discerned in Figure 2.3 for data from Western Australia. Rainfall predominantly occurs in winter, LAVPD is larger in summer and consequently minimum leaf water potential is lower in summer than winter.

Pre-dawn leaf water potential

Data on daily and seasonal patterns of water potential are best when they include values for **pre-dawn water potential.** Measurements of leaf water potential before dawn provide an estimate of the water potential of the wettest part of the soil profile that contains a significant amount of roots. It is assumed that leaf water potential equilibrates overnight to this water potential and therefore as the soil profile dries out, pre-dawn water potential will decline in parallel (Prior et al. 1997). Thus, pre-dawn water potential is a useful integrated measure of soil water availability to vegetation. However, differences in pre-dawn water potential are found between species growing at a common site even though they receive the same rainfall, solar radiation and evaporative demand and are growing in the same soil. These differences in pre-dawn potential reflect differences in plant attributes, especially rooting depth, canopy area per tree, stomatal conductance and hydraulic architecture (Fig. 2.4). Such differences in pre-dawn water potential are carried through to differences in daily patterns of water potential.

As pre-dawn water potential declines, the maximum stomatal conductance exhibited by leaves declines in order to reduce water use at a time of declining water availability. Therefore, interpretations of daily and seasonal patterns of tree water use and transpiration are best undertaken when changes in leaf water potential are available. We now consider patterns of tree water use.

Daily and seasonal patterns of tree water use and transpiration

Trees represent a significant pathway for the discharge of water from a catchment. Table 2.2 shows that, for a range of Australian and overseas woodlands and forests, total water use by forests is substantial. Indeed, vegetation water use can discharge almost all the annual rainfall at a site.

Table 2.2 A comparison of rates of daily, seasonal or annual water use by woodlands and forests around the world

Ecosystem/species	Daily, seasonal or annual water use	Location
Ash forest	Annual = 407 mm	UK
Beech forest	Annual = 344–393 mm	Belgium; UK
Oak forest	Annual = 151–340 mm	Denmark; France
Eucalypt savanna	Wet season = 124 mm Dry season = 190 mm Annual = 314	Darwin, North Australia
Paperbark swamp forest	Wet season = 240 mm Dry season = 269 mm Annual = 509 mm	Darwin, North Australia
Wet monsoon forest	Wet season = 183 mm Dry season = 385 mm Annual = 568 mm	Darwin, North Australia
Mountain Ash forest	Summer = c.1.9 mm d^{-1}	Yarra Ranges National Park, Victoria, Australia
Eucalyptus nitens	Summer = c. 2.6 mm d^{-1} unthinned; 1.4 mm d^{-1} thinned to 50%	Tasmania, Australia
E. grandis *E. camaldulensis*	Annual = 325 mm Annual = 303 mm	Victoria, Australia
E. globules	Late spring = 2.2 mm d^{-1} Summer drought = 0.33 mm d^{-1}	Victoria, Australia

Source: Data taken from the literature

Interestingly, across all the range of forest types and locations represented in Table 2.2, the range of tree water use appears to be relatively conservative (excluding periods of known drought), ranging from about 1 to 3 mm d^{-1}. Furthermore, daily tree water use shows consistent patterns around the world and across species. This simply reflects the universal importance of daily patterns in solar radiation and evaporative demand in determining vegetation water use.

Water use increases as soon as the sun rises and solar radiation is available to the canopy to drive transpiration. Increased leaf and air temperatures and increased LAVPD arise from the increase in solar radiation. Conversely, water use declines in the afternoon as the sun passes its zenith and solar radiation input declines, causing a decline in air and leaf temperatures and reduced evaporative demand. Figure 2.5 shows fluctuations in water flux from individual trees (measured with sapflow sensors) as deep but intermittent banks of cloud pass over three sites in northern Australia in the wet season. In contrast, the almost complete absence of cloud at Darwin and Newcastle Waters in the dry season results in tree water use varying little between 0900 h and 1700 h because sufficient radiation is available to drive an almost constant rate of transpiration.

Seasonal variation in maximum and daily total water use occurs because of changes in water availability (soil water content; see below), evaporative demand (LAVPD; see below) and leaf area (see below) and because of differences in solar inclination and hence solar radiation input. Figure 2.5 shows that peak water use per tree declined between the wet (summer) season and the dry (winter) season at all sites of northern Australia, despite LAVPD being much larger in the dry season than the wet season.

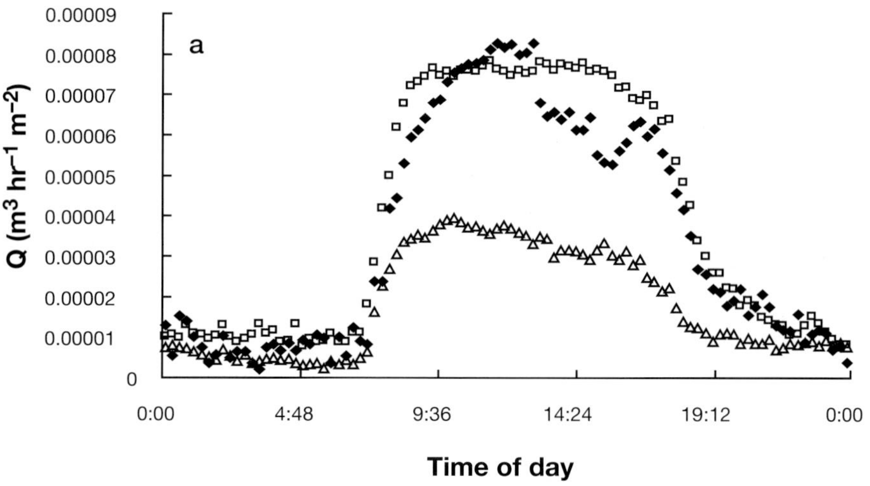

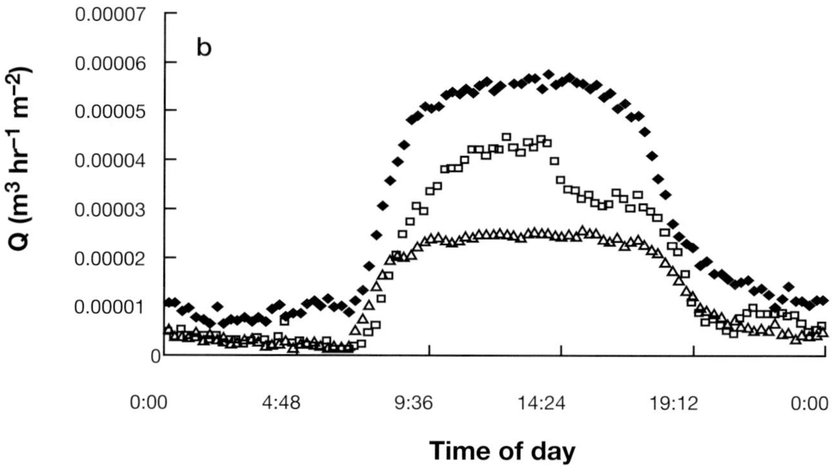

Figure 2.5 Diurnal patterns of transpiration at Darwin (diamonds), Katherine (squares) and Newcastle Waters (triangles). Data represent (a) the mean wet season and (b) dry season transpiration rates for the two dominant species at each site over a five day sampling period

Source: Redrawn from Eamus et al. (2000)

Scaling from tree water use to stand water use

Measurements of tree water use using sapflow sensors provide estimates of water use of individual trees. However, this information is generally less important than information about stand water use because it is stands of trees that influence the water balance of a site. Therefore, a way of scaling up from individual tree estimates to stand estimates of water use is required.

Sapflow sensors measure tree water use by measuring the velocity of water flow up the xylem. This is then multiplied by the area of the sapwood and the water content of the sapwood, to calculate the volume of water moving up the stem per unit time (Chapter 4). The challenge is how to use this single estimate of a volume flux moving up one tree, to provide a larger-scale estimate of stand water use.

Native forests and woodlands consist of trees of different species, different ages and different sizes. Even plantations, which are composed of a single species of a single age, have trees of

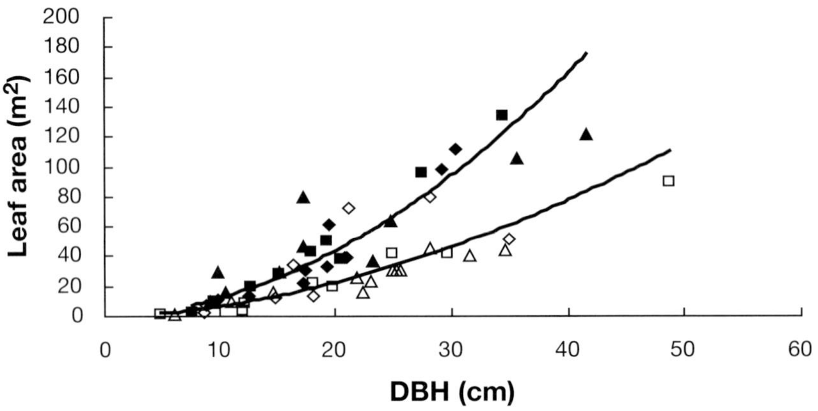

Figure 2.6 The relationship between leaf area on a tree and the DBH of a tree is significant, but varies seasonally in most tree species. This makes leaf area a difficult scaler to use when scaling rates of water use from individual trees to whole stands of trees. Different symbols represent three Northern Territory sites with differing rainfall. Open symbols represent dry season data, closed symbols wet season data

Source: Data from Eamus et al. (2000)

different size. Since it is not possible to put a sapflow sensor into every tree, a subsample of trees is measured and that estimate of tree water use is scaled up to calculate stand water use. Three measures are frequently used to scale water use from individual tree measures to estimates of stand water use: leaf area, sapwood area and diameter at breast height (DBH, usually 1.3 m above ground).

As trees grow from seedling to maturity, the total leaf area per tree, DBH and cross-sectional area of sapwood per tree increase. Thus, to take into account variation in tree size, water use from a range of tree sizes is measured and regressions of tree water use against leaf area, DBH and sapwood area can be plotted. An example of the relationship of DBH against leaf area for five species of trees growing in three sites in a savanna of the Northern Territory is shown in Figure 2.6. This plot highlights the problem of using leaf area as a scalar, namely that **leaf area varies seasonally**. Winter (i.e. low temperature induced) deciduous trees, drought deciduous trees and even forests of evergreen trees show seasonal maxima and minima of leaf area. Thus, in tropical Australia, average leaf area per tree declines in the dry season compared to the wet season, even for evergreen Eucalypts. Interestingly, in this study (Eamus et al. 2000) all five species appeared to have the same relationship between DBH and leaf area, which makes scaling much easier to accomplish as all trees can be treated identically, irrespective of species.

Sapwood area increases with increasing DBH and neither sapwood area nor DBH show seasonal fluctuations. However, estimates of sapwood area require drilling and coring of trees, while DBH is relatively easily measured either with a DBH tape or a Bittlich gauge. Sapwood-to-DBH ratios are highly site specific (Fig. 2.7) and therefore require calculation at all sites. Within a site, however, a strong correlation between sapwood area and DBH means that either can be used to upscale from water use by individual trees to water use by stands. DBH, being easier to measure in the field, is usually used. Strong correlations between water use and DBH are observed in all species studied, but the relationship varies with site and season and sometimes with species. Therefore **scaling** to water use by a stand from measurements on individual trees (replicated across a range of tree sizes and including all the major spcies within the site) requires a regression of DBH against water use for each season and each species.

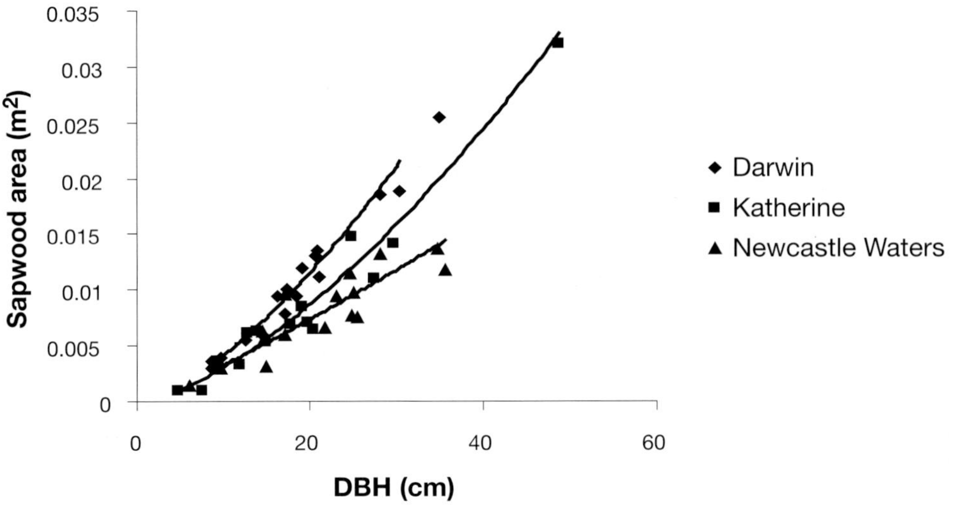

Figure 2.7 The relationship between DBH and sapwood area at three locations along the North Australian Tropical Transect at Darwin (diamonds), Katherine (squares) and Newcastle Waters (triangles). The relationship is site specific

Source: Redrawn from Eamus et al. (2000)

Factors determining transpiration rate, tree water use and stand water use

Tree water use (litres of water used per tree per day) and the related measure, transpiration rate, (mmoles of water transpired per square metre of leaf per second), vary over time-scales of tens of minutes to days to seasonal and decadal time-scales. Understanding what factors cause these fluctuations is important if we wish to interpret, and eventually manage, water use by vegetation at large scales (see Chapter 7).

Transpiration is under direct stomatal control. **Stomata respond to a large number of control factors**, but the most important from the perspective of tree water use are soil water content, atmospheric water content and solar radiation input. As soil water content declines (as revealed through the decline in pre-dawn water potential), stomatal conductance (G_s) declines, as shown in Figure 2.8a. Similarly, as atmospheric water content declines (as revealed by an increase in LAVPD), G_s declines (Fig. 2.8b). Although it is LAVPD that is manipulated experimentally, or vapour pressure deficit that is measured in the field and correlated with G_s, it is actually transpiration rate that regulates G_s, and not LAVPD per se. Soil and atmospheric water content can interact to influence G_s additively or synergistically. Solar radiation is usually not limiting to G_s except in the very early morning or very late evening.

The **principal mechanisms** by which stomata are controlled in response to reduced soil and atmospheric water content are outlined schematically in Figure 2.9. They involve the plant hormone abscisic acid, xylem sap pH, cellular water relations and xylem embolism. Stomatal responses to changes in atmospheric water content can occur within 10–20 minutes, while changes in soil water content around roots occur over a period of hours, days and weeks and therefore stomatal aperture adjusts over that time frame. Changes in stomatal conductance represent the first, the most reversible and the most finely tuned response of leaves (and hence tree water use) to changes in atmospheric and soil water content.

While it is true that G_s declines with increasing LAVPD, this is not necessarily indicative of a decline in tree water use because the decline in G_s is often insufficient to prevent an increase in transpiration rate and hence tree water use. Medhurst et al. (2002) show how tree water use for *Eucalyptus nitens* **increased over the initial increase in VPD** in a Tasmanian study.

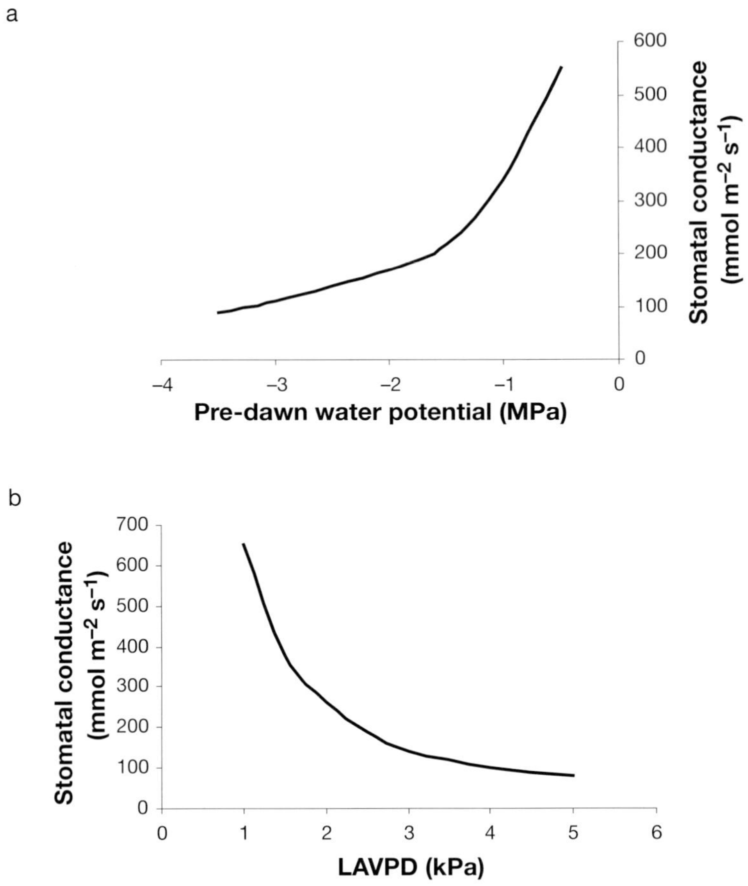

Figure 2.8 As (a) pre-dawn leaf water potential decreases or (b) LAVPD increases (indicating an increase in dryness of either the soil or atmosphere, respectively), stomatal conductance declines, linearly or, as in this example, curvilinearly

Source: Thomas & Eamus (1999); Prior et al. (1997b)

(Fig. 2.10). However, as VPD increased further, stand water use became regulated by stomata such that **water use became independent of VPD**.

Over the medium term (weeks and months), tree water use is regulated through changes in leaf area. Thus, seasonal changes in leaf area index (LAI; the ratio of total leaf area to ground area) are routinely observed in Australian ecosystems such that LAI declines in the dry season compared to the wet season.

Over the long term (years), **catchment water balance reflects tree height, canopy area per tree and tree density**, which in turn determine stand water use. Arid sites without access to groundwater have fewer trees per hectare (lower tree density) and the LAI is reduced (less leaf per unit ground area) compared to moist sites.

Table 2.3 compares stand water use for the wet and dry season at three locations along the North Australian Tropical Transect in northern Australia. This is a gradient of rainfall, ranging from 1730 mm of rain at Darwin to 520 mm rainfall at Newcastle Waters. The decline in water availability causes a decline in tree basal area, tree height (and hence canopy area per tree) and LAI, causing the observed decline in scaled estimates of tree water use. Large stores of water in the soil profile and a low rate of water use per tree allow trees in this seasonal environment to maintain significant water use in the dry season compared to the wet season.

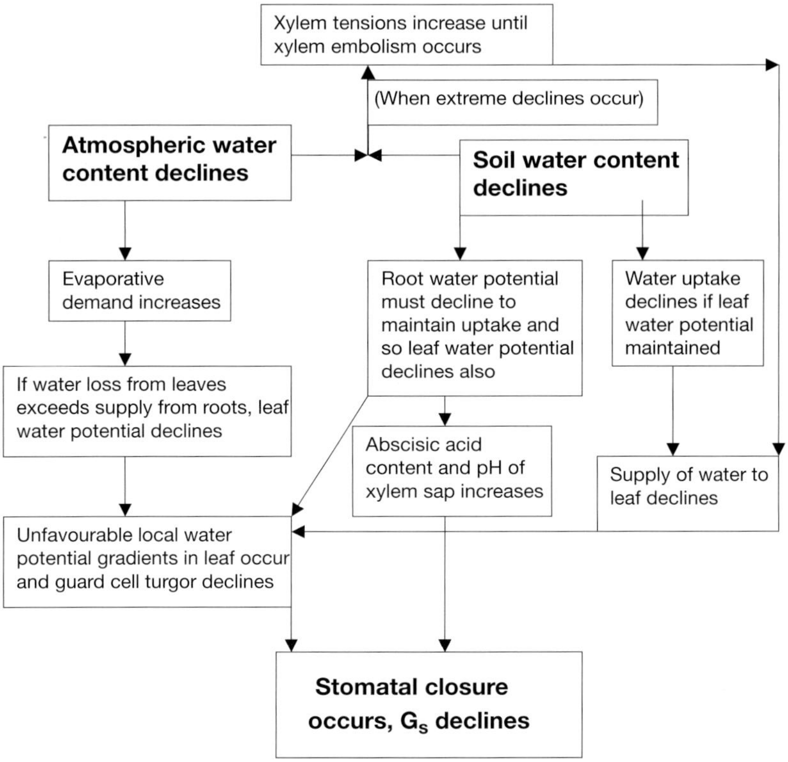

Figure 2.9 Schematic of the mechanisms linking reduced soil or atmospheric water content with reduced stomatal aperture and conductance (G_s)

Any consideration of the water relations of plants and vegetation must discuss water in soil since soil is the principal source of water used by vegetation. Although small amounts of water can be taken up from mist and fog in some species in some locations, and epiphytic (e.g. some orchids, bromeliads and ferns growing in tree canopies) and parasitic plants (e.g. mistletoes growing on tree branches) do not directly tap into soil water, the vast majority of plants extract water principally from soil. The following section discusses this topic.

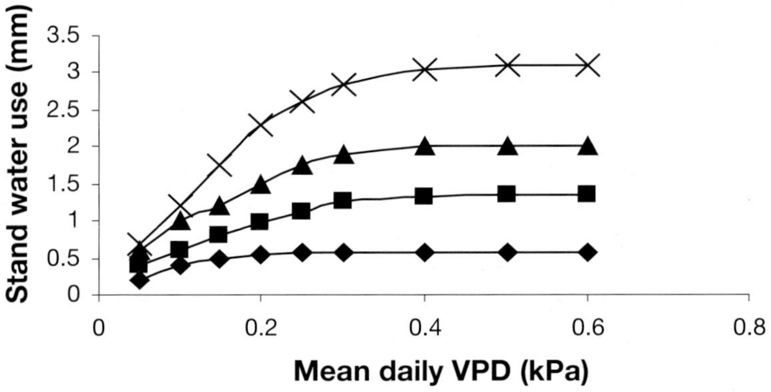

Figure 2.10 As mean daily vapour pressure deficit increased above four stands (crosses, triangles, squares and diamonds represent 1260, 600, 250 and 100 trees ha^{-1} respectively) of *E. nitens* in Tasmania, stand water use increased substantially but reached a plateau

Source: Redrawn from Medhurst et al. (2002)

Table 2.3 Relationship between rainfall, basal area, tree height, LAI and tree water use

Site	Rainfall (mm)	Basal area (m^2 ha^{-1})	Tree height (m)	LAI	Tree water use (mm d^{-1})
Darwin		10.0	15		
Wet season	1640			1.0	1.0
Dry season	90			0.7	0.9
Katherine		7.5	11		
Wet season	800			0.8	0.25
Dry season	70			0.4	0.20
Newcastle Waters		4.8	8		
Wet season	500			0.07	0.08
Dry season	20			0.06	0.06

Source: Data from Hutley et al. (2001)

Water in soils

Water in soils usually occurs as a dilute solution of salts, although in saline soils the concentration of the salt solution can be considerable (see Chapter 8). The presence of solutes in soil water causes the water potential of soil to be less than zero (because of the reduction in solute potential of the water). In addition, **matric** forces exist, resulting from the local interaction of water molecules with the surface of soil particles. Capillary (surface tension effects) and adsorptive forces between water and soil particles generate matric potentials (Ψ_m).

In saturated soils and groundwater, a positive pressure also exists due to the action of gravity pulling water downwards. Gravity also acts on water in unsaturated soils, of course, but a positive pressure is not generated in unsaturated soils because the water is too dispersed. Put another way, there is too much air in unsaturated soil for a positive water pressure to be developed. Soil water pressure head is therefore positive below the water table (i.e. within the saturated zone) and negative above it. In the unsaturated zone the negative pressure is also called the matric suction or tension. Matric potential is zero in saturated soil and becomes more negative as the soil water content declines because more and more work is required to extract water from drier and drier soils.

Thus, water potential of unsaturated soil is determined by the presence of solutes and matric forces, and is given by:

$$\Psi_w = \Psi_s + \Psi_m \tag{8}$$

Water, solute and matric potentials are negative. In saturated soil a pressure potential of soil water (which is absent in non-saturated soils) can be added, and this is positive.

Different soils exert different water potentials at the same soil moisture content (g m^{-3}) because of differences in particle size distribution. Sandy soils cannot hold a large volume of water because their large particle size allows rapid drainage of water due to gravity. In contrast, clay soils can hold a large volume of soil but their water potential declines quickly with small drops in water content. This effect is shown in Table 2.4. Sandy soils, with large particles, contain a much smaller volume of water than clay soils in both wet (water potential close to zero) and drier (water potential –1.5 MPa) soils. For a Handford sand, a decline in water content of only 3% causes a large drop in soil water potential, while for a Yolo clay soil the same drop in water potential requires an almost 20% decline in water content.

As soils lose water, their water potential declines and so does their **hydraulic conductivity**. Hydraulic conductivity is a measure of how easily water can move through a soil (or tree

Table 2.4 Relationship between soil texture and water holding capacity

Soil type (in decreasing order of particle size)	Water content as % of dry weight of soil	
	−0.03 MPa	−1.5 MPa
Handford sand	5	2
Yolo loam	13	7
Chino silty clay	41	22
Yolo clay	45	26

Source: Kramer (1983)

branch; see below). When soils are saturated, the (generally) downward movement of water is driven by gravity and occurs in small, medium and large pores. Large pores transmit large amounts of water at faster rates than small pores. Water movement is the result of differences in pressure (arising from the action of gravity) between upper and lower soil profiles (just as water pressure increases with depth below the sea surface). Hydraulic conductivity is large in wet soils because the largest pores have water in them, which can move freely.

As soils dry out, water content declines and water is lost first from the largest pores and then progressively more from medium sized pores; finally, the smallest pores dry out. During drying, soil water potential and hydraulic conductivity both decline (Fig. 2.11). Such unsaturated flow is influenced less by pressure differences (induced by gravity pulling water down) and more by matric forces. Matric forces tend to inhibit water movement due to the adsorption of water onto the surface of soil particles – hydraulic conductivity declines as the soil dries, making the movement of water through soils more difficult. Indeed, the flow of water to roots in dry soils can become limiting to plant water uptake (see below). Pressure gradients are minimal in unsaturated soils because of the presence of interconnected air spaces that allow equalisation of pressure throughout the soil profile. Solute potential gradients are also of minimal importance in driving water flow through unsaturated soils except in heavily salinised soils.

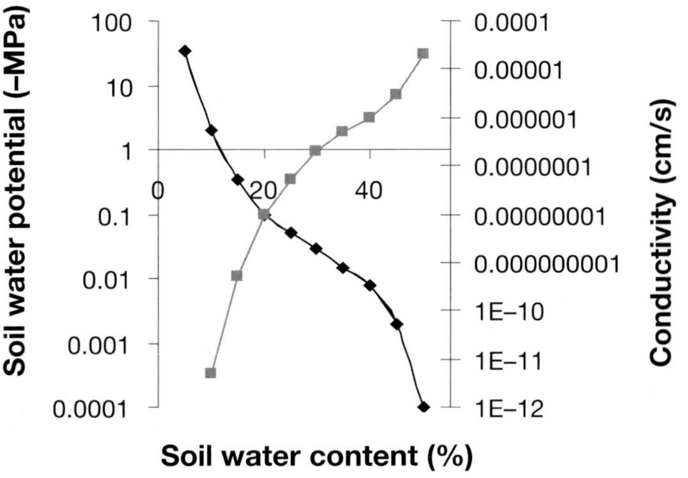

Figure 2.11 Soil water potential (diamonds; note log scale and negative value indicated in the units) is high (close to zero) when soils are wet and declines (gets more negative) as soils dry. Similarly, the hydraulic conductivity (squares; note log scale) is high in wet soils and declines as soils become dry

Source: Redrawn from Kramer (1983)

Hydraulic lift

Hydraulic lift (hydraulic redistribution) is the process whereby water from moist or wet layers of soil (usually deep in the soil profile) is moved into drier layers of soil closer to the surface via the roots of plants (Caldwell et al. 1998). This is a passive process and the movement of water is driven by differences in water potential between wet soil, roots and dry soil. About 30 species of Australian plants have been shown to exhibit hydraulic lift, including herbs, grasses, shrubs and trees. Hydraulic lift usually occurs at night when transpiration is minimal. The water released into the upper soil profile at night is then available for use (re-uptake) during the day. Re-uptake is favoured by the individual plant that released the water during the night, because of the close proximity of its roots. However, neighbouring plants have been shown to extract water lifted by another plant.

Hydraulic lift has three major benefits. First, it results in an increase in the volume of water available to plants during the day, because the water is moved from deep zones where root density is generally small, to upper layers where root density is larger. Second, hydration of the upper soil profile increases the availability of nutrients, which are taken up by roots in solution as part of the bulk flow of water to roots arising from transpiration. Finally, rehydration of the upper soil may aid in maintaining the viability of fine roots in the dry upper soil profile.

Although usually regarded as involving upward flow of water from deep to shallow soil layers, hydraulic 'lift' can occur laterally. The distinguishing feature of hydraulic 'lift' (i.e. upward or lateral flow) is that water is flowing in the reverse direction to that associated with transpiration. Thus water moves away from the root of the plant and out into the soil. Mark Adams and his colleagues at the University of Western Australia demonstrated hydraulic lift in *Grevillea robusta* and *Eucalyptus camaldulensis*. In addition, they showed that following the first wet season rains, when the surface soil layers are wetter than deeper layers, **reverse hydraulic lift** occurred and water moved from the upper profile into the deeper profile. Thus, **hydraulic redistribution** is a better description than hydraulic lift (Burgess et al. 1998). The movement of water from moist upper to drier deeper layers may also facilitate the initial growth of roots through dry deep layers in search of deeper wetter layers of soil (e.g. soil in the capillary fringe above groundwater).

Hydraulic redistribution has been demonstrated by three major techniques. First, measurements around roots have shown an increase in soil water content at night, especially in arid and semi-arid environments. Soil moisture content can be measured directly (gravimetrically) or by using one of a number of soil moisture probes (time domain reflectometry probes or soil psychrometers, see Chapter 4). Second, isotopes of water (usually deuterated water) can be injected into soil at depth, and if deuterated water appears in the xylem water of shallow rooted grasses around deep rooted trees redistribution has occurred. Finally, reverse flow in roots has been shown using sapflow sensors inserted into roots.

The volume of water made available by hydraulic redistribution in one night is now known to be a significant proportion of water transpired in the following day – some 14–30%. In the absence of hydraulic redistribution, when the upper soil profile (where there are more roots) is dry and the deeper profile (where there are fewer roots) is wet, the amount of root material at depth can limit the amount of water available for transpiration. Thus, the redistribution of water at night from a region with low root density to a region with high root density has an evolutionary advantage.

Determining the sources of water transpired by plants

Vegetation (especially trees because they are deep rooted) can use three sources of water. First, it may use recently arrived rain which is stored in the upper soil profile. Second, they may use stream or river water, where available. Third, they may use, where available, groundwater and

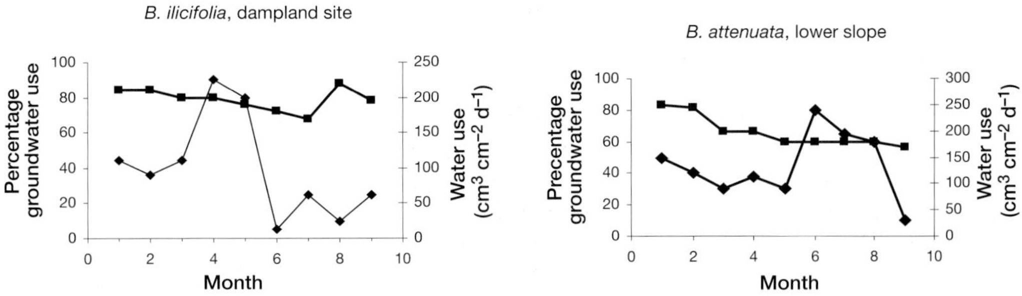

Figure 2.12 The percentage of groundwater used by vegetation is different at different times of year and in different species, and differs between location in the landscape. Closed circles are total water use and open circles are % groundwater use

Source: Data from Zencich et al. (2002)

the associated **capillary fringe** (the zone of soil that is wet due to the movement of water upwards, by capillary action, from the saturated groundwater zone). In some circumstances it is possible to determine which sources of water are being used by vegetation at different times by comparing the isotopic composition of water in xylem sap of plants to the isotopic composition of water in the upper soil profile, groundwater and stream water. The composition of the xylem water will most closely match the composition of the water source being most used by the plant. This method works only where differences in isotopic composition between soil water, groundwater and stream water are significant. The isotopes used in these studies are the stable isotopes of oxygen (^{18}O) and hydrogen (deuterium ^{2}H). For a detailed description of the methods, see Chapter 4. Stable isotope studies are invaluable in ascertaining whether vegetation is accessing groundwater. By comparing the extent of groundwater use seasonally, it is possible to determine the degree to which an ecosystem is **groundwater dependent**. Two examples illustrate the value of stable isotope studies in this area.

The Swan Coastal Plain of Western Australia is located in a region with a distinct Mediterranean climate (wet cool winter, hot dry summer). A transient unconfined aquifer underlies the Plain and provides a significant water resource for deep rooted *Banksia* woodland growing on the freely draining sandy (thus having low water storage capacity) soil. Deep rooted *B. prionotes*, *B. attenuata* and *B. ilicifolia* are **phreatophytes** (plants using groundwater) while the shallow rooted perennial shrub *Hibbertia hypercioides* does not use groundwater (Zencich et al. 2002). By comparing the $\delta^{2}H$ of groundwater and twig water, Zencich et al. (2002) showed that the *Banksia* species are facultative users of groundwater, with the proportion of groundwater use increasing during the dry summer and declining when winter rains recharge the upper soil profile (Fig. 2.12). For example, in 'dampland' locations with shallow depths to groundwater (and hence shallow unsaturated soil depths), almost 100% of the water transpired by *B. ilicifolia* was derived from groundwater in the summer dry season, but this declined to about 5% in the winter wet season. Therefore, this species is likely to be highly sensitive to anthropogenic extraction of groundwater and early tree death has been identified as an early indicator of negative impacts of groundwater extraction. In lower slope locations, where groundwater depth is larger than that at dampland locations, *B. attenuata* trees used more groundwater in autumn, compared with *B. ilicifolia* (Fig. 2.12) and maintained a more favourable water potential during summer. This difference probably reflects differences in rooting depth of the two species.

The riverbanks and many creeks of the floodplains of the River Murray – Australia's longest river – are dominated by *Eucalyptus camaldulensis* (River Red Gum). River red gums are tolerant of the repeated flooding to which they are subject, and indeed the flooding is necessary for seed germination and establishment. Thorburn and Walker (1994) examined the isotope composition of stream water, soil water and groundwater and xylem sap of trees growing close to a stream and 40 m away from a stream. A tree located 40 m from the stream used water from shallow (c. 0.2 m) soil and from groundwater. The proportion of groundwater use varied from 40% to more than 60% during the winter study months and the proportion increased as the upper soil profile dried. Creekside trees did not rely on creek water as a source and used shallow soil water at all times. Because the isotopic signature of the creek water and groundwater were similar, it was not possible to establish whether these trees were using one or the other source preferentially. However, Thorburn and Walker (1994) concluded that the creekside trees used soil water and groundwater only, despite the creek flooding the surface of the bank during the study period. They concluded that groundwater represented 0–60% of the total water used by creekside trees.

Having discussed the water relations of plants, and the evaporation of water from leaves and uptake of water by roots, it is important to consider how the plant's structure influences its behaviour and the movement of water within the plant. If trees can be considered as wicks linking the wet soil with the dry atmosphere, then the structure of the wick itself can be considered important in relation to its function. The following section explores this topic in more detail.

Hydraulic architecture

Hydraulic architecture describes the relationships among leaf area, sapwood area, hydraulic conductance (inverse of resistance) and conductivity of sapwood and xylem embolism. Thus it is a description of the structure and functioning of the water conducting pathway. A knowledge of hydraulic architecture allows prediction of the rates of water movement through various parts of a tree, the water potential gradients experienced by different parts of the tree under different climatic conditions and the vulnerability of the tree to xylem embolism. In addition, differences in hydraulic architecture are implicated in determining the limits to tree height, the rate of photosynthesis of a canopy and other aspects of the ecology of different groups of plants (e.g. gap colonisers vs late succession species).

Large amounts of water flow from the soil to the atmosphere through plants. The hydraulic architecture of plants determines the resistance to water flow in plants and also influences plants' responses to changes in environmental conditions. Changes in soil and atmospheric water content and solar radiation input occur during the day and also vary seasonally and between sites. Understanding how hydraulic architecture influences tree responses to these variables is central to successful use of vegetation in landscape-scale manipulations of catchment water balance.

Conductance, conductivity and Huber values

Fundamental variables that need measuring when assessing the hydraulic architecture of plants are the hydraulic conductance of a length of stem or branch, the hydraulic conductivity of the same, and the Huber value (see below) of the same. By knowing the hydraulic conductance/conductivity of the conducting path from root to leaf, it is possible to infer how water potentials will vary with time as a function of tree water use, or vice versa. An understanding of the Huber value provides insight to the relative allocations of plant carbon to light capture (through leaf area) and water conduction (through investment in sapwood).

The **hydraulic conductance** (Lp) of an excised length of tree stem or branch is defined as the rate of flow of water per unit pressure difference between the two ends of the segment of stem or branch. Thus, using a technique where water is sucked through an excised sample of stem or branch at different pressures, Lp can be calculated from the slope of a plot of flow rate (g s^{-1} or cm^3 s^{-1}) against pressure difference between the two ends of the stem or branch (kPa). In addition, Lp of intact trees in the field can be estimated by measuring the rate of a tree's water use (Q) using sapflow sensors (see Chapter 4) and the difference in water potential between root surface and leaf. Thus:

$$Lp = Q/(\Psi_r - \Psi_l) \tag{9}$$

where Ψ_r is root water potential and Ψ_l is leaf water potential.

In most cases root water potential is not measured as it is assumed to be adequately represented by the observed value of pre-dawn water potential and is assumed to remain constant throughout the day. This is probably an incorrect assumption in anything but very conductive wet soils but for comparative purposes, for example comparing species at a site, this method still provides some information of value.

Hydraulic conductivity (K_h) of an excised stem or branch is defined as the flow rate per pressure gradient ($\Delta\Psi_w/l$) where $\Delta\Psi_w$ is the difference in water potential between the two ends of the excised stem or branch and l is the length of the segment. Using conductivity allows us to remove the effect of path length as a determinant of water flow through the xylem and is preferred over conductance. Thus:

$$K_h = (Q/\Delta\Psi_W) \times l \tag{10}$$

Both Lp and are K_h best **scaled** using sapwood cross-sectional area or leaf area. This removes variation arising purely from differences in size of branches. To scale K_h to sapwood area, divide it by the sapwood cross-sectional area of the branch to calculate sapwood specific conductivity (K_s) ('specific' should strictly be used to express a value per unit mass). To scale to leaf area, divide K_h by the total leaf area attached distal to the point of excision to calculate leaf specific conductivity (K_l). Scale Lp using the same process.

According to **Poiseuille's law**, the conductivity of a pipe is proportional to the fourth power of its radius. Thus, while the K_h of wood is determined primarily by the number and the distribution of size classes of xylem conduits, it is also influenced by the size and number of the pores that link adjacent conduits (interconduit **pit pores**). The size of conduits and pit pores are important factors determining both the hydraulic conductivity/conductance of branches and stems and the sensitivity of branches and stems to embolism.

A final measure of hydraulic architecture is the **Huber value** (HV), which is defined as the ratio of sapwood cross-sectional area to leaf area distal to the point of excision. Clearly, HV = K_l/K_s. The Huber value is a measure of the investment of carbon in xylem tissue per unit leaf area supplied by that xylem. Measurements of HV should occur when leaf area is maximal. When comparing the behaviour of the water relations of different species at a common site, knowledge of HV provides significant interpretative insight to the different strategies employed by different species to survive in a given climate.

Hydraulic architecture of roots and shoots

Flow of water from soil to the atmosphere via a tree requires passage through several compartments. Water must traverse the soil-to-root interface, then move across the root (cortex, endodermis and stele) to the root xylem, then up the xylem of a stem to a branch and then through the branch xylem to a leaf and finally through the stomata of a leaf into the atmosphere. Of the total resistance to flow from wet soil to leaf, about half is located below ground. In moist or wet soils, the soil–root interface resistance tends not to be large and resistance to flow occurs principally in crossing the root. However, as the soil dries the soil–root interface resistance increases significantly because the hydraulic conductance of the soil is reduced. If this increase in resistance occurs during the day when evaporative demand is high and soil water content is moderate or low, it will cause a reduction in stomatal conductance (G_s). Stomatal conductance is a measure of how open the stomatal pores are on a leaf. A high stomatal conductance means the stomata are wide open and the flux of CO_2 into the leaf and H_2O out of the leaf will be large (assuming a warm sunny day). Conversely, low values of G_s mean small apertures and low values of CO_2 and H_2O flux, even on warm sunny days. A reduction in G_s as soils become drier occurs such that the rate of transpiration balances the rate of water uptake by roots. By maintaining this balance, excessive reductions in leaf water potential are avoided.

Most of the above-ground resistance to water flow occurs in minor branches, petioles and leaves. The K_l of minor branches is 0.1–0.001 of the value of major branches.

Xylem vessel size and K_l decrease from base to apex in most tree species, except where strong apical dominance is expressed, where K_l is relatively constant with height. The decline in K_l with height ensures that all parts of the canopy can receive water at the same rate (assuming the same water potential of all leaves in all parts of the canopy).

Ecological relevance of hydraulic architecture

Information about the hydraulic architecture of plants has been used in several debates concerning the behaviour, distribution and performance of different groups of plants. This section provides some examples of the application of information on hydraulic architecture in ecological studies.

Angiosperms may generally have a more efficient xylem transport system than gymnosperms. The **slow seedling hypothesis** proposes that, due to this difference in transport efficiency, seedlings of faster growing angiosperms will outcompete seedlings of slower growing gymnosperms in sites where water, temperature and nutrient supply are not limiting. Consequently, it is hypothesised, gymnosperms tend to be confined to sites where these factors are limiting and where the competitive advantage of the angiosperms is reduced.

The narrow, closed-end vessels of gymnosperms are generally less conductive to water than the larger diameter vessels of angiosperms and branches of gymnosperms tend to have sapwood of lower permeability than angiosperms. However, recent tests of the slow seedling hypothesis (Becker et al. 1999) in nine angiosperm and three conifer species were unable to find significant differences in hydraulic conductance of the two groups. Importantly, it was shown that while branch-scale, leaf-area-normalised conductance of angiosperms was larger than that of gymnosperms, at the whole tree scale there were no differences. Therefore the lower conductance of branches of gymnosperms is compensated for by alterations in other components of hydraulic architecture, especially sapwood–leaf area ratio.

A second example of the application of knowledge of hydraulic architecture to ecological studies derives from the observation that **whole plant conductance is strongly correlated with maximum transpiration** rate of a species and this is probably correlated with succession status and/or climate. Early colonisers, which exhibit high growth rates, tend to have larger conductance than late succession species. Consequently, early colonisers experience higher water potentials at a given transpiration rate (see equation 9) and this characteristic results in higher growth rates by maintaining higher leaf turgor. Similarly, gap specialists (i.e. trees that show rapid growth and take advantage of canopy gaps that periodically appear in closed forests) have significantly larger values of K_l than, for example, shade adapted species. This larger K_l is solely attributed to a high HV – a large investment in sapwood for a given investment in leaf area. Such a large investment in sapwood is required to supply the large transpiration rates that occur as an unavoidable consequence of the large stomatal conductances (and hence high rates of CO_2 influx) required to support the high rates of growth of gap specialists.

Table 2.5 Average values for Huber value and sapwood normalised hydraulic conductivity (K_s) of trees from several climates

	Huber value × 10^{-4} (m^2 sapwood/m^2 leaf area)	K_s (kg s^{-1} m^{-1} MPa^{-1})
Temperate angiosperms	2.6	4.1
Tropical vines and lianas	0.19	64.5
Tropical rainforest trees	4.8	10.9
Seasonally dry forests	1.7	1.4

Source: Data are the means of several species taken from the published literature (Eamus & Prior 2001)

A third example of the application of hydraulic architecture to ecological studies concerns the comparative ecology of tropical and temperate species and different life forms. Trees growing in tropical sites experience larger maximum evaporative demands than trees growing in temperate sites and, where water is readily available, exhibit larger transpiration rates. Given the fact that very low xylem or leaf water potentials must be avoided (to avoid xylem embolism and stomatal closure), high transpiration rates must be supported by large hydraulic conductance of the xylem. Eamus and Prior (2001) showed that this appears to be the case (Table 2.5).

Deciduous species tend to have larger K_s values than evergreen species. This is consistent with the observation that plants growing on wet sites or plants active only in wet seasons (drought deciduous species) tend to have larger xylem conduits. By restricting water use to periods when water availability is largest, the threat of xylem embolism is low and therefore large diameter conduits can be supported. From Poiseuille's law, small increases in diameter cause large increases in conductivity.

Vines and lianas generally have the largest conductivity, followed by tropical shrubs and rainforest species. Lianas have high K_s because they have large xylem vessels. Vines and lianas do not invest much in structural wood because they are supported by the tree they are growing up. Trees with low rates of water use (e.g. savanna trees and conifers) have the lowest conductivities. Presumably this is because trees with low rates of water use occur where water availability is limiting, at least for some part of the year. When water availability is limiting, xylem embolism is a recurrent threat. **To avoid uncontrolled embolism**, xylem vessel diameters and pit pores need to be narrow because small diameter vessels are more resistant to embolism than large diameter vessels. Narrow pores prevent the movement of gas bubbles between adjacent vessels. It has been argued that the low and constant rate of transpiration of north Australian savanna trees is because of the need to avoid runaway embolism in the dry season (Eamus et al. 2000).

Hydraulic architecture and tree height

All trees eventually stop growing taller. Trees on a mesic site are usually taller than trees of the same species on xeric sites, and use more water. It has been proposed that both observations can be explained by differences in hydraulic architecture of tall and short trees. This is the **hydraulic limitation to tree height** theory (Ryan & Yoder 1997).

Ryan and Yoder (1997) proposed that taller (and hence older, if comparing trees of the same species at a single site) trees have a lower whole tree hydraulic conductance than shorter (and younger) trees of the same species at the same site. This does not mean that taller trees always have a lower conductance than shorter trees if comparing different sites or different species.

The total resistance to water flow increases (conductance declines) as the length of the path over which the water has to flow increases. In addition, gravity exerts a drag on water moving vertically up trees (0.01 MPa m^{-1}). To compensate for this, as trees grow taller xylem is produced with increasing permeability. Thus, as the diameter of a tree increases, for a constant tree height the conductance of the stem increases. However, for a constant tree diameter (and hence a constant sapwood cross-sectional area), as trees get taller stem conductance declines because of increased path length. Thus, the increase in permeability and the increase in sapwood area observed in taller trees are not sufficient to totally compensate for the decline in whole tree conductance.

The decline in total tree hydraulic conductance as trees get taller explains why tall trees have a lower stomatal conductance than young trees and why old trees have a lower rate of transpiration and photosynthesis. It can be shown that:

$$G_s = (K_s A_s)(\Delta\Psi_W)/DA_l l \tag{11}$$

Therefore, since K_s declines with age (height) of the tree and assuming the evaporative demand (D) and gradient in water potential is constant, G_s must decline, as is commonly observed. This decline in G_s also explains why the **net primary productivity** (total C fixed by photosynthesis minus rates of respiratory loss of C by vegetation) of old forests is smaller than that of young forests. Recent work showed that both stomatal conductance and assimilation rate are linearly related to hydraulic conductance in a large number of angiosperms and conifers and that these are lower in older than in younger trees. Similarly, a linear relationship between average leaf area specific conductivity of stems (k_l) and mean quantum yield was observed (in 7 conifers and 16 angiosperms) (Fig. 2.13; Hubbard et al. 1999; Brodribb & Field 2000).

The same reasoning can be applied to water limited ecosystems. It is possible that conditions in the dry season of northern Australia (high evaporative demand, low soil water availability) determine the hydraulic architecture of trees so that the rate of water use in the wet season (lower evaporative demand and higher soil water availability), is limited by the hydraulic architecture and not by the availability of water. This would explain how water use per tree remains constant in northern Australia, despite large differences in soil and atmospheric water content. The low constant rate of water use throughout the year, despite large changes in soil and atmospheric water content, is limited by tree hydraulic architecture that has evolved to prevent runaway xylem embolism. See Chapter 6 for a discussion of the ecohydrology of north Australian sites.

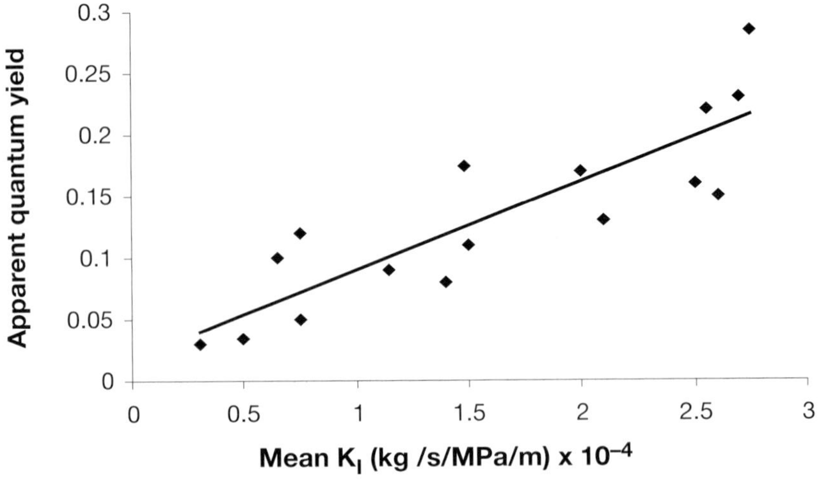

Figure 2.13 There is a strong correlation between average leaf area specific conductivity of stems (k_l) and mean quantum yield, for a number of angiosperm and coniferous species

Source: Data redrawn from Brodribb & Field (2000)

The hydraulic limitation theory was questioned by Becker et al. (2000), who proposed that changes in leaf area–sapwood area dominate in determining whole plant conductances. They also proposed that changes in soil-to-root hydraulic conductances as soils become drier are important, possibly more important than any changes in conductance of stems as trees get taller.

Hydraulic architecture, water flux and water potential

It is possible to integrate the factors determining the balance between gradients in water potential and transpiration. The rate of sap flow (Q) can be calculated thus:

$$Q = EA_l \tag{12}$$

where E is transpiration rate and A_l is leaf area.

Transpiration rate (E) is determined by the conductance of the stomata (G_s) (we ignore the influence of boundary layer conductance here) and the concentration gradient of water between the inside and outside of the leaf (D). The classic analogy is with Ohm's law, where the flow of electrical current (equivalent to the flow of water in transpiration) is determined by the voltage and the resistance of the pathway. Thus, $E = G_s \times D$ (i.e. the rate of water flow through stomata is equal to the stomatal conductance (G_s) multiplied by the concentration gradient of water across the stomata (D; expressed as a mole fraction)). Since $K_s = K_h/A_s$, where A_s is sapwood cross-sectional area, then equation (11) can be rewritten as:

$$K_s A_s = G_s / D / A_l l / \Delta\Psi_W \tag{13}$$

By rearrangement:

$$\Delta\Psi_W = G_s / D / A_l l / K_s A_s \tag{14}$$

Plants regulate their rates of water loss and hydraulic architecture so as to maintain the gradient of water potential from root to shoot within narrow limits, because too large a gradient in water potential leads to xylem embolism. This equation highlights the strategies available to achieve this. Thus, plants can adjust stomatal conductance in the short term (tens of minutes), leaf area in the medium term (days to weeks) and sapwood area and sapwood conductivity in the long term (weeks to months). In northern Australia the large increase in evaporative demand of the dry season, plus the impact of drying of the upper soil profile, is exactly matched by a decline in leaf area per (evergreen) tree. Consequently (evergreen) tree water use remains constant all year, despite large seasonal fluctuations in supply and demand for water. In contrast, *Pinus sylvestris* growing at warmer drier sites show a homeostatic reduction in the ratio of leaf area to sapwood area, in comparison to trees growing at cooler moister sites (Mencuccini & Grace 1995).

Xylem embolism

Patterns of water use by vegetation can be dramatically altered by drought. Understanding how drought influences vegetation water use is important to effective management of catchment water yield. Many forested catchments are managed to maintain the source of water for towns and cities (see Chapter 7). Droughts lasting 3 months and 18 months are likely to have different scales of impact on water supply and understanding the mechanisms by which this occurs can contribute to modelling the response of catchment water yield to drought. Xylem embolism is one response to drought that can be modelled.

Xylem conduits are rigid pipes conducting water that is under tension (the pressure of water in the xylem is less than atmospheric pressure). The water potential of xylem sap can be as low as –1 to –5 MPa. The hydrogen bonds of water allow water to remain meta-stable despite large tensions existing in the water column. However, the water column can snap if tensions are generated that are too large to be sustained. Once snapped, the column is said to be **embolised** and is no longer able to conduct water. A gas bubble forms within the xylem conduit, breaking the continuity of the water column between root and leaf. Thus xylem embolism results in a loss of hydraulic conductance and a reduced ability to maintain a favourable canopy water balance. Consequently G_s declines to reduce the rate of water use by the canopy. By balancing water loss to water uptake, excessively large declines in leaf water potential are avoided and xylem embolism is minimised.

Xylem embolism can be caused by drought and by freezing. During freezing, bubbles form within xylem conduits as a result of freeze thawing of xylem sap. During drought, air seeding of the xylem occurs when air from surrounding tissue is pulled into the xylem through pores of the xylem wall. Air is pulled first through the largest pores in the xylem. The size of the pore determines the pressure difference required to pull air into the xylem. The larger the pore, the smaller the pressure gradient required to pull air in. Consequently plants growing in arid zones have smaller xylem diameters but also, more critically, smaller pore diameters, to make the xylem less sensitive to embolism. Conduit length is also important in influencing vulnerability to embolism.

A narrow pit pore diameter, plus a narrow conduit diameter and short conduit length, make xylem **resistant to embolism**. However, these characteristics also increase the resistance to flow. Species growing in cool moist habitats tend to have wider conduits, larger pit pore diameters and longer conduits than those growing in hot arid environments because the probability of emboli developing in cool moist habitats is less than that in hot dry environments. The degree of embolism (expressed as a percentage of the loss of total conductivity due to embolism) can range from zero at night in cool humid seasons to more than 60% in unusually hot and dry conditions.

Vulnerability curves

A key question in the study of plant water relations is the vulnerability of the hydraulic pathways to xylem embolism. Related questions are whether species adapted to different environments/climates have different sensitivities to embolism, and whether differences among species can be correlated to habitat preferences of different species.

Figure 2.14 shows a vulnerability curve for two species. Vulnerability curves relate the water potential of xylem to the degree of xylem embolism. The lower the water potential required to induce a loss of 50% of conductivity, the less vulnerable the xylem to embolism. Generally, the vulnerability of a species correlates well with the xylem potentials routinely experienced by that species in its natural habitat. For example, species growing in Dipterocarp forest, which have high water holding capacity soils and therefore rarely experience low leaf water potentials, experience a 50% loss of conductivity at approximately –0.6 MPa. In contrast, species growing on heathland, with a lower soil water storage capacity, exhibit a 50% reduction in conductivity at -0.8 MPa. Species growing in seasonally dry forests, where a pronounced dry season occurs, do not lose 50% of their conductivity until approximately –1.8 MPa. It is likely that species operate at the 'edge' of embolism formation. Thus, the water potentials experienced by plants in the field are only slightly higher (closer to zero) than those required to induce large amounts of embolism, and stomatal regulation is used to balance uptake and loss of water so that xylem potentials do not fall to levels that can induce large rates of embolism formation.

Drought avoiders are plants that avoid the development of high levels of embolism by dropping their leaves at the start of the dry season. **Drought tolerators** use stomatal regula-

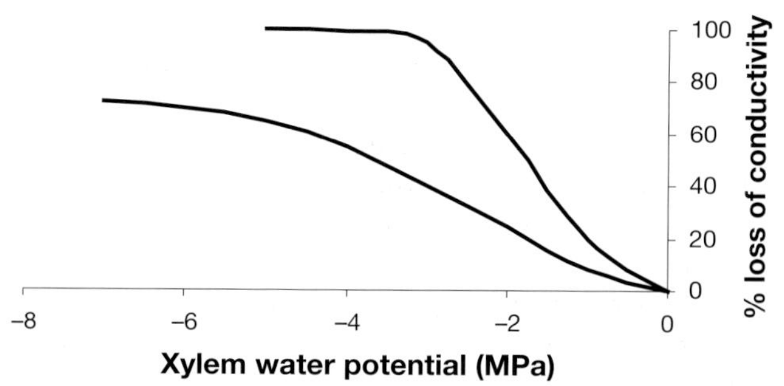

Figure 2.14 The vulnerability curve of two hypothetical species, differing in vulnerability to embolism. The upper curve represents a species with a larger vulnerability to embolism than that represented by the lower curve

tion, deep roots and a hydraulic architecture that is relatively resistant to embolism to survive the dry season with a leaf canopy that transpires through much of the day.

Recovery from embolism

Recovery from xylem embolism is an important process that can repair most or all of the loss of conductance of xylem following embolism. At night, when xylem sap pressure can increase from negative to positive values (because water uptake continues at night in the absence of large rates of transpiration), emboli are dissolved (the gas bubbles redissolve in the xylem sap). Such removal of emboli by root pressure is adequate for plants less than 10–20 m tall, where root pressure is sufficient to pressurize xylem to these heights. How emboli can be removed from the tops of tall trees, where root pressure is generally considered insufficient to reach, remains unknown. In temperate ring-porous trees, the formation of new xylem conduits in spring is one mechanism by which functional xylem can become available, in the absence of embolism repair.

Conclusion

The water status (water potential) of a plant is determined by several factors. These include:

- radiation input (shortwave and long-wave);
- soil and atmospheric water content;
- hydraulic architecture of the plant;
- stomatal behaviour.

The flux of water through a plant is similarly determined by the same range of factors and leaf water potential and plant water use are linked, principally through stomatal behaviour and hydraulic architecture. The hydraulic architecture of a tree influences the gradients of water potential that develop within the plant as transpiration rate varies in response to changes in solar radiation and soil and atmospheric water content. When the balance between uptake and loss of water is upset and stomatal regulation is insufficient to preserve leaf water status, xylem embolism occurs and the conductance of the sapwood declines, further exacerbating the decline in leaf water status that occurs during drought. Understanding the daily and seasonal patterns of water status and water use of vegetation is an important prerequisite for understanding the ecohydrology of a catchment. The next chapter examines the basics of hydrology as they relate to ecohydrology, with a particular emphasis on features of Australia that influence the hydrology of Australian catchments.

Further reading

Andrade JL, Meinzer FC, Goldstein G, Holbrook NM, Cavelier J, Jackson P & Silvera K (1998). Regulation of water flux through trunks, branches, and leaves in trees of a lowland tropical forest. *Oecologia* **115**: 463–471.

Eagleson PS (2002). *Ecohydrology: Darwinian expression of vegetation form and function.* Cambridge University Press.

Roberts J (1999). Plants and water in forests and woodlands. In *Ecohydrology* (eds AJ Baird & RL Wilby). Routledge, London, pp. 181–236.

Tyree MT (1997). The cohesion-tension theory of sap ascent: current controversies. *Journal of Experimental Botany* **48**: 1753–1765.

Tyree MT (1999). Water relations of plants. In *Ecohydrology* (eds AJ Baird & RL Wilby). Routledge, London, pp. 11–38.

Chapter 3

Basics of hydrology

In the previous chapter the fundamental language, theories and concepts of plant water relations, plant water use and the soil-plant-atmosphere continuum were described. This chapter discusses the terms, concepts and knowledge base required to understand hydrology. Working hydrologists might omit this chapter, but for plant scientists it will provide the foundations needed to progress through the rest of the book.

The chapter starts with a description of the water cycle and a water balance, at both local and regional scales. Above (overland flow, or surface flows) and below ground processes (subsurface flows, groundwater discharge and recharge) are discussed. Aquifers (both confined and unconfined) are defined and recharge and discharge processes described. Movement of water, through above ground (rivers and lakes) and below ground (aquifers) water bodies are discussed, as is the connection between these water bodies. Finally, the dynamic behaviour of water bodies is described.

After reading this chapter readers should be familiar with the following:

- hydrologic balance;
- inter-annual variability in rainfall;
- local and regional water balances and the principal components of such water balances;
- surface and subsurface flows of water;
- aquicludes and aquitards;
- the meaning of the terms disconnected, connected, gaining and losing streams, discharge and recharge lakes and flow through lakes;
- the importance of salt in the Australian landscape.

Introduction

> 'All streams flow into the sea, yet the sea is not full, though the streams are still flowing' (Ecclesiastes 1: 7) – a very early reference to hydrology.

The focus of this book is the linkage between terrestrial vegetation (mostly trees) and hydrology of the Australian continent. Yet, as a text aimed at understanding Australian ecohydrology, the basics of hydrology must first be considered. Only after we understand water and solute balances of landscapes can we appreciate the importance of ecohydrology to an interpretation of landscape function.

Australian hydrology is subject to the same general mathematical and physical treatment as hydrology worldwide, but there are properties of the Australian climate and geomorphology that lead to a distinctive Australian ecohydrology. In particular, Australian rainfall is highly variable between years, the climate is characterised by high evaporative demand for much of the year across much of the continent and Australia is basically flat (low relief). The hydrological consequences of this combination are profound.

Hydrologic cycle

The hydrologic cycle is the movement of water between the various stores of water that exist on the earth. Figure 3.1 is a representation of the hydrologic cycle and indicates the pools and fluxes of water. Major pools (stores) of water include:

- oceans and seas;
- rivers and lakes;
- soil and groundwater;
- atmospheric water;
- frozen water (glaciers, polar ice caps, mountain tops).

The major **fluxes** (movement across a unit surface per unit time) of water are well known – they are precipitation (including rain, snow, mist, fog and hail) and evapotranspiration, i.e. the combination of evaporation from wet surfaces (lakes, oceans, wet canopies and wet soil) and transpiration from leaves. The amount of rainfall across Australia was discussed in Chapter 1. The magnitude and importance of inter-annual variation in rainfall is discussed below.

Evaporation from oceans accounts for about 0.5 m of water per year in polar regions and 1.5 m of water in tropical regions, giving an annual average of about 1 m per year from all oceans. Tropical regions evaporate more water than polar regions because there is more solar radiation available to drive evaporation (Chapter 2) in the tropics.

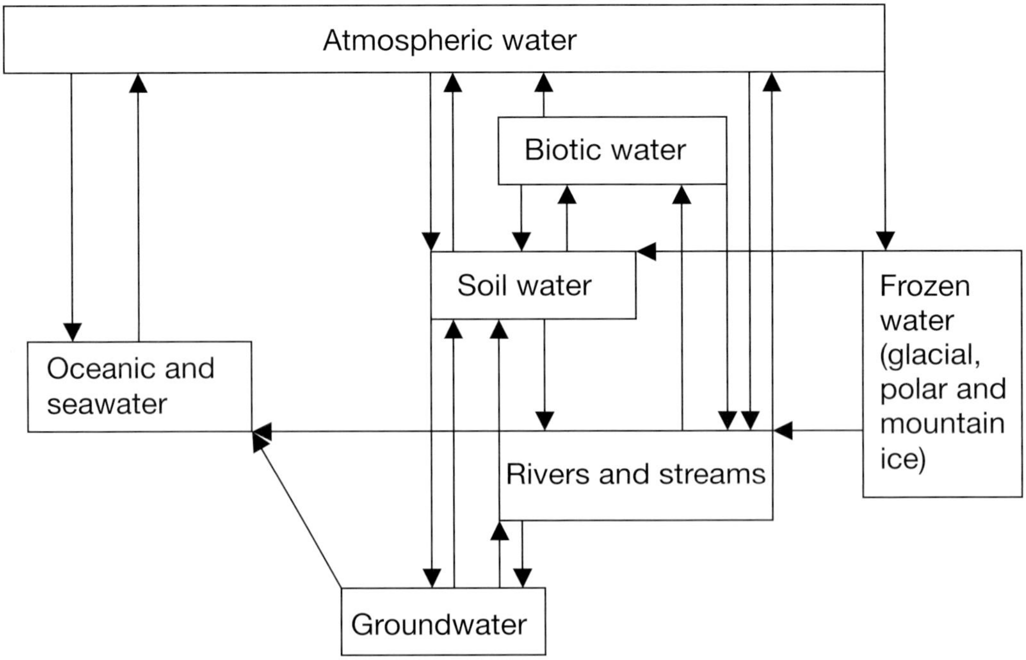

Figure 3.1 The hydrologic cycle consists of the movement of water (represented by arrows) between several major stores of water (represented by the text boxes)

Water evaporated from oceans stays in the atmosphere for only 2–4 weeks before it falls as precipitation, mostly on oceans again. Most precipitation occurs when large, warm moist masses of air come into contact with cold dense air. This occurs by movement from tropical to temperate regions (such as tropical monsoon storms in northern Australia), by orographic cooling (when air is lifted aloft by the presence of mountains) or by a warm surface causing the lifting of warm moist air to higher cooler regions of the atmosphere.

The movement of water from terrestrial surface water bodies (rivers and lakes) to groundwater stores (called groundwater recharge) is an important flux. Without recharge, groundwater stores would become depleted. Groundwater recharge is discussed later.

Rainfall variability

Rainfall is the most important input of water to the hydrologic balance of many (but not all) ecosystems. **Hydrologic balance** refers to an input–output model (or equation) of a region. The region could be a small catchment or an entire continent. Forests generally receive most of their water supply from rainfall falling on the catchment, but some lakes and wetlands and some mound springs (see Chapter 6) receive most of their water input as overland flow or groundwater input. However, at this point, we consider rainfall as the principal source of water.

Australia's climate was described in detail in Chapter 1. However, the high inter-annual variability in rainfall and seasonal drought, leading to high hydrologic variability, is worth reiterating. Australia's mean annual rainfall (350–400 mm y^{-1}) is lower than that of all other continents and the mean annual solar radiation inputs are very large (hence mean temperatures are high for most of the mainland). This results in very low mean annual run-off in Australia (approximately 70 mm), which is significantly lower than in any other inhabited continent – the values for Africa, Asia, Europe, North America and South America are approximately 180, 215, 220, 280 and 440 mm, respectively. Also, the variability of streamflow is far higher in Australia than in the rest of the world, with the possible exception of South Africa (McMahon et al. 1992).

Australia is the only continent where the overwhelming influence on climate is non-annual (i.e. seasonal) climatic variation (Flannery 1994). In the rest of the world, seasonal patterns of temperature and rainfall are highly predictable but in Australia inter-annual unpredictability is the most predictable feature of climate.

The general aridity of much of Australia (Fig. 1.3a, Colour Plate 3) and the lack of predictability of rainfall (Fig. 1.3b, Colour Plate 3) has major implications for the evolutionary adaptations of Australian biota and the productivity of Australian ecosystems (Hatton & Nulsen 1999). Furthermore, there is now some indication that human-induced climate change may be a feature of recent weather patterns in at least part of the continent. Climate change is likely to increase the frequency and severity of extreme events, such as drought and flood.

To summarise the discussion of climate in Chapter 1, the seasonality in climate of the southern part of the continent tends toward either a Mediterranean (cool wet winters) climate or a fairly uniform distribution of rainfall. Rainfall in the southern parts ranges from 200 to 1200 mm annually. Potential evaporation exceeds rainfall in most months even in higher rainfall zones and certainly exceeds it in all months in the lower rainfall zones. Northern Australia, in contrast, has summer dominant rainfall (1500–2500 mm y^{-1}, averaging about 1760 mm y^{-1}) and dry winters.

Geomorphology and river flows

Irrespective of the variability and generally low levels of rainfall that characterise much of Australia, the fate of rainfall is partly dependent on the geomorphology of the landscape. Furthermore, the geomorphology of the landscape influences whether rivers cascade rapidly through steep gorges or meander slowly across broad floodplains. The morphology of land also influences where rain itself is formed as mountains can force warm moist air to rise, leading to condensation and rain and a rain shadow on the leeward side (**orographic rain**). Therefore some broad knowledge of the geomorphology of Australia is valuable in ecohydrological studies.

Australia can be viewed as a vast plateau with a concave centre (and hence slightly raised edges). The Great Dividing Range extends along much of the eastern coast. Slopes on the eastern edge are steep and rugged but the western edge slopes more gently to foothills and a low plain. A lowland extends from the eastern edge of the Dividing Range to the coast. This land is more fertile than that in the arid interior. On the western side of Australia a large plateau extends, ranging in elevation from 300 to 500 m. Two ranges (the MacDonnell and Musgrave Ranges) rise in the middle of this plateau to a maximum height of about 1500 m. The arid interior contains the Great Sandy Desert, the Gibson Desert and the Victoria Desert. The centre of the continent consists of a vast sandy and stony low lying plain which receives very little rain.

Clearly, the geomorphology is characteristically flat, with only occasional relief (variation in altitude). Deeply weathered regolith arising from ancient rock covers the continent. Australia's geology has been relatively quiet over the past 60 million years; in some cases, any uplifting that has occurred has actually restricted drainage. Little renewal of surface material means that soils are old and nutritionally poor, profiles are deeply weathered, and geomorphologic units result from differential erosion of an ancient surface (MacArthur 1993). This geologic history has resulted in the flattest continent on earth with generally **low hydraulic gradients** (slopes or pressure gradients down which water flows) and low transmissivities (ease with which water can move through a substrate). Water and solutes generally do not move quickly through the Australian landscape. The sluggish movement of water, combined with deeply weathered regolith and aridity, makes groundwater–biota interactions a singularly prominent feature of this continent. For example, fissured limestone in the Barkly Tableland and Nullarbor Plain cause the landscape to be highly permeable and hence overland flow is extremely rare.

Australia's low topographic relief, low rainfall and high **evaporative demand** (the ability of the climate to evaporate water is high) leads to a sparse network of rivers and lakes. Run-off is extremely low, less than 10 mm y^{-1}, over much of the continent. Areas with significant run-off (>100 mm y^{-1}) are minimal and restricted to the tropical monsoonal north, where rainfall is typically 1500–2500 mm y^{-1} and the east coast (New South Wales and Queensland), the Blue Mountains and the Snowy Mountains, where orographic rainfall is significant. Mean annual run-off from the Malgrave, Johnston and Tally Rivers is about 1600 mm y^{-1} while run-off from the high rainfall zones of northern Australia is in the range 100–500 mm y^{-1}. In many rivers, groundwater inflow maintains river flow during periods of zero or very low rainfall. This represents **baseflow** for the river and indicates that many aquatic systems are dependent, for at least part of the year, on groundwater flows. Baseflows are the flows of water in rivers that are derived from groundwater discharge into the river. Such discharge may occur at a few localised points or extend along many kilometres of river bank.

The Murray-Darling Basin is the largest Australian river system. It originates from the slopes of the Dividing Range then spreads through a vast plain that is almost flat. Most of the water of the Murray-Darling Rivers is used for irrigation or it evaporates – very little reaches the sea. In its middle reaches flooding is common and an important part of the recharge of groundwater in the region.

Having established the meaning of the hydrologic cycle and discussed the broad-scale importance of rainfall and geomorphology to key fluxes within this cycle, we now discuss the concept of the water balance. Understanding water balances is primary to the science of hydrology. As previously stated, a water balance is an input–output model, or equation, for a defined area of land.

Water balance

Rainfall is the principal immediate source of water to most ecosystems in Australia (but exceptions exist – see Chapter 6). The geomorphology of the landscape influences how this rainfall moves across the landscape. Resolving exactly the fate of all the water in a region is a key activity for hydrologists and is summarised by a water balance. Indeed, hydrology is concerned with resolving the **water balance** – the accounting of water inflows, outflows and changes in storage that occur within a defined volume (e.g. within a soil pedon, a dam, a catchment). We can describe the water balance as simply:

$$\text{Inflow} = \text{outflow} + \text{change in storage}$$

Although inflow is often solely considered as rainfall, it should include other inputs such as irrigation diversions, surface run-on and groundwater inflow.

Outflows include streamflow, deep groundwater discharge, drainage loss, soil evaporation, plant transpiration and evaporation of rain directly from leaves (canopy interception loss). Change in storage can include unsaturated soil moisture storage, deep groundwater level changes and changes in dam storage.

The hydrologic system can be viewed at two different levels of complexity, both relevant to understanding the interactions between water and biota. First, there are systems in which the water balance is constrained by the rain that falls directly on the local land surface (a localised water balance); second are systems in which lateral redistribution of water is a significant component of the water balance (a regionalised water balance).

Localised water balance

If we assume that a plot of ground receives no additional water other than the rain which falls directly upon it, then the local water balance can be written:

$$P = E + T + RO + \Delta S + D \tag{1}$$

where P = rainfall/precipitation, E = evaporation, both soil and canopy interception, T = transpiration, RO = surface run-off, ΔS = change in soil moisture storage and D = vertical drainage below the plant root zone.

The schematic diagram shown in Figure 3.2 recognises losses of water (run-off, deep drainage, evaporation and transpiration). It is, in most locales and applications, strictly incorrect: at any scale, there is probably some lateral redistribution of water. However, emergent ecohydrological properties of many Australian ecosystems imply that this approximation is quite robust and that 'imported' water does not always play a significant role in determining the character of the associated biota.

Regionalised water balance

A full treatment of the water balance of a landscape or large catchment requires an account of lateral processes such as the movement of water overland as run-on into a region or the discharge of groundwater into streams (Fig. 3.3). A more complete water balance for larger scales is therefore:

$$P = E + T + O + RO + G + \Delta S + \Delta w \tag{2}$$

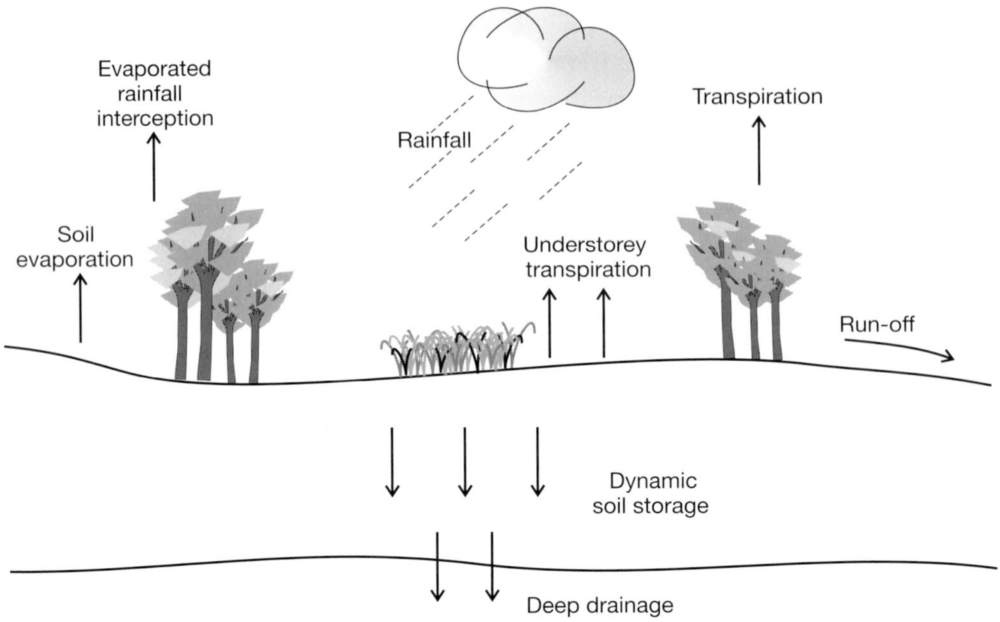

Figure 3.2 Conceptual model of a local soil water balance

where E = evaporation, both soil and canopy interception, T = transpiration, O= surface run-on, RO = surface run-off, G = groundwater discharge, ΔS = change in soil moisture storage and Δw = change in groundwater storage.

Consideration of groundwater

In the localised water balance discussed above (equation 1), a vertical drainage, D, is included. Once water is past the root zone of the vegetation in the localised water balance, it is considered to be 'outside the system' and no further consideration is given to it. However, in the regionalised water balance, D is replaced by an explicit treatment of the groundwater components and is implicitly the **recharge** to the groundwater system. In a defined groundwater system below the surface soil, without groundwater inflow, all drainage of water beyond the roots gets to the groundwater system and so this volume represents groundwater recharge. Thus the water balance for a groundwater system can be written as:

$$D = G + \Delta w \qquad (3)$$

where G is groundwater outflow (or discharge) that includes deep groundwater flow as well as discharge to the surface as seeps and springs, and direct discharge to streams; Δw is the change in water stored.

When water is added to or removed from groundwater, the **groundwater level** changes substantially more than the equivalent quantity of water. This means that a 20 mm input of water to an aquifer raises the level (depth below the surface) of the aquifer much more than 20 mm. Note that 20 mm of water actually means that 0.02 m^3 of water moving across 1 m^2 of ground area (0.02 m^3/1 m^2 = 0.02 m = 20 mm).

The reason that 20 mm of water entering an aquifer raises the aquifer by much more than 20 mm is because solids occupy much of the volume of the saturated regolith. The relationship between fluctuations in water table depth and the actual quantity of water leading to the changes in water table is described by the expression:

$$\Delta H = \Delta h / S_y \qquad (4)$$

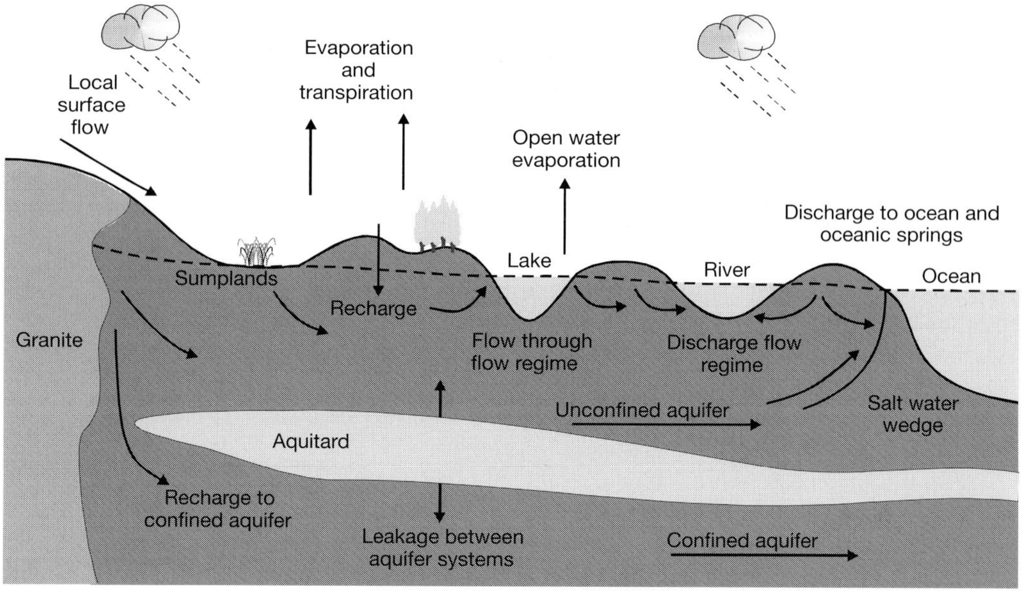

Figure 3.3 Conceptual model of a regionalised water balance

where ΔH is the change in the height of the water table, Δh is the added water in depth equivalents, and S_y is the **specific yield**.

A static description of the sizes of the various pools of water in a description of a catchment water balance is an insufficient description of the water balance. Water moves between these pools at rates that vary spatially and temporally. The measurement of these fluxes of water and an understanding of the driving forces causing water movement is also required. It is the flux of water down a river (including its volume, timing and duration) for example, that very much determines the ecology of the river. The following section addresses the flux of water through the landscape.

Flow of water through the environment

Precipitation clearly represents a major flux of water from the atmospheric pool of water to soil store. Other fluxes, such as groundwater recharge and discharge, are also important. Understanding flows of water within a landscape is central to the study of ecohydrology and central to the management of water, vegetation and landscapes. This section provides a description of the movement of water in the landscape.

In environments like those found across Australia, the accumulation of water in excess of local rainfall due to lateral hydrologic processes is of paramount importance to the biota. These relatively well watered places usually have distinctive species assemblages, higher productivity, and an importance to the overall ecosystem far in excess of their size relative to the surrounding landscape. It is worth exploring the diverse processes by which water accumulates and flows through ecosystems, for these processes differ in their relative rates, water qualities and natural controls.

Although rainfall is clearly a flow process, it has been separated from other flow processes because it is largely beyond human control or manipulation. In contrast, flows of water that occur after rainfall are amenable to some degree of manipulation and are affected by small- and large-scale human activities.

The first process to consider after rainfall is that of infiltration. An understanding of this process, and what happens when the rate of rainfall exceeds the capacity of a site for infiltration, is important to understanding all other water flows. The second flow process is run-off (also called overland flow), which occurs when the rate of rainfall at a site exceeds the capacity of the site to absorb water. Run-off is a significant pathway for the movement of water, especially across landscapes with significant slope.

Infiltration

Infiltration of water is the passage of water into soil by forces of gravity and capillarity. A soil's infiltration capacity depends on the properties of the soil and the moisture content. Once the water has entered the soil, further movement is known as percolation. Infiltrating water replenishes soil moisture deficiencies but infiltration should not be confused with groundwater recharge. Without this vital process, plants would soon deplete soil water stores and groundwater recharge would cease. A very rapid initial rate of infiltration into dry soils is followed by a drop in infiltration rate as soil water content increases to saturation. Infiltration rate now becomes gravity driven and is termed **percolation.**

Infiltration rate is determined by several factors, including the depth and composition of the litter layer, soil particle size, soil pore size and the amount of water and organic material in the soil. The maximum rate of infiltration is called the **infiltration capacity**, and this maximum rate is attained when the rate of supply of water (rain intensity) exceeds infiltration capacity. When this occurs, surface run-off occurs.

As water moves through a dry soil profile a **wetting front** is established. The wetting front is the boundary between the saturated layer of soil in the upper soil profile and the dry soil below it, through which water is infiltrating from above. The movement of the wetting front through the soil profile is dependent on the supply of water from above. The maximum rate of water flow through the pores in soil is described by **Poiseuille's law**, which essentially says that the maximum rate of water flow is proportional to the fourth power of the radius of the pore:

$$Q = (\Pi \times r^4 \times \Delta P)/(8 \times \eta \times L) \quad (5)$$

where $\Pi = 3.142$; r = the radius of the pore; ΔP = the pressure gradient over length L and η is the viscosity of water (which increases as temperature decreases).

Poiseuille's law explains why water flow through coarse textured soils (sands) is very much faster than through fine textured soils (silts). Sands have far larger average pore sizes than silts.

Overland flow

When the rate of supply of water to the soil surface exceeds the rate of infiltration, surface **run-off** or **overland flow** occurs. Water that runs off the surface of a catchment generally ends up in stream flow, although in some cases **sheet flow** moves into wetlands without passing through a channelised stream. Alternatively, surface run-off can infiltrate downslope and feed into subsurface lateral flow or groundwater recharge. Run-off increases on steep slopes (compared to flat terrain) and is lower across sandy soils than soils with a high clay or silt content.

Rainfall can generate overland flows by two distinctive processes: **saturation excess** and **infiltration excess**. With saturation excess, the soil is already at or near saturation due to conditions antecedent to the rainfall, and further rain cannot enter the soil but runs off. This phenomenon is typical of low-lying portions of catchments (e.g. riparian areas), which can wax and wane in size over the wet season and are therefore referred to as **variable source areas.** By contrast, **infiltration excess** occurs when the rate of rainfall exceeds the rate of infiltration of surface water into the soil. This latter process is characteristic of heavy textured soils or non-wetting sands, and tends to decrease with increasing vegetation and surface litter that detains the ponded water for longer periods of time. Overland flow ceases relatively quickly after rain by either mechanism. The following equation links the three components determining overland flow:

$$\text{Overland flow} = \text{precipitation} - (\text{surface retention} + \text{infiltration})$$

The amount of overland flow that occurs at a site is determined by several factors, including the intensity and duration of precipitation, the type of vegetation or other ground cover, the slope of the land and the permeability of the overland flow. Increased duration or intensity of precipitation and increased slope increase the amount of overland flow. A decrease in vegetation cover or soil surface permeability also increases the amount of overland flow. **Stream discharge hydrographs** (plots of stream height or stream volume flow over time) can record the impact of overland flow into streams in the early stages, prior to increased groundwater base flow that may arise sometime after the start of precipitation (Figs 3.4–3.7).

Subsurface flow

Water that has infiltrated into soil on sloping ground will tend to move both downwards and laterally. While such movement occurs in a very limited way when soil water is in an unsaturated state, upon saturation water can move quickly, depending on the slope gradient and **hydraulic conductivity** (given the symbol K), which depends on the pore size within the soil. Pores are of similar size to the soil particles. These can range from small rocks and pebbles, with a size greater than 2 mm, to sand (0.02–2 mm), silt (0.002–0.02 mm) and clays, which have particles and pore sizes less than 0.002 mm. The hydraulic conductivity of sands may be several metres per day, while saturated flow through heavy clay is usually very slow (<1 mm d^{-1}). In soils with a strong texture contrast between adjacent layers (e.g. a sand layer over clay), a **perched water table** can develop during the wet season. A perched water table is one that is found above the permanent water table whenever the infiltration rate exceeds the saturated hydraulic conductivity of a layer. This usually transient hydrological feature can give rise to substantial lateral **subsurface flows** (also termed **throughflow**) on hillslopes in winter in southern Australia and summer in the north. In most soils, subsurface flows are negligible for slopes less than 2%.

Consideration of subsurface flows must include Darcy's law. Thus, the flow of water through saturated soil is described by Darcy's law:

$$Q = K_s \times (\Delta H / \Delta x) \tag{6}$$

where Q is the rate of flow, K_s is the saturated hydraulic conductivity of the soil, and $\Delta H/\Delta x$ is the pressure gradient within the saturated soil, which is the same as the slope of the water table. Δx is the distance over which the pressure gradient is measured. The flow rate, Q, is the volume of flow that passes through a unit area cross-section per unit time, and has units of velocity (m d^{-1}). In perched hillslope water tables, the pressure gradient, $\Delta H/\Delta x$, is typically similar to the land surface slope, which in agricultural areas is often less than 5%. K_s is rarely more than 1 m d^{-1}. The symbol for saturated hydraulic conductivity unfortunately is the same (K_s) as the symbol for sapwood specific conductivity. However, the context of the symbol should provide a clue to its meaning in each instance.

Water flows much more slowly through the soil matrix than through cracks and channels such as old root channels and earthworm holes. Such channels, called **preferential flow channels** or **preferred pathways**, can cause large variation in rates of water flow through soil across the landscape. A small 2 mm diameter (cross-sectional area 0.000003142 m^2) root channel can transmit as much water as 1.3 m^2 of clay matrix, and a 4 mm channel as much as

21 m^2. Such a rapid rate of flow occurs only when the channel is in contact with saturated soil, such as under waterlogged conditions. This hydrologic feature can greatly increase K_s over the value expected from bulk soil texture. Accounting for such preferred pathways when measuring the K_s of soils in the field is very difficult.

Shallow subsurface flow discharges water into the environment much more slowly than overland flow, and is important in sustaining streamflow and seeps and springs from local hillslope hydrological systems. Discharge takes place at a point in the flow path where either $\Delta H/\Delta x$ or K_s decreases sufficiently to bring the water table to the land surface. A decrease in the former quantity is often termed 'break-of-slope' and is a common discharge feature in uplands. If the rate of delivery is very slow, evaporation or transpiration can discharge this water to the atmosphere. At higher rates, water will flow freely as a spring and perhaps contribute to streamflow.

Using hydrographs to measure water flow

Overland flow frequently ends up in streams and rivers. A record of the volume of flow in streams and rivers over time is called a hydrograph. While specific use of hydrographs is covered later, a general understanding of hydrographs is provided now as it can assist in understanding the hydrology of a catchment and the links between rainfall, surface flow, infiltration and soil moisture.

Four scenarios can be used to highlight the principles underlying interpretation of a hydrograph. In all scenarios, rainfall intensity is uniform across time and space. Soil moisture deficit here is the difference between the amount of water currently in the soil and the amount of water that the soil could hold at field capacity. The field capacity of a soil is the maximum amount of water held in a soil before it starts to drain downwards due to gravity.

In the first scenario, rainfall intensity is less than the rate of soil infiltration and the volume of water infiltrated is less than the soil moisture deficit. In this case, overland flow is minimal and because the volume of rain infiltrated is less than the soil moisture deficit, lateral flow of water in the soil to streams is minimal. Consequently the rate of stream discharge increases only very slightly and for a short period, and both of these are due to rainfall within the stream itself (Fig. 3.4).

In the second scenario, rainfall intensity is still less than the rate of infiltration, but the volume of water infiltrated is larger than the soil moisture deficit. Consequently a larger peak and a long sustained increase in stream flow occurs (Fig. 3.5). This is because in addition to rainfall within the stream, a lateral subsurface flow of soil water and groundwater to the stream occurs.

In the third scenario, the difference between the lower dashed line and the upper solid line (Fig. 3.6) reflects the fact that the volume of water entering the soil is less than the capacity of the soil to retain that water but the intensity of rainfall exceeds the rate of infiltration and consequently overland flow to streams occurs.

In the final scenario, when rainfall intensity is larger than the rate of infiltration and the volume of water arriving as rainfall is larger than the soil moisture deficit, then overland flow, subsurface flow, groundwater flow and in-stream rain all contribute to increased stream discharge (Fig. 3.7). Groundwater flux to the stream is not altered by rainfall. Once overland flows have stopped, stream discharge returns to the level set by groundwater baseflow.

Groundwater flow

Overland flows of water include surface run-off and river and stream flow. These are conspicuous flows and have received the attention of hydrologists, ecologists and managers for many years. In contrast, flows of water to and from groundwater occur predominantly out of sight and

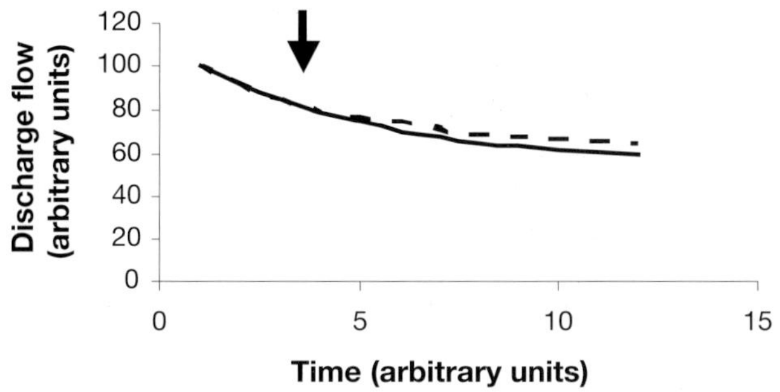

Figure 3.4 Only small increases in stream discharge occur when the intensity of rainfall is less than the rate of infiltration and the volume of rainfall infiltrated is less than the soil moisture deficit. The dotted line shows the new flow resulting from rainfall (indicated by the arrow) across the catchment

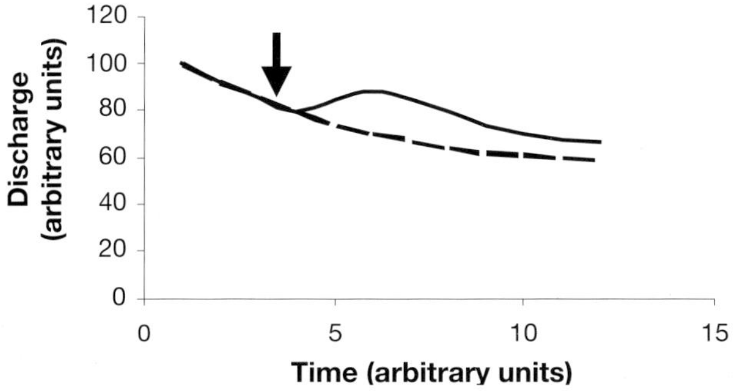

Figure 3.5 When rainfall intensity is still less than the rate of infiltration but the volume of water infiltrated is larger than the soil moisture deficit, a larger peak and a longer sustained increase in stream flow occurs. The solid line shows the new flow resulting from rainfall (indicated by the arrow) across the catchment

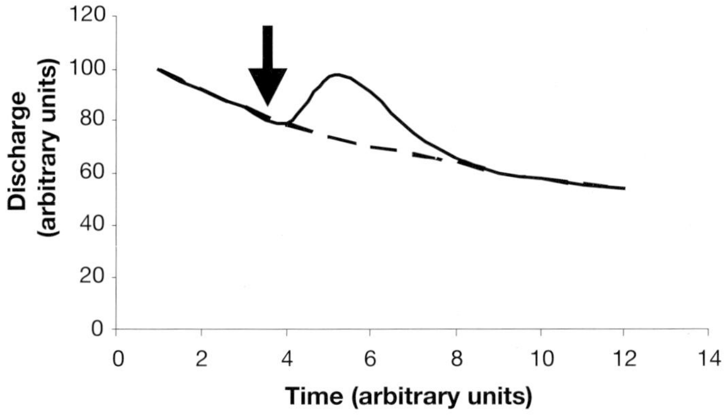

Figure 3.6 When rainfall intensity is larger than the rate of infiltration to the soil but the volume of water infiltrated into the soil is less than soil moisture deficiency, surface run-off is a major cause of the increased discharge from the stream. The solid line shows the new flow resulting from rainfall (indicated by the arrow) across the catchment

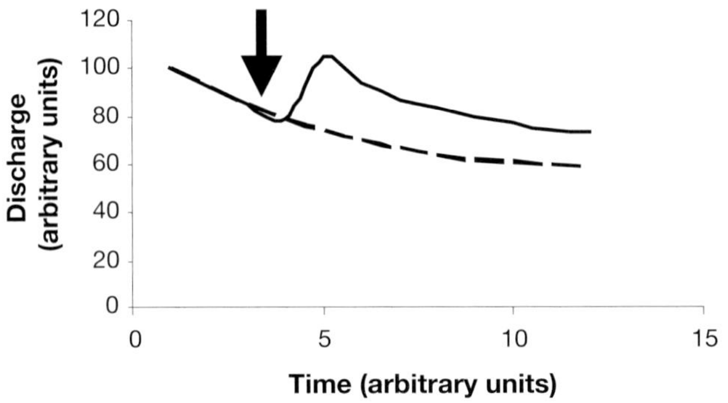

Figure 3.7 When rainfall intensity exceeds infiltration capacity, and the volume of water infiltrated exceeds the volume of the soil moisture deficit, surface run-off, groundwater influx and lateral subsurface flows are the main causes of increased stream discharge. The solid line shows the new flow resulting from rainfall (indicated by the arrow) across the catchment

at much slower rates. Consequently they have received less attention than surface flows. However, as will be seen in later chapters on groundwater dependent ecosystems and salinity, the importance of groundwater flows is now being realised and receiving increasing research interest.

A groundwater hydrograph from a highly transmissive sand aquifer on the Swan Coastal Plain in Western Australia is shown in Figure 3.8. The annual cycle of recharge and discharge under the seasonal rainfall regime (Mediterranean climate) is apparent. The interpretation of such hydrographs can yield an understanding of recharge rates and their relationship to rainfall, land use change and abstraction. For example, with knowledge of the specific yield (say, 0.1) we can estimate the seasonal recharge rate. In this case, a 1 m rise translates to 10 mm of recharge. Healy and Cook (2002) provided a comprehensive treatment of methods for using groundwater levels to estimate recharge.

The slowest flows of water through the environment are associated with deeper pathways of water through the regolith, conventionally termed **groundwater aquifers**. An aquifer is a satu-

Water pressure and groundwater levels

Many near-surface aquifers are unconfined, meaning that they can receive recharge from directly above, as the unsaturated zone above them is fairly permeable. The upper surface of these aquifers is marked by the water table. When drilling into these aquifers the rest water level in the borehole is the same as the depth at which saturated water is first encountered. When water is pumped from a borehole intercepting an unconfined aquifer, the water level drops (the pumped or dynamic water level) and the aquifer close to the borehole is de-watered as the level drops.

Confined aquifers are overlain by low permeability strata. This disconnects them from the surface and they do not receive direct recharge from the land above the confined part of the aquifer. The low permeability layer maintains a pressure head in the aquifer. The rest water level in a borehole intercepting the aquifer will be higher than the top of the aquifer, where water was first 'struck' during drilling. The pressure head may be higher than the land surface, in which case an artesian borehole is formed which will flow freely due to natural pressure gradients. Water levels in confined aquifers also fall due to pumping. As the response is a pressure response and not a dewatering response, it may be propagated quickly through the confined part of the aquifer and observed in boreholes far from the pumping borehole.

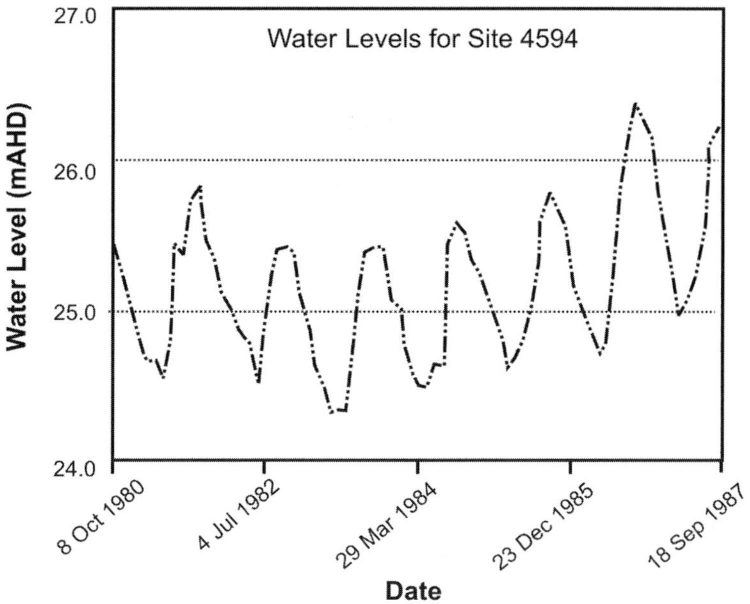

Figure 3.8 A groundwater hydrograph from a highly transmissive sand aquifer on the Swan Coastal Plain in Western Australia

rated, permeable geological structure or strata that can store and transmit significant quantities of water under relatively natural pressure gradients. An **aquiclude** is a saturated, low permeability geological structure or strata that does not transmit appreciable water flow under such natural gradients. An **aquitard** is a geological structure which functions somewhat between an aquifer and aquiclude, in that water can flow through it but flow is slow. The notion of an aquitard is a relative one, in that it implies a lesser permeability than a neighbouring aquifer.

A **confined aquifer** is one that is bound between two aquicludes or aquitards. It is separated from the surface soil, and water flows in it in response to pressure gradients from remote sites. The water in a confined aquifer is often under significantly higher pressure than the water perched above its top boundary. An **unconfined aquifer** is one in which the upper surface is the water table. In unconfined systems, the pressure of the water (P) at a point is a function of the height of the column (h_p) of water above that point, the density of the water (ρ) and the force of gravity (g). Thus:

$$P = \rho g h_p \tag{7}$$

Confined aquifers (or **artesian aquifers**; Fig. 3.9) contain water under pressure. The pressure of the water (P) at a point in a confined aquifer can be calculated with the same equation as above, but in that case h_p is replaced by the height that the water would rise to if it were allowed to, e.g. if a piezometer was installed thereby breaking the seal in the confining layer.

Artesian groundwater systems

A groundwater system is said to be artesian if the local pressure level in the aquifer is higher than the level of the land surface. Such situations arise where the aquifer is confined over a distance such that the levels at the intake (recharge) end of the aquifer are higher than the ground level some distance away.

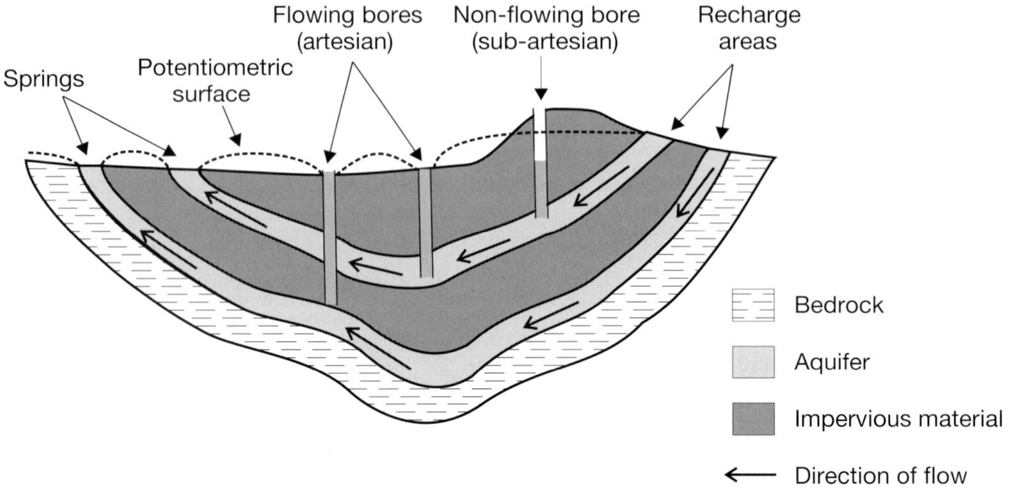

Figure 3.9 A bore sunk through the impervious aquiclude will flow freely at the surface if it is located lower than the elevation of the intake zone of the aquifer it accesses

The **total pressure head** in a confined aquifer is the sum of the head of pressure measured at a point in the aquifer, plus the pressure head between that point and an arbitrarily chosen reference point below the aquifer.

It is the difference in total pressure head between two points in an aquifer that creates the hydraulic gradient which causes water to move from one point to another. In unconfined aquifers, the height of the water in observation bores gives a direct measure of the gradients in water pressure and hence can be used to predict directions of movement. In confined aquifers, the piezometric or potentiometric (pressure) surface defines the height to which water would rise if a well was drilled into the confined aquifer. When the land surface is lower than the potentiometric surface, water flows unaided from the aquifer through a well that has been drilled (**artesian bores**).

Transmissivity, storativity, porosity and specific yield

The volume of water in an aquifer, how quickly that water can move and the rate at which water can be pumped out of an aquifer without significant reductions in flow rate to the pump well, are important features of a groundwater store (aquifer). Big aquifers with large volumes of water, and aquifers that can be pumped out at high rates to supply water for use as irrigation without suffering large reductions in the rate of supply of water to the well, are highly prized commodities for industry and water managers alike. In order to discuss these issues, we need to define transmissivity, storativity and other related terms.

A key hydrological property of an aquifer is its permeability, the measure of how quickly water can flow through the saturated geological medium (K_s). Open fissures in limestones and large voids in coarse grained gravels allow rapid flows of groundwater, between 1 and 10 km d^{-1} in karst limestone and around 150 m d^{-1} in gravels. Whereas fine grained silts and clays have low permeabilities of 2 m d^{-1} and 0.00002 m d^{-1}, respectively. Permeability may be the same in all directions (isotropic) or higher in one direction, for example horizontally compared to vertically (anisotropic). When all else is constant, thicker aquifers contain a larger volume of water. **Transmissivity** of an aquifer is the amount of water that can flow laterally through the aquifer when a hydraulic head (or head of pressure) of 1 m m^{-1} is applied. Thus:

$$T_r = bK_s \tag{8}$$

where T_r = transmissivity ($m^2\ d^{-1}$), b = saturated thickness (m) and K_s = saturated hydraulic conductivity ($m\ s^{-1}$).

When the hydraulic head of a saturated aquifer changes, water will either be discharged from the aquifer or more water will be stored in the aquifer. The **storativity** (capacity) of an aquifer is the volume of water that can be stored or discharged from a saturated aquifer per unit surface area per unit hydraulic head. The **specific yield** of an aquifer is the ratio of the volume of water that can drain from an unconfined aquifer under gravity to the total volume of the aquifer, and is always less than porosity.

Water in aquifers does not exist as a lake of free-standing water. It exists within the pores (voids) that exist between soil and rock particles. Water in aquifers saturates these pores and so aquifers occur within the **saturated zone** of the subsurface. Above this zone is a **capillary zone**, where water has moved upwards because of the action of capillary forces (attractive forces) between water and soil particles. This zone might be up to 1 m thick, depending on the soil particle size; sands have a thinner capillary zone than silt because average pore size is larger in sands and so the capillary forces are smaller. Above this region lies the unsaturated or aerated zone of the subsurface. This is where most roots occur, although some roots may penetrate the capillary fringe to access groundwater.

The **porosity** of an aquifer is a measure of the proportion of the total aquifer that is taken up by pores (or voids). Thus:

$$\text{Porosity} = (100V_v)/V_t \qquad (9)$$

where V_v = the void volume in an aquifer and V_t is the total volume of the aquifer. It is usual to determine this for a small subsample of the aquifer.

Porosity is influenced most by particle size. In an aquifer composed of coarse sand, porosity might be 15%. For a clay filled aquifer, porosity might be 50%. The effective porosity is a measure of the pore volume through which water can actually flow (pores that are interconnected). Highly connected pores with large pores sizes results in an aquifer that is highly permeable and whose water can be extracted faster than from an aquifer with low permeability.

Groundwater recharge and preferential flow

Groundwater recharge occurs when a catchment is said to be 'leaky'. Leakiness is deemed to occur when water percolates out of the root zone down towards the water table. The balance between rainfall, evapotranspiration and overland flow determines the rate of deep percolation (leakiness). It is difficult to measure and highly variable in space and time. Leakiness is an important concept when discussing the value of trees in landscapes to reduce the extent, or prevent the development, of dryland salinity. This is discussed in Chapter 8.

Recharge of groundwater is a highly complex process and can happen through a variety of routes. These include direct recharge through the unsaturated soil (vadose zone), indirect recharge where there may be some lateral flow before the water reaches the water table (e.g. recharge from surface water in a river) and localised recharge, where water collects (but not in a water course) before percolating down into groundwater (De Vries & Simmers 2002). In high rainfall environments most recharge will be direct but as the rainfall decreases the importance of the other mechanisms increases. **Preferential flow** occurs where infiltrating and percolating water finds routes where vertical flow rates can be orders of magnitude higher than hydraulic conductivities through the soil, for example through old root channels (Beekman & Xu 2003). Opportunities for preferential flow are greatest where there are interfaces between soil types or soils and rocks, natural cracks or openings in the soil, or old or existing root channels (Scott & Le Maitre 1998), but may also occur in highly homogenous soils. Shallow soils over bedrock, or exposed rock which is highly faulted or karstic may also permit high rates of percolation of recharge through the rock joints, fractures and solution passages. Studies in the Kalahari

Table 3.1 Indicative K_s values for various geological materials, and the sensitivity of discharge rates to changes in the parameters in equation (10)

	K_s (m d^{-1})	Δh	A (m^2)	Q (m^3 d^{-1})
Shale	10^{-7}	1	1000	10^{-4}
		10	1000	10^{-5}
Sandstone	10^{-3}	1	1000	10
		10	1000	10^{2}
Coarse sand	10	1	1000	10^{4}
		10	1000	10^{5}

region of Botswana and other areas have estimated that preferential flow may account for about 50% of the total recharge, and up to 79–90% in some situations. These kinds of values indicate that preferential flows, often via indirect and localised recharge, are the norm rather than the exception in semi-arid landscapes, and need to be considered when attempting to assess recharge and quantify its temporal and spatial variability (Edmunds & Tyler 2002; Scanlon et al. 2002).

The rate at which groundwater moves through an aquifer (Q in m^3 d^{-1}) can be calculated from:

$$Q = \Delta_h K_s A \tag{10}$$

where Δ_h is the hydraulic gradient, K_s is the saturated hydraulic conductivity of the aquifer (in m d^{-1}) and A is the cross-sectional area of the aquifer (m^2). Some indicative values for K_s for various geological materials, and the sensitivity of discharge rates to changes in the three parameters in equation (10), are given in Table 3.1. For a similar gradient and aquifer cross-sectional area, a coarse sand aquifer can discharge one hundred million times the volume of water than a typical shale aquifer, directly reflecting the huge range in K_s among geological materials and its direct control of potential discharge rates.

Groundwater **discharge** is the rate at which water is lost from an aquifer. This can include upward movement of water through the unsaturated zone in response to evaporation or transpiration, or flow to rivers, wetlands or the sea. Discharge from one aquifer may become recharge to another in a layered multiple aquifer system.

The rate of discharge (the rate at which water is lost from the aquifer; it includes discharge through vegetation as transpiration, lateral flow to rivers, wetlands and the sea and movement to overlying or underlying aquifers) is limited by the location in the aquifer where the product of $(\Delta_h \times K_s \times A)$ is a minimum. Where recharge exceeds discharge, groundwater levels increase.

In contrast to shallow perched hillslope systems, the hydraulic gradients of aquifers are usually less than the surface topography. These lower gradients, combined with the fact that the K_s of aquifers is often lower than that of surface soil, means that the flow velocities of groundwater are far slower than overland and subsurface flows. Like shallow subsurface flows, groundwater discharges at the surface either to the atmosphere (through direct evaporation or transpiration) due to a decrease in K_s, aquifer cross-sectional area, or gradient.

Water at the surface of the landscape (rivers, dams) is often assumed to be isolated from water in aquifers. This mistake leads to many errors when allocating water to various consumptive users, as it is possible to count the water twice. Because aquifers discharge and are recharged, there is a link between surface and groundwater. The following section discusses this further.

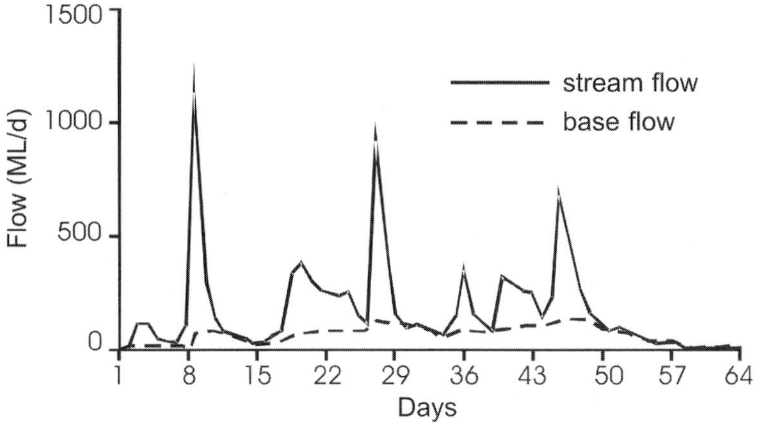

Figure 3.10 Example of the separation of the base flow component from total stream flow

Surface–groundwater interactions

The various processes that can generate and transmit water flow can operate in the same landscape, and give rise to characteristic curves describing the relationship between water flow and time, where the time interval can be days or weeks or months. Given the practical complexity implied in trying to independently characterise each process and pathway for water flow, the hydrological response of stream flow to rainfall is usually treated as a bulk response. One useful characterisation of stream discharge response is to deconstruct the stream hydrograph into **quick flow** and **base flow** components. The base flow component is the contribution to total stream flow derived from groundwater influx. Quick flows are the components of total stream flow that arise from increased input to the river from surface and subsurface flows and rainwater intercepted directly by the river (Fig. 3.10). This separation is only a statistical concept, but such an analysis illustrates the seasonality (or perenniality) of stream flow and the peak discharges expected following storms. Moreover, in natural (unregulated) streams, the base flow component is largely comprised of groundwater discharge. These qualities of quick flow and base flow relate directly to the biotic response. For example, some species (animal in particular) require a constantly flowing (base flow) stream for survival. At the other extreme, floodplain vegetation may need only the most occasional flooding (associated with a quick flow component of flow) to survive either an extended dry season or even a period of several years between rainfall events.

A complication for understanding and managing the dependence of ecosystems on environmental water that can arise more frequently in Australia than on continents with more topographic relief, is where the groundwater catchment has different flux boundaries (limits to flow) than the (partially) overlying surface catchment. Surface catchments may 'import' water from outside their topographic boundaries if the discharge area of a groundwater system (perhaps whose recharge waters arise remotely from that catchment) lies within them. Conversely, streams may lose water as recharge into a local groundwater system far downstream.

Hyporheic zone

The saturated zone of mixing between groundwater and surface water below rivers and streams is known as the **hyporheic zone** (Woessner 2000). This zone represents an ecotone between the surface environment characterised by light, high dissolved oxygen and temperature fluctuation, and the groundwater environment characterised by darkness, less oxygen, stable

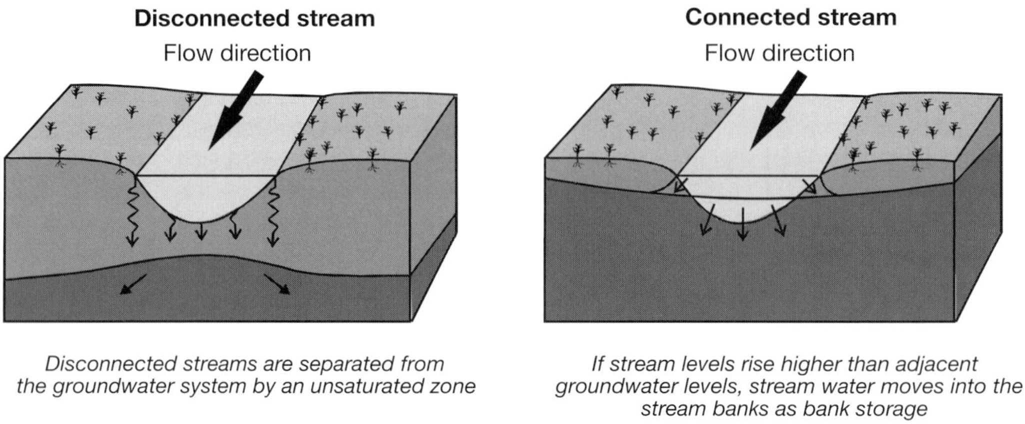

Figure 3.11 Conceptual diagram of a disconnected and connected stream

temperature (Hayashi & Rosenberry 2002), and interstitial spaces. Some of the species found in in-aquifer ecosystems removed from surface water, are linked to hyporheic zone ecosystems.

The hyporheic zone provides a number of **ecologically important services**. These include (Gardner 1999; Hayashi & Rosenberry 2002):

- thermal, temporal and chemical buffering of water moving between groundwater and surface water;
- providing habitat for micro-organisms, macro-vertebrates, fish and wildlife;
- sustaining of refugia for aquatic species that rely on a particular chemical or temperature regime for their survival;
- supplying nutrients and inorganic ions to surface water systems through groundwater discharge.

It is expected that aquifer discharge and mixing with surface water will play an important role in controlling the physio-chemical environment of the hyporheic zone, especially in middle and lower catchments, where surface flow is slower.

Disconnected, connected, gaining and losing streams and lakes

The water balance treatment thus far has implicitly assumed that water in the landscape continues to accumulate as it moves downstream in the form of streams and rivers and it all eventually debouches into the sea or lake. In fact, surface water and groundwater can interact in a variety of ways with profound implications to the biota. This section discusses these different interactions.

The first example of surface–groundwater interactions is the contrast between those water bodies (e.g. streams or lakes) whose bases are either in direct hydraulic contact with groundwater (**connected**) or are perched above an unsaturated zone (**disconnected**). Disconnected streams or lakes can slowly *lose* water into an underlying groundwater system, but at a rate insufficient to raise the water tables to the base of the stream/lake (Fig. 3.11). With connected streams or lakes, the surface water body can exchange water with the underlying aquifer, including the recharge of the streamside or lakeside land (**bank storage**).

When surface water bodies are connected to groundwater systems, groundwater can discharge into the surface water. When this occurs into a stream they are called **gaining streams/lakes**. This occurs when the surrounding water tables (or pressure levels) are higher than the

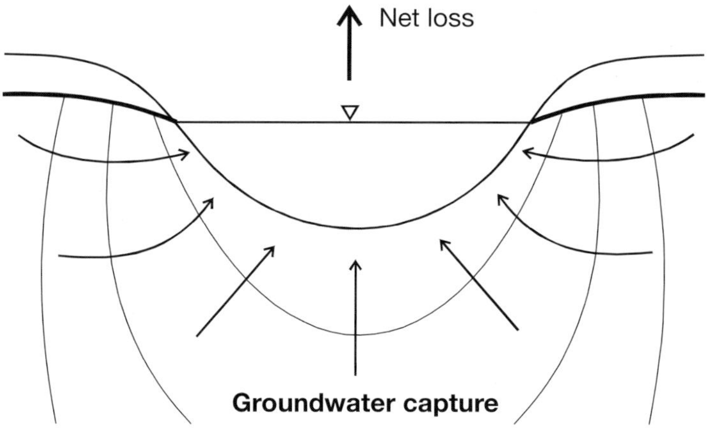

Figure 3.12 A lake in a groundwater discharge regime (a gaining lake)

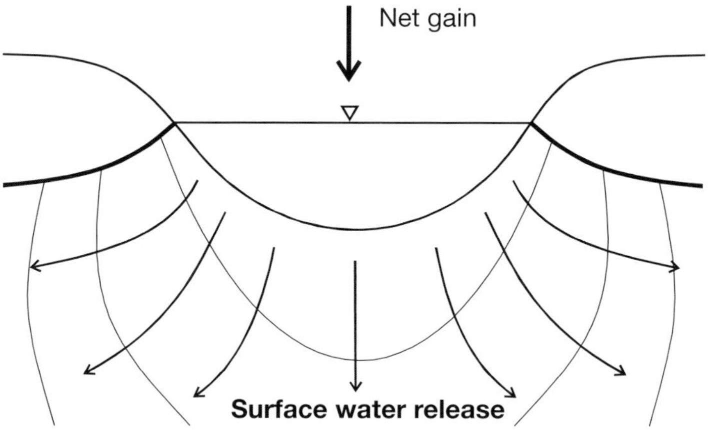

Figure 3.13 A lake in a groundwater recharge regime (a losing lake)

stream surface (Fig. 3.12). When the stream surface is higher than the surrounding groundwater levels, surface water can leak into the groundwater system and the stream is then called a **losing stream/lake** (Fig. 3.13).

Discharge lakes, recharge lakes, flow through lakes

The surface–groundwater interactions of lakes can be characterised similarly to streams. In the case where a lake is largely driven by groundwater discharge, and evaporative losses or engineered withdrawals exceed inflow and rainfall inputs, the surrounding groundwater levels can exceed the lake level and groundwater inflow will occur, known as a **groundwater discharge regime** (Fig. 3.12). When a lake has inflows or rainfall inputs that raise its level above that of the local groundwater surface, it is in a **groundwater recharge regime** (Fig. 3.13). Where lakes occur along a relatively steep groundwater gradient, i.e. groundwater levels are higher on one side of the lake that the other, then groundwater passes into the lake on one side and is lost from the lake on the other. This is known as a **flow through regime** (Figs 3.14, 3.15). These latter systems have a characteristic capture zone upstream and a dispersed flow of surface water on the down gradient side of the lake (Fig. 3.15).

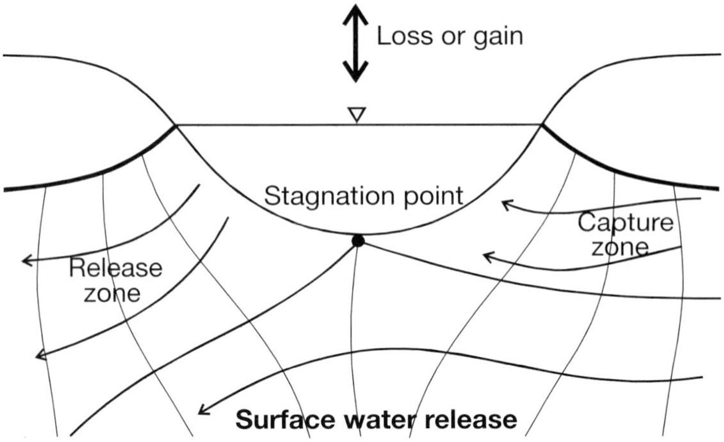

Figure 3.14 A lake in a groundwater flow through regime. There is a capture zone upstream of a local stagnation point, with a downstream release zone

Regime dynamics

The particular surface–groundwater regime of a lake or stream can vary with season and with relative rates of rise or fall in the groundwater and surface water body (lake or stream). Figure 3.16 illustrates four cases. In the first case, wet season lake levels rise faster than those of the groundwater, and a recharge regime is established. In the second case, wet season groundwater levels rise faster than those of the surface water body, and a discharge regime develops. The third and fourth cases can occur in the dry season. If the surface water levels fall faster than those of the groundwater system, a discharge regime can occur; if groundwater levels fall faster than the surface water levels, a recharge regime develops. These regimes are major determinants of the water balance and water quality in an ecosystem.

In addition to groundwater interactions with lakes and streams, some consideration must be given to fluctuations in groundwater depth (as revealed by hydrographs) arising from other causes. Changes in groundwater depth may result from (a) very large (spatially and total amount of rainfall) rainfall events arising, for example from cyclonic activity or other large

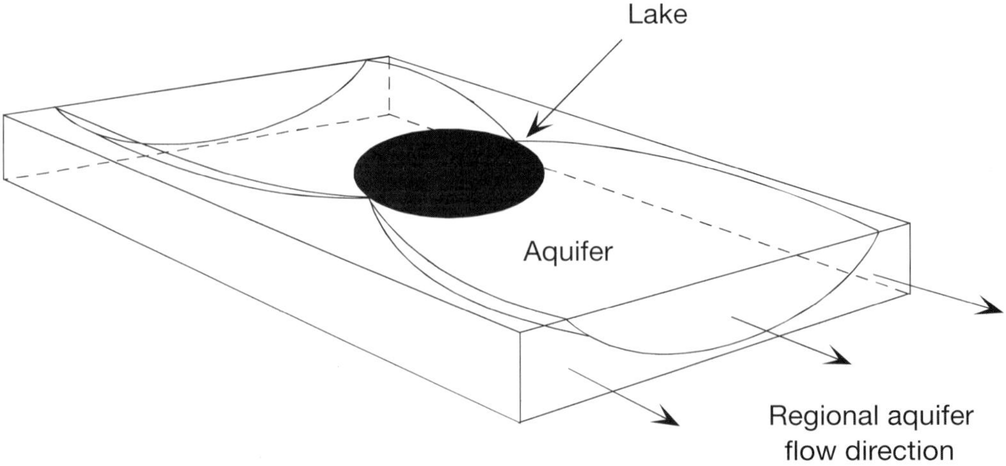

Figure 3.15 A lake in a groundwater flow through regime, showing the capture and release zones in plan view

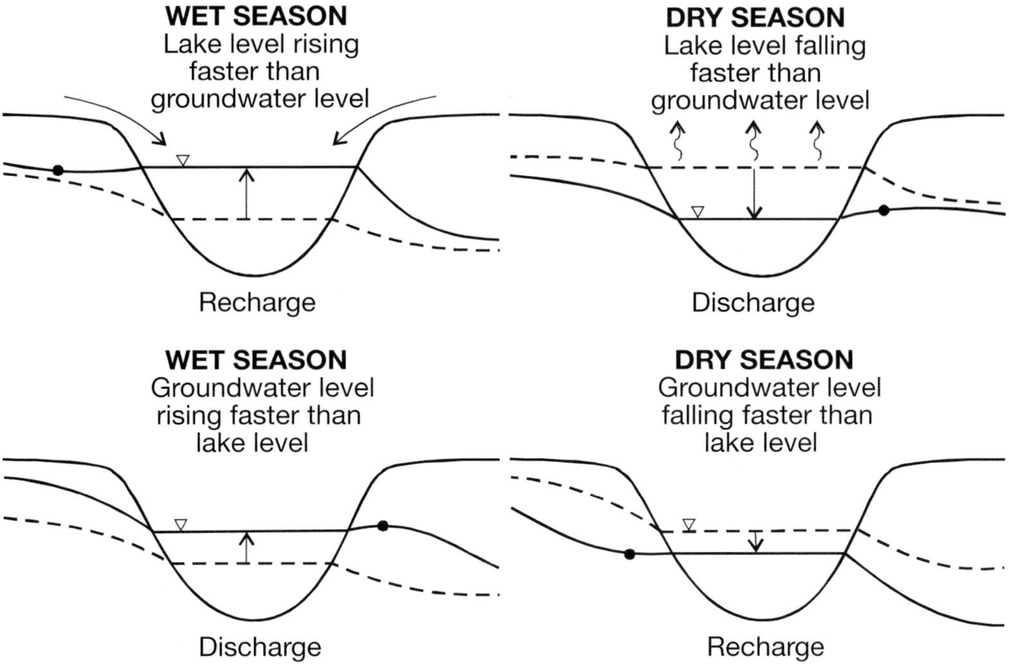

Figure 3.16 Conceptualisation of surface–groundwater regime dynamics showing seasonal dynamics and dependency on the relative rates of rise or fall of the two hydrological systems

storm events; (b) changes in land use, for example replacing large areas of woodland and forest with crops and pastures; (c) seasonal changes in climate (especially rainfall) and vegetation function (water use); (d) extraction by pumping. These changes occur over different time-scales – days and weeks for (a), weeks and months for (c) and (d) and years to decades for (b) and (d). Some changes in groundwater depth may be considered to represent natural changes ((a) and (c)) or anthropogenic changes ((b), (c) and (d)). For natural patterns of change, such as the seasonal patterns observed in the Northern Territory (see Fig. 6.6 for an example), vegetation function adapts. For example, seasonal changes in leaf area index in north Australian savannas are partly driven by this change in groundwater depth (Chapter 6). In contrast, increased depth of groundwater arising from long-term changes in rainfall and large-scale changes in land use, as has occurred in Western Australia (see Fig. 6.17), has started to influence vegetation function and vegetation structure (discussed in the Gnangarra Mound case study in Chapter 6).

Pumping wells, drawdown and cone of depression

Where groundwater is sufficiently shallow, the roots of vegetation can access it for transpiration. This vegetation may show temporary or even permanent dependence on this source of water. The topic of groundwater dependent ecosystems is extensively discussed in Chapter 6. At this stage, we merely wish to show that by extracting water from an aquifer, it is possible to alter the local water table. The impact of this change is discussed in Chapter 6.

Wells are routinely used to access groundwater. Wells can be dug manually to shallow depths (to 10 m, perhaps) or to deeper depths using drilling rigs, which can reach hundreds of metres. In both cases a lining is required to prevent the well from collapsing inwards. The

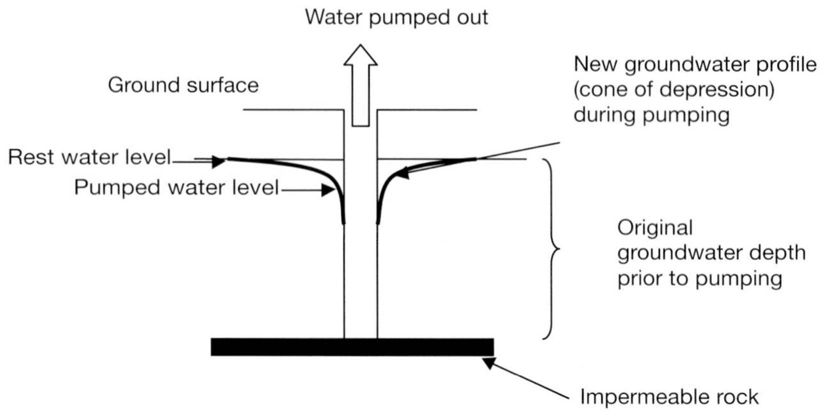

Figure 3.17 A cone of depression is formed in the height of the water table when pumping occurs from an unconfined aquifer

Source: Redrawn from Brooks et al. (1997)

lining must be perforated at the lowest end to allow groundwater to enter the well. A fine mesh screen is usually used to allow water to enter and keep soil out. In an artesian well, water will reach the ground surface freely without pumping because of the pressure within the aquifer. For other wells, a pump is required.

It is important to determine the **drawdown** of a well following pumping. The drawdown is a measure of how much the groundwater level has dropped immediately around the well following an extended period of extraction (pumping) from the aquifer. There is no value in attempting to extract water from the well by pumping at a rate that exceeds the aquifer's ability to supply water. To establish the drawdown, the water level in the well must be measured without pumping, then a constant pump rate is established. After a sufficient time the water level will be established at a new lower depth, which must be measured. The difference in water depths is the drawdown. Too large a drawdown for shallow aquifers can be expected to influence vegetation around the well. If too many wells are located too close to each other so that drawdown of adjacent wells overlap (interference effects), and the aquifer is shallow, large-scale impacts on vegetation can occur.

The ratio of the rate of pumping (m^3 per second) to the drawdown is defined as the **specific capacity** (m^3 per sec) of the well and its surrounding aquifer. The specific capacity is a measure of the rate at which water can be extracted from the well and indicates the purpose for which the well may be used. Pumping the well for prolonged periods generates a **cone of depression** around the well, a dimple-shaped lowering of the water table (drawdown). When wells extract water at low rates from high yielding aquifers the cone of depression is small and shallow. In contrast, wells that extract very large volumes from aquifers of a low yield can generate cones of depression that extend for several kilometres from the well (Fig. 3.17). Pumping wells that are too close together result in interference between the cones of depression, and the water table is reduced much more than if the wells were further apart. A cone of depression can influence the vegetation and stream flows associated with the surface above the cone and therefore it is important to know the areal extent and magnitude of the cone to assess potential negative impacts on the ground surface.

Solutes in the Australian landscape

This chapter has been concerned with the fundamentals of hydrology, with an emphasis on the hydrology of Australian landscapes. Of particular importance in Australia is the impact of solutes (mostly salts) on the ecology and hence hydrology of Australian landscapes. Conversely, changes in tree cover and land use have altered the hydrology of landscapes and resulted in the development of dryland salinity (see Chapter 8 for a detailed description). Therefore, any treatment of Australian ecohydrology must explicitly consider the influences of solutes in conjunction with water balance. In places, there is more than one tonne of salt per square metre of soil (measured to beyond the root zone of trees). Most of the salt in Australia's soil derives from three main sources. These are, first, as a remnant of the original parent rock material; second, as old sea bed material left from when the land was covered by ocean; and finally but most importantly, accumulated salt within the regolith due to atmospheric inputs (deposition) and low rates of leaching over millennia. Rain carries small quantities of salt of oceanic origin. When plants take up water they exclude the salt from their roots, and it accumulates in the soil. This is not true for mangroves and halophytes, which are plants that grow best in salty substrates. However, for the majority of terrestrial native ecosystems, such plants are of limited importance and are not considered further.

Over long periods (millions of years), huge quantities of salt can build up in soil. Most of this salt is deep in the soil but it is highly soluble, and when water passes through the salt bearing soil it is easily mobilised. If the hydrologic regime changes in favour of increased recharge, then the water level rises through the salt bearing soil and the salt is dissolved and brought closer to the surface. When this salty water table reaches stream beds, the streams run more frequently and with higher salt content. Where the soil surface is within about 2 m of a saline water table, surface evaporation drives capillary action bringing water upwards from the water table, leading to ever increasing salt accumulation at the surface.

In places, this process of salt accumulation at the surface due to the discharge of salty groundwater and its evaporation (or transpiration) is entirely natural. Salt playas were a natural feature of Australia before human induced changes in landscape hydrology led to the widespread salinisation observed today. Streamside terrestrial vegetation can have a natural dry season dependency on saline groundwater, relying on periodic flooding of fresh surface water to change the local regime into a recharge mode that leaches accumulated salts, thereby decreasing the concentration of salts in the root zone (Chapter 8).

Conclusion

Hydrology is concerned with resolving the water balance of a catchment or region. In order to do that, hydrologists and ecohydrologists need to understand the principles underlying the hydrologic cycle and the difference between localised and regional water balances. The flow of water, from a hydrological perspective, includes movement as overland flow, subsurface flow, groundwater discharge and recharge and river flow, as well as transpiration through leaves. Streams and lakes can receive or discharge water and these flows have implications for the ecology and hydrology of streams and lakes (see later chapters). The techniques to measure these flows are covered in Chapter 4. It is clear that Australia's low rainfall and its high inter-annual variability, plus the low relief over most of the country, has influenced the hydrology of Australia and is the reason why there is so little river flow in Australia.

In places in Australia, the underlying water balance is sufficiently localised that the potential influence of lateral flows can be disregarded when considering plant–water interactions. However, the particular nature of the Australian landscape and climate is such that lateral hydrological processes are crucial in determining the character of the biota. These flow processes operate over a variety of time and spatial scales. The surface and groundwater that accumulate interact in a variety of ways that can change over time. These interactions have profound implications to the biota in a land characterised by long dry seasons and high inter-annual variability. The key measures or properties resulting from these water balances and interactions that determine the nature of the biotic response include the:

- total amount of rainfall and its temporal distribution throughout the year and over years;
- depth to groundwater and its seasonal and inter-annual dynamics;
- quality of the groundwater, especially the salt content;
- duration and volume of stream base flow.

These emergent hydrological properties of Australian ecosystems are fundamental to understanding the distribution, productivity and functionality of plant and animal communities. These issues are discussed in the remainder of this book.

Further reading

Brooks KN (1997). *Hydrology and the management of watersheds.* Iowa State Press, Iowa.

Burch GJ, Bath RK, Moore ID & O'Loughlin EM (1987). Comparative hydrological behaviour of forested and cleared catchments in south-eastern Australia. *Journal of Hydrology* **90**: 19–42.

Cech TV (2005). *Principles of water resources.* John Wiley & Sons, New York.

Eagleson PS (2002). *Ecohydrology: Darwinian expression of vegetation form and function.* Cambridge University Press.

Chapter 4

Techniques in ecohydrology

In the two preceding chapters the concepts, language and paradigms of plant physiology, ecophysiology and hydrology were discussed. However, the practice of ecohydrology in the field requires some knowledge of the techniques that can be applied to answer ecohydrological questions. Such questions include:

- What is the water status of the soil, vegetation and atmosphere at a site and how do these change daily and seasonally?
- What is the rate of water movement through vegetation, and the transpiration rate of individual plants and canopies of plants?
- From where is vegetation extracting water?
- What is overland flow and how might we measure it. If vegetation cover changes, what are the likely impacts on overland flow, streamflow and groundwater recharge?
- What is the rate of movement of water through soils and aquifers and how does this influence stream flow. How do we measure stream flow?

An understanding of the techniques used in a discipline contributes significantly to the development of an understanding of that discipline. This chapter outlines some of the key methodologies used in ecohydrology and relates them to specific questions that can be addressed using each method. The first half deals with plant ecophysiology (plant function in the field), the second half deals with hydrological methods. After reading this chapter readers should be able to describe the appropriate application of a large number of techniques used in the study of plant ecohydrology.

Introduction to plant ecophysiological techniques

From a plant ecophysiological perspective, the principal techniques of concern in studies of vegetation and water can be divided into four main groups plus one 'other' group. These are:

1. techniques to measure the amount of water in soil, including gravimetric and volumetric techniques, soil thermocouple psychrometers, time domain reflectometry and neutron-probes;
2. techniques to measure the amount of water in the plant, including leaf thermocouples, the Scholander pressure bomb and measurements of relative water content;
3. techniques to measure the amount of water in the atmosphere, including wet and dry bulb thermometers, capacitance probes, cooled mirror hygrometers and infra-red gas analysers;

4 techniques to measure the fluxes of water through components of the soil–plant–atmosphere continuum. Such techniques include sapflow sensors, open-top chambers, eddy covariance and Bowen ratio techniques;
5 other techniques that contribute to ecohydrological studies. This includes methods to assess plant functioning in the field, such as using stable isotopes to determine from where water is being absorbed in the soil profile, root distribution, measurements of leaf area index and tree hydraulic architecture.

This chapter reviews some of the major techniques for these five measures. Thermocouple psychrometry is applicable to both soil and plants and is therefore considered first.

Thermocouple psychrometry

Soil and leaf water potential can be measured in the lab or field using **thermocouple psychrometers**. Thermocouple psychrometers consist of a thin wire of one type of metal joined to a thin wire of another type of metal. This thermocouple junction is enclosed in a volume of air that is allowed to come into equilibrium with either the soil sample or leaf sample being studied. When a small electrical current is passed through the wire, the junction is cooled. It can be cooled until it is below the **dewpoint** temperature of the surrounding air. The dewpoint is the temperature at which water condenses from the air. When the junction is colder than the dewpoint temperature a small drop of water condenses from the air. The current is then switched off and the water drop evaporates. This cools the junction and a small electrical current is generated. The rate of evaporation of the water drop is determined by the water vapour content of the air around the junction. In very dry air, the water drop evaporates quickly and a larger current is generated than if the air is more moist. Calibration of the thermocouple psychrometer is undertaken by using filter paper discs soaked in a solution of known water potential.

By measuring the current produced when the drop of water evaporates after equilibration with a leaf or soil sample of unknown water potential, it is possible to determine the water potential of soil and leaves. **Soil psychrometers** are buried in soil and the thermocouple is protected by being encased in a small ceramic or steel woven cup about 10 mm long and about 5 mm diameter. The cup allows water vapour to diffuse through it and so the air around the thermocouple equilibrates to the water potential of the soil immediately around the cup. Soil psychrometers can be used during wetting and drying cycles of soil to follow changes in soil water potential but can only measure the water potential of soil immediately around the thermocouple. **Leaf psychrometers** have the thermocouple in a small hollow in a metal mount. The metal mount is pressed firmly onto a leaf and the air around the thermocouple equilibrates to the water potential of the leaf. By cooling the thermocouple the water potential can be determined as for the soil psychrometer. A leaf thermocouple can be attached to intact whole leaves to measure changes in leaf water potential over a diurnal or seasonal period or during a drought experiment.

Thermocouple psychrometers can measure water potentials, solute potentials and turgor potentials over the entire range encountered in most vegetation studies (leaves mostly) and in very dry soils (0 to -4 MPa). In some models a leaf disc is placed within a chamber and its leaf water potential measured. The disc is then frozen with liquid nitrogen and allowed to thaw. The solute potential of the sample is then determined as described above for the measurement of water potential and the difference between the water and solute potential gives the turgor potential of the sample. Slight errors are introduced as the solute potential of the **symplast** (the contents of the cell inside the cell wall) is diluted by about 6% by mixing with water in the

apoplast. The apoplast is the volume of a plant organ that lies outside the plasmalemma and is essentially the cell walls plus intercellular air spaces.

Thermocouple psychrometers cannot be attached to data loggers and require manual operation. They are moderately cheap but very sensitive to temperature gradients between sample, chamber walls and thermocouple. Contamination with dirt and salt crystals is a constant hazard in field work, as is damage to the very thin wires of the psychrometer.

Measuring soil water content

Gravimetric and volumetric water content and water potential of soil

Gravimetric (by weight) **water content of soil** is a measure of the mass of water contained in a known mass of soil. It is perhaps the easiest and least demanding of all measurements of soil water content. A sample of soil is collected from the field and sealed in a container with minimal air space above the sample (to reduce evaporative losses). The soil is weighed in the lab and then dried (at 105°C) in a shallow tray to constant weight (i.e. all the water has been removed) and reweighed. Gravimetric water content is expressed in gH_2O (g soil)$^{-1}$. Large numbers of replicates and sampling at different depths are possible with this technique. However, it does not tell us about the amount of water available to the plant because the relationship between soil water content and water potential (**moisture release curves;** see below) is different for different soils. However, if a moisture release curve is available for the soil being studied, gravimetric values are very valuable.

Volumetric water content is the volume of water present in a soil sample expressed per unit volume of soil. Sampling a known volume of soil is not as easy as sampling a known weight of soil. The soil is dried as described above. Volumetric water content = gravimetric water content multiplied by **bulk density** of the soil. Bulk density of soil is defined as the mass of dry soil per unit bulk volume. The bulk volume is determined before the soil is dried.

Moisture release curves define the relationship between the water content (often expressed gravimetrically) of a specific soil and the water potential of the soil. As the soil dries, soil water potential declines non-linearly and the curve is specific for a given soil type. While time-consuming to measure, the development of a moisture release curve is very useful in ecohydrological studies because it allows rapid conversion of gravimetric water contents (which can be measured easily and cheaply) to water potential, thereby allowing gravimetric water contents to be compared with plant water potential. Moisture release curves use a **pressure plate extractor,** whereby undisturbed samples of soil are placed in a pressure chamber above a porous ceramic plate. The pressure above the soil sample is increased incrementally and the volume of water that is squeezed out through the plate is measured at each pressure.

Soil water potential can be measured using thermocouple psychrometers, as described above, or by using the **filter paper method** of Greacen et al. (1989). In this method, three layers of filter paper are placed inside a soil core in a temperature controlled environment and the change in weight of the middle layer (which is not physically in contact with soil and therefore does not become contaminated with soil) between the start and end of the equilibration period is recorded. The change in weight (which is due to water uptake from the soil) is closely correlated with the matric potential of the soil. Thus, for one particular study using one particular type of filter paper (O'Grady 2000):

$$\Psi_s\,(\mathrm{RWC{:}}\ 0.278 - 0.453){:}\ \ \Psi_s = -\exp(12.27 - 17.93RWC)$$
$$\Psi_s\,(\mathrm{RWC{:}}\ 0.453 - 1.784){:}\ \ \Psi_s = -\exp(5.55 - 3.11RWC)$$

where Ψ_s (kPa) is the matric potential and RWC ($g\ g^{-1}$) is the gravimetric water content of the filter paper. These equations are not universally applicable.

Time domain reflectometry

Time domain reflectometry (TDR) provides a means of measuring bulk soil water content over moderate distances (tens of centimetres). Such integrated measures overcome problems associated with small-scale spatial variation in soil moisture content.

Time domain reflectometry exploits the fact that the **dielectric constant** of a soil varies with soil water content. The dielectric constant of a medium is the squared ratio of the speed of propagation of an electromagnetic pulse in a vacuum relative to that in the medium. A waveguide is inserted into the ground and the speed of propagation (v) of an electromagnetic pulse sent down the waveguide is calculated from the time response of the system. The speed of propagation is determined by the soil bulk dielectric constant (E_b) thus:

$$E_b = (c/v)^2 = (ct/2L)^2 \quad (1)$$

where c is the speed of light in a vacuum (3×10^8 m s^{-1}), and t is the time taken for the pulse to travel down to the end of the wave guide and back to the detector. Soil bulk dielectric constant (E_b) is mostly determined by the dielectric of liquid water E_w (which is about 81 (a dimensionless value) because the dielectric constants of other soil components are much smaller, e.g. soil minerals E_s = 3 to 5, and air E_a = 1. These large differences in dielectric constants between liquid water and other soil components makes this method relatively insensitive to soil composition and texture and thus a good method for liquid soil water measurement.

The waveguides are usually solid steel or brass rods. Pairs of rods are inserted into the soil profile, either directly from the surface (and hence vertically) or horizontally into a soil profile exposed after digging a trench. Rods can be up to 1.5 m long and can be 3 mm or more in diameter. Commercial TDR probes are readily available and are robust, cheap and portable and can be left in situ for long periods of time. The measurements are relatively insensitive to soil texture, soil density and soil temperature and can be used to provide estimates of soil salinity.

Application of TDRs in stony ground is very difficult, especially if using narrow rods. Similarly, insertion of rods into dry hard ground is difficult as it is important for the pair of rods to remain parallel throughout their entire length. However, miniature TDR probes (approximately 10 cm long) have been developed which can be buried for many months, or can be moved around and used to determine the water content of the upper 10 cm of soil. Miniature TDRs buried at different depths allows the determination of the rate of movement of a wetting front through soil and can be used to determine soil water stores at depth.

Neutron moisture probes

Neutron moisture probes provide an integrated measure of the water content of a small volume of soil (a sphere of soil with a diameter of about 20–40 cm). Access tubes to depths of many metres mean that soil water content can be measured to these depths repeatedly over time.

When neutrons are emitted from a point source, and that source is located below ground, the neutrons have a finite probability of hitting a hydrogen nucleus (a proton) which is part of a water molecule. The more water in the soil, the higher the likelihood of a neutron hitting a proton in a water molecule. When a neutron hits a proton it bounces back with reduced energy because some of the energy is transferred to the proton during the collision.

In **neutron moisture probes** (NMPs) a source of fast neutrons is slowly lowered down an access tube which has been lined with a metal or PVC tube. Below the neutron source is a detector that counts the number of neutrons bouncing back. The more water in the soil, the larger the number of neutrons bouncing back to the detector.

The NMP has been used in many ecosystems to make repeated measures (repeated over time) of soil water content at a number of replicated sites within the ecosystem. Very sandy soils tend to make it difficult to insert the liner (access tube) because the hole collapses before the tube is inserted. Very stony or rocky soils make it even harder, but not impossible, to insert access tubes.

Calibration of the NMP for each soil type is required because particle size distribution and mineral composition of the soil affect the counts. In addition, soils of high organic content, soils with high root densities and soils with pronounced layers require calibration for each layer, which can be difficult. Measurements near the soil surface are not possible because the NMP measures over a sphere of about 20–40 cm radius. When the source is within that distance from the surface, neutrons are emitted into the air above the soil and cannot bounce back in response to hitting soil water molecules. Different material for the access tube can also affect the output and therefore this should be standardised. Calibration can be undertaken in the field by making measurements immediately after rain and then sequentially for many days after rain, with samples of soil taken back to the lab for determination of soil water content. Calibration in the lab, using large containers (eg. a 200 L oil drum), can be a problem since soil structure and root content are altered when removed from in situ. In the field, considerable care is required to backfill the access tube so that large airspaces are not created. In addition, care is required to prevent the access tube itself (either inside or outside) acting as a conduit for preferential water movement from the surface into the soil.

The NMP presents a significant safety hazard because of the radiation source. Training and correct equipment are vital for safe use. A licence is usually required for operation of an NMR system. However, the system is portable and once the access tubes are in place (and covered to prevent filling by rainwater) a single NMP can be used at many sites, at many times of year. Access to considerable depths is possible if mechanical drilling rigs are used.

Tensiometers

Tensiometers are commercially available and are cheap and robust and can be left in situ for months or years, requiring only the routine addition of water into the reservoir. Because they are cheap many tensiometers can be located in the field for estimation of spatial variability in soil moisture content. They measure only the matric potential of the soil, not the solute potential component, which makes them unsuitable for saline soils. They also work best in the wetter end of the soil water potential spectrum, from about -0.005 to about -0.08 MPa, and fail when soils become too dry (less than -0.2 MPa matric potential).

Tensiometers work by allowing the soil solution to come into equilibrium with a reference pressure indicator (a vacuum gauge) through a permeable ceramic cup which is pushed into the soil to the required depth. Tensiometers provide point measurements and have a relatively slow response time (hours). If air bubbles enter the system, the tensiometer does not work.

Soil moisture blocks (gypsum blocks)

Gypsum blocks are cheap, robust and commercially available, making them suitable for multiple placement around a site to obtain information about spatial variability in soil moisture content. Gypsum blocks consists of gypsum ($CaSO_4$) with two electrodes embedded. The electrical resistance of the block changes when the water content of the gypsum changes. This occurs as the gypsum equilibrates with local water content relatively quickly. These can be used in low or moderately saline soils. A single calibration of each block is required, irrespective of the soil type in which the block is located. Gypsum blocks will dissolve over time and are therefore replaced after 2–3 years.

Soil moisture sensors can be used to address such questions as:

- how soil moisture varies with time, depth and position in the landscape;
- whether there is a relationship between root distribution and changes in soil moisture over time.

Measuring plant water status

Pressure bomb and pressure volume analyses

The pressure chamber or **Scholander pressure bomb** is used for measuring leaf or branch xylem water potential, also referred to as xylem pressure potential. It consists of a steel chamber connected to a source of high pressure (usually a cylinder of compressed gas) (Fig. 4.1). A removable lid allows the petiole of a leaf or a thin (typically 1–5 mm diameter) branch with attached leaves to be placed in the chamber with the cut end of the petiole/branch protruding through the lid. A pressure gauge is connected to the pressure chamber. Once a leaf/branch has been excised it is immediately placed in the chamber, with the petiole of the leaf or a length of branch protruding through a hole in the chamber lid. The lid is attached to the chamber after a screw assembly has been used to compress a neoprene seal onto the petiole/branch. The pressure inside the chamber is increased slowly and the cut surface of the petiole/branch observed with a magnifying glass. The pressure inside the chamber is noted at exactly the point in time that xylem sap appears at the cut surface. Pressure is indicated on a pressure gauge attached to the pressure chamber. This pressure is equal and opposite in sign (i.e. the pressure in the chamber is positive, leaf water potentials are negative) to the average water potential of the leaf or branch. Exactly what water potential is being measured when a branch with several leaves is attached is open to debate.

The pressure chamber is expensive, but robust and portable (though heavy, especially since bottles of compressed gas are required to pressurize the chamber). Commercial pressure chambers can measure water potentials as low as -8 MPa, which certainly covers the range of water potentials encountered in most studies of woody vegetation. Pressure chambers cannot be automated and require manual operation.

There is a very slight risk of the seal around the petiole/branch failing. If the seal fails the leaf/branch invariably shoots directly upwards at very high speed. Therefore safety glasses must be worn when examining the petiole. It is not acceptable to look down from above at the petiole/branch; always observe from the side.

Pressure volume analysis (PV analysis) is a technique for establishing the relationship between the water content of a leaf (expressed as relative water content) and the inverse of water potential of the leaf. This is known as a type II transform PV curve and is represented in Figure 4.2a. From such a PV curve the relationships among water, solute and turgor potential of the leaf can be derived by following the procedure described in detail below. A graph of turgor, solute and water potential as a function of relative water content is called a **Hofler** diagram (Fig. 4.2b).

To derive a PV curve the following process is undertaken.

1 Excise a leaf or branch with attached leaves from the plant and immediately place the cut end of the petiole or branch under water in a closed container overnight, in the dark. This allows the leaf/branch to fully rehydrate. The previous practice of floating the leaf on the surface of water is to be avoided as this tends to artificially rehydrate the airspaces of the leaf.
2 Following rehydration, weigh the leaf accurately and quickly, taking care to ensure the leaf surface and petiole or branch surface is dry. This represents the turgid weight (TW).
3 Immediately measure leaf water potential using the pressure bomb (it should be very close to zero). This is Ψ_1.

Figure 4.1 Use of the pressure bomb. A fine needle valve (in the operator's right hand) is used to gradually increase the pressure in the chamber (indicated on the large pressure gauge, centre) while the operator observes the end of the petiole or branch with a magnifying glass to determine when xylem sap first appears at the surface

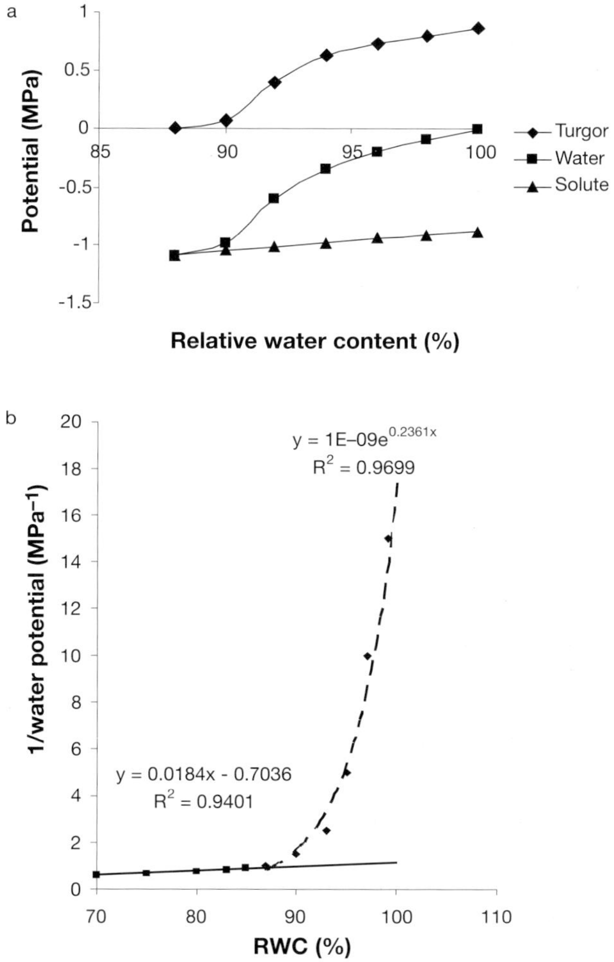

Figure 4.2 (a) An example of a type II transform of pressure volume data, from which (b) a Hofler diagram can be derived (see text for explanation)

4 Allow the leaf to lose water for 2 minutes, under lights.
5 Reweigh the leaf/branch and measure its new leaf water potential. Both should have declined a little.
6 Repeat steps 4 and 5, typically 10 times, gradually increasing the time between subsequent measurements.
7 At the end of step 6, place the leaf (not the branch) in an oven at 65–80°C for 3 days and determine leaf dry weight (DW).
8 For each value of leaf weight, calculate relative water content (RWC) from the following:

$$RWC = [(TW - DW)/(FW - DW)] \times 100$$

9 Construct a table of results and derived values as follows.

Leaf fresh weight (g)	Leaf water potential (MPa)	Inverse water potential (MPa^{-1})	Relative water content (%)
3.0	**–0.1**	10	99.5
2.8	**–0.18**	5.556	97.0
etc	**etc**	etc	etc

10 Plot the inverse of water potential (y-axis) against RWC (x-axis) to yield a plot that looks like Figure 4.2a (type II transform).
11 Regress the linear portion of the curve (for the lower values of relative water content). This linear portion of the curve corresponds to the data set where turgor potential is zero and hence water potential is equal to solute potential (see Chapter 1). The curved part of the curve corresponds to all values of water potential where turgor is positive.
12 Extrapolate the linear portion as shown in Figure 4.2a. For suitable values of relative water content (in this example, between 87% and 99%), calculate the solute potential using the linear regression equation, remembering to note that the regression will yield the inverse of solute potential. Using the power regression for the curved part of the data set, calculate the water potential. By difference, turgor potential is calculated.
13 Plot a Hofler diagram, as shown in Figure 4.2b, which shows the relationship between water, solute and turgor potential as a function of relative water content. As the water content of the leaf (and hence, by inference, the cells of the leaf), declines from full hydration, turgor declines (the cell deflates) and solute potential declines due to the increasing concentration of solutes in the remaining volume of cell water. Hence water potential declines as the sum of both. At some point, turgor potential is zero and water potential now declines only as a function of the decline in solute potential, thus the abrupt reduction in slope at the point of zero turgor in the type II transform PV curve (Fig. 4.2a).

In response to water stress (drought), many plants **osmoregulate** (accumulate solutes within the cell) to reduce solute and water potential, thereby maintaining the cells' ability to cause water to move into them from outside. Osmoregulation can be determined by examining a PV curve of leaves before and after drought. Osmoregulation has occurred when the solute potential of the leaf after drought is lower, at a given water potential, than a leaf prior to drought. The elasticity of cell walls can also be adjusted in response to drought. The slope of the relationship between cell pressure and cell volume is proportional to the elasticity of the cell wall.

Pressure volume analyses are useful in the following studies:

- comparative studies of water relations of species growing at a site;
- determination of osmoregulation responses of different species in response to drought;
- determination of relative sensitivities of leaves of different species to drought.

Humidity, atmospheric water content and temperature

Humidity is a commonly used measure of atmospheric water content. However, the terms relative humidity (RH) or humidity should be avoided. Relative humidity is so highly temperature dependent (Fig. 4.3) that it rarely provides a useful measure of atmospheric water content in the field. A simple example will illustrate this.

Table 4.1 shows how the absolute **saturated water content** of air increases non-linearly as temperature increases. Atmospheric water content can be expressed as concentration (grams of water per cubic metre of air) or as water vapour pressure. At 30°C, 50% RH corresponds to a water content of either 15.19 g m^{-3} or 2.12 kPa. However, at 20°C, 50% RH corresponds to a substantially different atmospheric water content (8.65 g m^{-3} or 1.17 kPa), despite the air being at the same RH. Thus, although the RH of the air at these two temperatures is the same (50%), the evaporative demand imposed by the air on a leaf or on the surface of a lake is very different. It is the difference in concentration of water between the inside and outside of a leaf that drives diffusion of water through stomata of a leaf. Leaf and air temperatures are often different, making the use of RH even more problematic. When leaf and air temperatures are different it is possible for the RH of the air inside and outside the leaf to be identical but there still exists a large concentration gradient of water driving diffusion out through the stomata.

Measuring the amount of water in the atmosphere

The amount of water in the atmosphere is a key measure in ecohydrology as it strongly influences the rate of water evaporation from a wet surface and influences stomatal conductance. Dry air increases the rate of evaporation from a wet surface and induces stomatal closure. Consequently, measuring atmospheric water content in conjunction with stomatal conductance assists us in understanding how vegetation water use might respond to changes in atmospheric conditions. Several techniques are now discussed. However, first, it is important to understand that humidity, as a measure of atmospheric water content is a poor expression of the water content. This is explained in the box text above.

Thin film capacitance humidity sensors

Thin film capacitance probes are fast, cheap, portable systems for measuring atmospheric water content that are suitable when high levels of precision are not required. They consist of two metal electrodes, separated by a thin hygroscopic polymer film. The first electrode is used

Table 4.1 As air temperature increases, the saturated (100% RH) water content of air (expressed as a concentration or as water vapour pressure) increases non-linearly

Air temperature (°C)	Saturated water content (g m^{-3})	Saturated water vapour pressure (kPa)
0	4.85	0.611
5	6.80	0.872
10	9.40	1.227
15	12.83	1.704
20	17.30	2.337
25	23.05	3.167
30	30.38	4.243

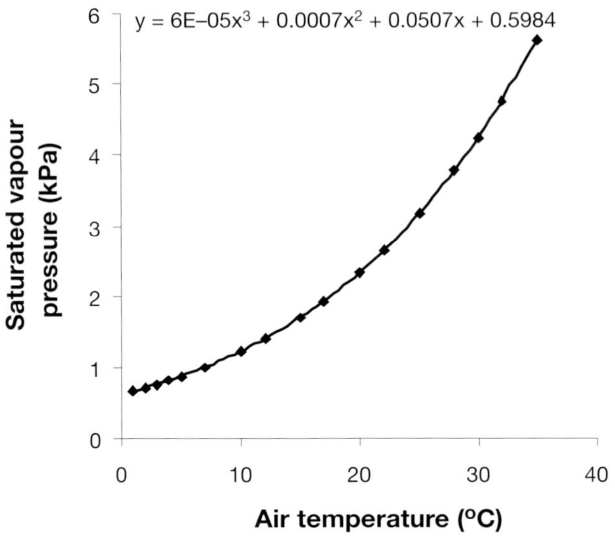

Figure 4.3 An empirical relationship between air temperature and atmospheric saturated vapour pressure

to coat a small glass slide which acts merely as a physical support. The thin (typically 1 μm) hygroscopic polymer is then applied and the second electrode coats this polymer. The second electrode is very thin (<0.05 μm) and therefore allows water vapour molecules to diffuse into and out of the polymer. As the polymer gains or loses water the capacitance of the film changes; this is measured electronically. The internal electronics of the meter convert the capacitance of the probe into a relative humidity value. A low supply voltage (DC) of around 7 V is required to operate the instrument and power consumption is very low.

These sensors are relatively cheap, physically robust (but should not come into contact with liquid water, salt, atmospheric pollutants such as sulphur dioxide and ozone, or dust) and lightweight. They can be attached to data loggers for remote use and can be logged at 2 minute intervals if required, or hourly or daily. Response time is a few seconds at moderate temperatures (20°C) and intermediate humidities (25–85%). However, at low or high humidities the response time is longer and accuracy diminishes. There is a temperature dependence of the sensor, and most humidity sensors require a measurement of temperature at the same time as humidity (because humidity is itself influenced by temperature). A temperature sensor is often included as standard.

There are two major problems with these sensors: **hysteresis** and drift over time. When exposed to high or low humidity, the response of the sensor tends to take several hours to subsequently show accurate readings at intermediate humidity. This lagging effect is known as hysteresis. Drift over time requires periodic calibration, which is easily achieved in the lab but is less easy in the field.

Capacitance sensors are commonly used in portable gas exchange systems because of their small size, ease of use and stability in moderate humidity ranges.

Wet and dry bulb temperatures

Wet and dry bulb temperatures are temperatures recorded using dry sensors and wet sensors at the same time and same place. These are used when both air temperature and the water content of the air are being measured.

When a pair of highly matched thermometers (mercury filled or, more commonly, platinum resistance thermometers) are ventilated by a wind and one of the thermometers is covered by a thin layer of water, the wet thermometer will be cooler than the dry thermometer. The difference between the **wet and dry 'bulb' temperatures** is determined by the water vapour pressure of the air moving over them. Hand-held wet and dry bulb thermometers are available – the two thermometers are held in a sling which is manually rotated to achieve the minimum speed (of greater than 3.6 m s^{-1}) of airflow over the two bulbs. A fine cotton wick keeps one bulb wet via connection to a small reservoir of pure distilled water. The presence of any salt in the water invalidates the reading. In some systems a battery powered fan can be used for automated measurement with a platinum resistance thermometer (PRT), providing an electrical current proportional to temperature.

The wet and dry bulb temperatures can be used in tables to determine the ambient water vapour pressure. Alternatively, the following equation can be used:

$$e_a = e'_w - (A\rho)(T_d - T_W) \qquad (2)$$

where e_a is the ambient water vapour pressure (kPa), e'_w is the saturated water vapour pressure at the wet bulb temperature, A is the psychrometric constant (approximately 0.000666 Pa K^{-1} for temperatures above zero), ρ is atmospheric pressure (approximately 100 kPa) and T_d and T_w are dry and wet bulb temperatures respectively.

Saturated water vapour pressure can be approximated from the following empirical regression:

$$e_s = 0.00006T^3 + 0.0007T^2 + 0.0507T + 0.5984 \qquad (3)$$

where e_s is saturated water vapour pressure at temperature T. The T in this equation could be the wet or dry bulb temperature depending on application.

Wet and dry bulb thermometers are very cheap, robust and reliable. The PRT version can be attached to a data logger and used remotely. They can be used in subzero temperatures and are most reliable when the water content of the air is not close to saturation. Salt in the water coating the wick must be avoided while the difference in temperature between the wet and dry bulbs is itself influenced by temperature.

Cooled mirror hygrometers

Cooled mirror **dewpoint hygrometers** consist of a small (metal) mirror bonded with a special heat conductive polymer to a metal plate that can be cooled or heated slowly. The mirror is housed in a small (typically 30–100 cm^3 volume) light-tight housing with an air inlet and outlet. A small light source is angled above the mirror and a detector angled to catch the reflection of the light. As air is passed over the mirror the plate is cooled gradually. When the temperature of the mirror attains the dewpoint temperature of the air, a fine mist of water condenses on the mirror surface and the amount of light reflected suddenly drops. The temperature of the mirror is recorded automatically with a temperature sensor embedded in the mirror. The dewpoint of the air is directly proportional to the water content of the air – a low dewpoint indicates a low water content.

Using equation (3), substituting the dewpoint temperature for T in the equation yields the water vapour pressure of the air being sampled.

Cooled mirror dewpoint hygrometers are moderately expensive, but reliable. Particulate matter in the air stream can give false readings, as can any salt or dirt build-up on the mirror. They are mostly used in the laboratory but can be used in the field if the air stream is filtered. They can be used in cold, temperate and tropical environments. Cooled mirror dewpoint hygrometers can be used remotely and attached to a data logger and their output is entirely independent of ambient temperature.

Infra-red gas analysers to measure concentrations and fluxes

Infra-red gas analysers (IRGAs) are used to measure the concentration of both water vapour and CO_2 in a volume of air. When measuring the concentration of CO_2 they can be used to measure the rate of photosynthesis of leaves and canopies (see below). When measuring the concentration of water vapour they can be used to determine the rate of transpiration of individual leaves or, after suitable scaling, evapotranspiration of stands of vegetation.

IRGAs consist of an infra-red radiation source (a hot wire) at one end of a gas-filled chamber and an infra-red detector located at the other end. The detector is designed to measure the change in absorbance of infra-red radiation that occurs due to absorbance of *either* CO_2 *or* H_2O vapour in the chamber. Infra-red gas analysers used in plant ecohydrology are invariably part of a larger system that incorporates mass flow controllers to accurately measure rates of air flow, and temperature sensors (thermocouples) to measure leaf and air temperatures. An IRGA is generally attached to a computer that converts the absorbance into meaningful values of changes in H_2O or CO_2 concentration. Rates of photosynthesis and transpiration are calculated from rates of air flow, changes in H_2O or CO_2 concentration, and leaf area.

Two types of IRGAs are used in ecohydrology. The first is a portable hand-held system that measures the rate of transpiration, photosynthesis and stomatal conductance of a leaf enclosed in a hand-held cuvette (chamber). The rate of air flow through the chamber, the amount of light incident on the leaf, leaf temperature and differences in CO_2 and H_2O concentration of the inlet and outlet air streams of the leaf cuvette are measured and recorded every second by an inbuilt computer. Air is stirred rapidly within the leaf cuvette, to minimise boundary layer resistance, and a transparent window allows light to fall on the entire leaf surface enclosed within the chamber. Rates of photosynthesis, transpiration and stomatal conductance are displayed on the digital display and recorded in the computer. Sampling of replicate leaves in full sunlight, on replicate trees within replicate plots, throughout the day is a useful sampling strategy.

The second type of IRGA is an open path analyser that measures the CO_2 and H_2O concentration in a volume of air that is not enclosed in a chamber. These analysers are used in **eddy covariance** studies (see below) and typically sample at 5 or 10 Hz (5 or 10 times per second).

All IRGAs require daily calibration. Both zero drift (offset of the calibration line from a zero/zero intercept) and span changes (changes in the absolute value recorded for a specified calibration gas) occur over time, especially when ambient temperature fluctuates. Therefore calibration with gases of known concentration (e.g. zero and 400 μmol CO_2 mol^{-1}) is essential. The air stream in the portable hand-held analysers must be filtered to remove particulate matter. These analysers are expensive but portable, reliable and moderately robust. The hand-held system requires a human operator but the open-path system can be used remotely for continuous monitoring over periods of several weeks in the field.

Portable IRGAs for measuring leaf scale rates of photosynthesis can be used to address questions such as:

- how the rate of photosynthesis varies between species growing at a common site;
- how changes in light flux density, water vapour pressure deficit, soil water content and temperature influence photosynthesis in the field.

Application of the second type of IRGA (the open path system) is considered in the section on measuring gas fluxes from canopies.

Sensors used to measure atmospheric water content can be used to address questions such as:

- what are the daily and seasonal patterns of water vapour pressure at a site?
- how does water vapour pressure vary with height above a canopy?
- how does stomatal conductance of plants at a site vary as the water content of the atmosphere changes.

Measuring fluxes of water through components of the soil–plant–atmosphere-continuum

Measurements of fluxes of water vapour (and CO_2) are extremely important in ecohydrology. The discharge of soil water and groundwater through vegetation is a major pathway for the movement of water into the atmosphere. Understanding the patterns and controls of such fluxes is central to understanding landscape-scale water balances. These processes can be monitored at different scales, from leaf to individual tree to canopies.

Sapflow techniques

Sapflow is the movement of water through the xylem of a plant, usually measured as litres of water per day. Tree water use is measured by determining the rate of sapflow up the stem. The rate of movement of water up the stem of a tree in the xylem is, to a very good first approximation, equal to the total rate of water use of that tree. Thus, liquid flow of water up the stem equals, in the short term, the vapour flow of water from the canopy. In the longer term there is some storage of water in a canopy as leaves grow. There is also some potential for diurnal storage in stems of some species of tree.

Three principal methods are used to measure the volume of water moving up the stem, per unit time. All require heating the transpiration stream and determining the velocity of the heat pulse or the rate of transfer of heat up the stem, from which rate of water movement is calculated.

The **heat pulse method** requires an input of a pulse of heat, typically of 1–2 seconds duration, directly into the transpiration stream. The time taken for this pulse of heat to reach a pair of temperature sensors located in the transpiration stream above and below the source of heat is used to calculate the rate of flow of transpiration. Installation of the heater and temperature sensors involves drilling three parallel narrow holes into the sapwood of a tree. The upper and lower holes house the upper and lower temperature sensors. The heater element is inserted into the middle hole.

When a heat pulse is generated by the heating element, some heat will move to the lower temperature sensor by conduction through the sapwood, against the direction of flow of the transpiration stream. The upper sensor receives heat in the transpiration stream and via conduction. The lower temperature sensor is closer to the heating element than the upper temperature sensor. As the lower sensor is warmed, the difference between the upper and lower temperature sensor increases. At some point (typically 10–100 seconds) the upper temperature sensor is warmed by the heat pulse and the difference in temperature between the upper and lower sensor begins to decrease. The time taken (t) for the upper and lower temperature sensors to reach the same temperature is the critical value recorded by the data logger attached to the system.

The velocity of the sapflow (s) (mm s^{-1}) is calculated as d/t where d is the distance between the heater and temperature sensor. Thus, $s = d/t$.

Heat pulses are usually given every 15 minutes so that rate of tree water use can be estimated at 15-minute intervals. These can be averaged to give half-hourly or hourly estimates of tree water use.

To **convert velocity to volume** of water moving up the stem (tree water use) it is necessary to know two additional items of information – the cross-sectional area of the sapwood in the stem, and the volumetric water content of the sapwood. The cross-sectional area of sapwood can be estimated by taking increment core samples from two or more places around the circumference of the tree and measuring the width of the sapwood, or by cutting the tree down and measuring the width of the sapwood directly. The junction between sapwood and heartwood is often discernible by a distinct colour change between cambium and sapwood and between sapwood and heartwood (see Fig. 7.7, Plate 7).

To estimate the **volumetric water content of the sapwood**, a sample of sapwood is obtained from an increment corer and its fresh weight (W_f) determined quickly. The sapwood sample is then rapidly placed in pure water so that the entire sample is immersed. The weight of the water with and without the wood sample is determined quickly, so that very little time is allowed for the sample to absorb water. The difference in weight is the weight of water in the sapwood (W_w). The wood sample is then dried in an oven (60–70°C) for three days and the sample reweighed (W_d).

The volume fraction of water (V_h) is then given by:

$$V_h = \left(W_f - W_d\right)/W_W \tag{4}$$

The volume fraction of wood (V_w) is then given by:

$$V_W = W_d/\left(1.53W_W\right) \tag{5}$$

The factor 1.53 is the average specific gravity of wood of many species.

The volume of water moving up the stem per second is obtained by multiplying the velocity of sap by the cross-sectional area of sapwood and taking into account the fraction of sapwood that is water.

The second type of sapflow sensor uses the **thermal dissipation method** and was developed by Granier in the 1980s (Granier et al. 1996). The principal difference between this method and the method described above is that in the Granier method xylem sap is continuously heated rather heat applied as a pulse. Two access holes are drilled into the tree. The lower hole is for a probe containing a heating element and a thermocouple. The upper hole (downstream of the lower probe) contains only the thermocouple. The temperature difference between the lower and upper probe is a function of the sapflow density passing the lower probe. Sapflow (F) is calculated thus:

$$F = 0.428S_a\left[\left(T_m - T_d\right)/T_d\right]^{1.231} \tag{6}$$

where F = sap flow, S_a = sapwood area at probe height, T_m = maximum temperature difference obtained when sapflow is zero (obtained at night or with the canopy removed), and T_d = actual temperature difference at any sampling time.

Granier type sensors can be purchased or constructed in the lab, and are cheap, robust and reliable. However, for some suppliers, the probes tend to be 'one-use' only as they can't easily be removed from the trees. Also, T_m requires determination for each species as the thermal conductance of each species tends to be different. Finally, good thermal insulation from the effects of fluctuations in solar radiation input to the stem is required as this causes artifacts in the value of T_d recorded.

The third type of sapflow sensors uses an external heat source and does not require drilling into the xylem. This system, known as the **stem heat balance** method, continuously applies heat externally to a stem or branch. The temperature of the stem/branch below and above the heated section is measured and the flow of sap determined by the following equation:

$$F = \left(Q_h - Q_r - Q_v - Q_s\right)/C_s\Delta T \tag{7}$$

where F = sapflow (in g per second), Q_h = heat input to stem, Q_v = heat loss vertically, Q_r = heat lost radially, Q_s = heat stored in the stem, C_s = heat capacity of the sap, and ΔT = temperature difference between top and bottom of the heated section. For a full discussion of this method see Smith and Allen (1996).

The advantage of this third method is that it can be used in herbaceous (i.e. non-woody) species. However, ensuring good thermal contact between heater and stem can be a problem and sustained use may kill the plant.

The choice of heat-pulse versus thermal dissipation is a personal choice, although there is a developing consensus that the thermal dissipation method is better able to measure low flow rates (during the night, early and late in the day).

Use of sapflow sensors has had a remarkable, important and rapid impact on the study of tree water use, catchment hydrology and ecophysiology in the past 20 years. The sensors are commercially available at a moderate price or can be home-made. They are robust, portable and can be left running continuously for weeks. Replication of trees within a species, sampling of several species within an ecosystem, and sampling for many days within each major season, is required to generate accurate assessments of the behaviour of trees with respect to their water use.

When starting a new project using sapflow sensors to examine water use of a new species, it is important to ensure that:

- the radial **depth profile** of water flux has been established;
- the minimum number of sensors per tree for a sufficiently accurate estimate has been established using a **Monte Carlo** simulation study;
- **wound width** has been established for each species.

For the radial depth study, a large tree of the study species is selected and 16 sensors inserted into the stem. Four sensors are inserted into each of four quadrants of the bole of the tree, at different depths. The depth of each sensor within each quadrant is different, but the four sets of depths are the same. In this way it is possible to establish how sap velocity varies with depth and thereby establish the depths to which the sensors must be placed in all subsequent experiments.

The **Monte Carlo** study is a way of establishing how many sensors at a given depth are required to generate a sufficiently accurate estimate of sap velocity. A plot of the coefficient of variance against the number of sensors is used to estimate how many sensors per tree are required. It has been generally found that four sensors (two probe sets) per tree is sufficient to get the coefficient of variance to less than 25%. Similarly, the number of trees required can be estimated from plots of coefficient of variance against number of trees sampled. Typically, four trees per species will give a reasonable estimate, but obviously the more the better. The importance of having an accurate estimate of wound width cannot be overstressed. One method to estimate wound width is to apply a water based food dye, for many hours, to a deep cut several centimetres below the sap probes. After a day or two, the wood around the probes is chiselled out. The food dye will not be found in the xylem immediately around the insertion holes because of damage and the development of a wound response. The diameter of this wound can be measured accurately with a dissection microscope by measuring the diameter of the sapwood around the holes that is without dye.

There are several sources of error in using sapflow sensors. These include:

- error in estimating sapwood area;
- error in determining the width of the wound that develops around the heater and temperature sensor – the default value frequently used is often wrong;
- deviation of the probe spacings during insertion;

- spatial variability in sapwood velocity, both as a function of sapwood depth and compass position around the stem;
- trees that have non-symmetrical stems.

The use of sapflow sensors allows us to address such questions as:

- what are the daily and seasonal patterns of water use by trees;
- how tree water use varies with changes in atmospheric and soil water content;
- are there any relationships among rates of tree water use, hydraulic architecture and climate;
- can we estimate catchment water use for multi-species woodlands and forests using tree-based estimates of water use?

Open top chambers

Open top chambers (OTCs) are useful for measuring the rate of evapotranspiration from low-lying vegetation. They sum the water lost from vegetation and soil within the chamber and have been used in studies of evapotranspiration and pollution (especially gaseous pollution such as ozone and SO_2) for the past 40 years. They are conceptually similar to leaf chambers of laboratory based gas exchange systems. They consist of a (usually) round cross-section of clear plastic (occasionally glass) with a tapered top (Fig. 4.4). Round cross-sections are preferred as this minimises the volume of stagnant (stationary) air trapped in corners of square-section chambers. Cost, required degree of robustness, and transmissibility to solar radiation affect choice of material for the walls of the chamber. Glass is heavy and makes the OTCs less portable, but it has very good transmission of photosynthetically active radiation. Mylar has good optical properties but is expensive. Perspex is cheap, but has poorer transmission and scratches easily.

Figure 4.4 An open top chamber under construction in the laboratory. This chamber can be rapidly assembled and disassembled in the field. Postgraduate students are not normally kept inside the chamber as this interferes with measurements of water flux from vegetation

Open top chambers can be as small as 0.5 m in diameter and 1 m tall, or as large as 5 m in diameter and 20 m tall, depending on the vegetation type being studied. The smaller chambers are semi-portable and can be placed on and off a patch of low-lying vegetation as required. The large ones are semi-permanent and remain in place for weeks or years.

Air is blown into the base of the OTC at one or several points around the base and at one or several heights up the side. Typical flow rates through an OTC are such that the total volume of air in the chamber is pumped through once or twice per minute. The slower the rate of air movement through the chamber, the larger the difference in water vapour concentration between the inlet and the outlet of the chamber (and hence the larger the signal strength). However, the slower the rate of air movement, the larger the temperature build-up within the chamber, which influences transpiration significantly. Too fast a flow rate results in too small a signal for measurement.

Light, temperature and water vapour concentration are measured within and outside the chamber. Ideally (but unrealistically) there should be no difference in light and temperature between inside and outside the chamber. The rate of air flow into the chamber must also be measured very accurately, using either mass flow controllers or vane anemometers. Air flow should be sufficient to cause the leaves in the chamber to move slightly, thereby reducing boundary layer resistances to diffusion of water from the leaf.

Transpiration from the vegetation within the chamber is calculated from the following:

$$E = \left((\Delta e)(F_a)\right)/(A_l) \tag{8}$$

where E is transpiration rate (mol m^{-2} s^{-1}), Δe is the difference in water vapour concentration (mol m^{-3}) between the inlet and outlet of the OTC, F_a is the rate of air flow through the chamber (m^3 s^{-1}) and A_l is the total leaf area (m^2) within the chamber at the time of measurement.

Measuring evapotranspiration of canopies

Two principal methods for measuring total evapotranspiration from large areas of vegetation are commonly used. The first technique requires the determination of the **Bowen ratio**. The second is called the **eddy covariance** technique. These methods use a **micro-meteorological** approach and provide estimates integrated over areas of up to 1–2 km^2 of ground area. They can provide data at 15-minute resolution (or less) and can be used over short (grasses) and tall (rainforest) canopies. These techniques cannot separate fluxes of water vapour from bare earth, the understorey and the overstorey. Hence, they measure total evapotranspiration from all surfaces within the **footprint** (1–2 km^2) of the technique. The footprint is the area of ground that each technique is able to integrate (average) over during the measurement period.

Canopy energy balance and the Bowen ratio method

The energy balance of a canopy is:

$$R_{n,c} = G + H_c + \lambda E_c \tag{9}$$

where $R_{n,c}$ = net canopy radiation, G is soil heat flux (temperature of the soil increases or decreases if there is a net receipt or net loss of heat), H_c is sensible heat flux density from the canopy and λE_c is canopy latent heat flux. Canopy latent heat flux is the transpiration rate of the canopy, since we can convert the evaporation of 1 g of water into an equivalent energy flux (as discussed on p. 36). Metabolic heat flux (heat produced from metabolism), and changes in temperature of the vegetation are ignored because they are so small on an hourly or daily scale.

Calculating evapotranspiration rates using the **Bowen ratio** method requires measurement of all the components of a canopy or stand-scale energy budget. Thus, calculating evapotranspiration rates using the Bowen ratio method requires measurements of fluxes of energy and mass (water) above a canopy. The Bowen ratio (β) is defined as:

$$\beta = H/\lambda E \tag{10}$$

Now:

$$H/\lambda E = \gamma((K_h \Delta T)/(K_e \Delta e)) \tag{11}$$

where H is sensible heat flux, λ is the heat required to evaporate water (latent heat of vaporization 2.45 kJ g^{-1}) and E is the rate of evaporation, γ is the psychrometric constant, K_h and K_e are eddy diffusivities (a measure of the ease of diffusion through air) of heat and water vapour respectively, ΔT is the difference in temperature between the two sensor heights and Δe is the difference in water vapour concentration between the two heights. λE is the latent heat of vaporization of water (= 2.45 MJ kg^{-1}) multiplied by the rate of transpiration.

Differences in temperature and water vapour concentration can be measured relatively easily, but K_h and K_e are not easy to determine. The Bowen ratio, β, can also be written as:

$$\beta = -(C_p \rho_a K_h \Delta T)/(\lambda K_e \Delta e) \tag{12}$$

Solving equation (9) and equation (10) for E yields:

$$E = (R_n + g)/(\lambda(\beta + 1)) \tag{13}$$

The Bowen ratio technique is not readily applicable to tall vegetation such as forests because of the small size of the temperature gradient that exists above a forest and because of the large length of fetch required. The **fetch** is the length (distance) of uniform vegetation upwind of the sensor required to allow the wind flowing across the vegetation to have interacted sufficiently to have reached equilibrium. This distance, for a 20 m tall woodland, can be 2 km or more. Changes in soil temperature (and hence energy store) are also required when making Bowen ratio measurements.

In order to calculate the Bowen ratio, measurements of net radiation, sensible heat flux and water vapour pressure at two points above the canopy, soil heat flux and air temperature are also required. A radiometer measures net radiation (R_n) above the vegetation. As mentioned in Chapter 2, net radiation is the balance between incoming shortwave plus incoming long-wave radiation minus outgoing long-wave radiation plus reflected shortwave radiation. In addition to measuring net radiation, sensors are used to measure changes in heat content of the soil to determine the heat flux to/from the soil. Perhaps most importantly, water vapour concentration and air temperature are measured at two points above the canopy. For a low grass sward, these sensors might be 1.5 m and 2.5 m above the ground. From these measurements it is possible to calculate total evapotranspiration.

With automated instruments and data loggers, Bowen ratio systems can be used continuously and remotely for weeks or months. The biggest problem with this technique is getting paired temperature and humidity sensors that are calibrated and sensitive enough to detect the small differences that occur above the canopy. To overcome this problem, many automated systems routinely interchange the upper and lower pair of sensors every 20 minutes or so. Measurements of temperature and water vapour concentration are made every second and averaged over longer time intervals.

Eddy covariance

The technique known as **eddy covariance** relies on the fact that air flow above and through a canopy is **turbulent**. Thus, above a canopy, wind at any single point in space and single point in time can be moving in any direction because air travelling over a canopy consists of a large number of eddies (small chaotic currents of wind).

Figure 4.5 The top of an eddy correlation tower above a savanna woodland in the Northern Territory of Australia. Located at the top of the tower is a 3-dimensional sonic anemometer, a fast response Krypton hygrometer, a radiation sensor and the inlet tube for sampling air for CO_2 concentrations

Photo courtesy of L Hutley, Charles Darwin University

An eddy covariance system consists of three major components, all located several metres above the canopy (Fig. 4.5). First, an **open-path infra-red gas analyser** (or a **hygrometer** if measuring only water vapour fluxes) measures the concentration of water vapour and CO_2 at a single point. Second, a **3-dimension sonic anemometer** measures the velocity of (in theory) the same eddy of air that was sampled by the gas analyser. The 3-dimensional anemometer determines the velocity in the x, y and z planes (directions: vertical, horizontal 1 and 90° to horizontal 1). It does this by measuring the time taken for a pulse of ultrasound to move from three sources to three detectors, arranged at 90° angles to each other. Third, a temperature sensor measures the temperature of the air being sampled. In addition, a radiometer is placed above the canopy and soil heat flux plates are installed into the ground in a representative patch of vegetation.

Measurements of wind speed, temperature and CO_2 and H_2O concentrations are made at a frequency of 10 Hz (10 times per second). Large amounts of data are generated in short periods of time.

CO_2 and H_2O concentrations in air change as air moves over and through a canopy. During sunlit hours, the canopy photosynthesises and therefore air that is moving down towards the canopy from the bulk air above will tend to have a higher concentration of CO_2 than air moving up from the canopy into the bulk air stream. At the same time, during sunlit hours, air moving up from the canopy will tend to have more H_2O vapour than air moving down because of transpiration by the canopy. It is therefore possible to derive values for the net flux of water vapour from the canopy and CO_2 flux to the canopy. This is achieved by integration of all the 10 Hz data supplied by the 3-d anemometer into upward and downward components of air movement and the corresponding concentrations of water and CO_2 for each anemometer record. Typically, rates of transpiration and CO_2 uptake are calculated for 15-minute averages of the 10 Hz data.

Eddy covariance (EC) remains the only direct measure of gas flux to and from a canopy. As with the Bowen ratio method, the EC technique cannot separate fluxes from soil, understorey and upper-storey (Fig. 4.6).

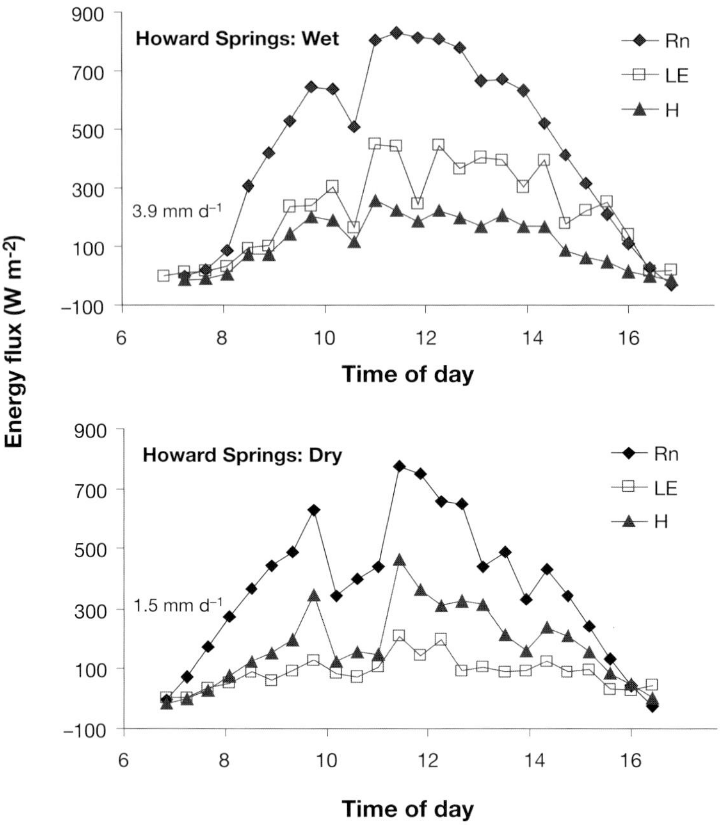

Figure 4.6 Examples of the energy budget of a tropical savanna in northern Australia. In the wet season, net radiation peaked at about 900 W m^{-2} and latent energy peaked at about 450 W m^{-2}. Fluctuations occurred because of cloud cover. Sensible heat flux peaked at about 200 W m^{-2}. Water use for this day was 2.9 mm. In contrast, in the dry season, despite similar net radiation levels latent energy flux was much reduced (peak value of 180 W m^{-2}) and sensible heat flux increased (peak values of 450 W m^{-2})

Source: Redrawn from Hutley et al. (2000)

The EC technique provides an integrated measure of the behaviour of a patch of vegetation approximately 1 km square when placed several metres above a woodland or forest. This means that it is the **average** or bulk behaviour of the ecosystem that is being measured. The technique cannot discriminate how the trees, shrubs, grasses and bare soil are behaving independently. The technique requires flat terrain and a relatively uniform canopy structure for a significant distance upwind of the equipment tower. For a low canopy (e.g. a wheat crop) uniform vegetation 500 m upwind is required. For a taller rougher canopy such as a woodland or forest, a fetch of 1–2 km is required. This distance is necessary so that the wind travelling over the canopy has time to equilibrate with the canopy and therefore measurements of eddies reflect an equilibrium value.

Several sources of error exist in the use of the EC technique. These include:

- hilly terrain that is too close to the sensors so that the sensors do not measure air that has equilibrated with the canopy;
- terrain that is too spatially variable and contains a range of ecosystems too close to the measurement tower;

- the tower is located too close to open water bodies within the fetch;
- there is insufficient wind speed (especially a problem at night) and vertical stratification of temperature within the lower atmosphere. This leads to errors in the sampling of air, which is assumed to be travelling at a sufficient speed to allow adequate mixing. Stable stratification of the volume of air within the canopy can act as a large buffer, with CO_2 being stored at night within the canopy and released in the early morning;
- instrument errors arising from poor positioning of the anemometer and gas analyser;
- turbulence created by the presence of the tower structure;
- time lags between anemometer and gas analyser not being corrected.

Despite methodological complexities and high set-up and maintenance costs, EC has proven to be invaluable in addressing the following questions:

- the daily and seasonal patterns of ecosystem water and CO_2 flux of boreal, temperate, tropical rainforest and savanna ecosystems;
- how temperature, rainfall, soil and atmospheric water content and radiation influence canopy gas exchange;
- the water and CO_2 budgets of catchments.

Penman-Monteith equation

The most famous equation linking climate factors (temperature, solar radiation, vapour pressure), plant factors (stomatal conductance) and transpiration is the **Penman-Monteith** equation. It is used for calculating the rate of evapotranspiration from a uniform canopy. Monteith (1965) and Atwell et al. (1999) provide the following version of the P-M equation (there are many variations):

$$\lambda E = \left[sR_n + (\rho C_p)(\delta_e)(g_a)\right] / \left[s + \gamma(1 + g_a/g_c)\right] \tag{14}$$

where:
λ = the latent heat of vaporization of water = 2.45 MJ kg^{-1}
E = rate of evapotranspiration (mm d^{-1})
s = slope of the curve relating saturation water vapour pressure and temperature (kPa K^{-1})
R_n = net radiation (W m^{-2} or MJ m^{-2} d^{-1})
ρ = air density (g m^{-3}; typically 1.22 kg m^{-3})
C_p = specific heat capacity of air at constant temperature (J kg^{-1} K^{-1}; typically 1012 J kg^{-1} K^{-1})
δ_e = ambient (bulk) air water vapour pressure (kPa)
g_a = aerodynamic conductance to water vapour (m s^{-1})
g_c = canopy (or surface) conductance (m s^{-1})
γ = psychrometric constant (kPa K^{-1}, typically 0.0662 kPa K^{-1}).

Many of these variables (net radiation, ambient water vapour pressure) can be measured directly. Other parameters in the equation are constants, assumed to be constant or it is known how they vary with air temperature (λ, s, C_p, air density). However, estimation of g_a and g_c has proven very difficult. Total **aerodynamic conductance** (g_a) is the conductance to water vapour of the diffusion path from an imaginary evaporation surface in the vegetation canopy to an imaginary reference point in the bulk atmosphere above the canopy. The main determinants of g_a are wind speed, vegetation structure, leaf density and the distance between the imaginary evaporating surface and reference surface above the canopy. Aerodynamic conductance can be calculated thus:

$$g_a = k^2 u / \{\{\ln[(z - d)/z_o]\}^2\} \tag{15}$$

where z = height of measurement of wind speed, d = displacement height, which is the distance between the actual height of canopy (h) and a point within the canopy where wind speed (u) tends to zero, and z_0 is the roughness length of the canopy. Roughness length is a descriptor of the wind speed profile that develops over any surface as air moves across it and is the height above the ground surface at which wind speed approaches zero due to the effects of vegetation. k = von Karman's constant = 0.41, u_z is the wind speed in the vertical direction (as opposed to either of the two horizontal directions, u_x and u_y).

The roughness length of the canopy (z_0) and the displacement height (d) can be adequately approximated for forests and row crops using the following estimates:

$$z_o = 0.076\ \text{h (forests)}$$

and

$$d = 0.78\ \text{h (forests)}$$

or

$$z_o = 0.123\ \text{h (row crops)}$$

and

$$z_o = 0.67\ \text{h (row crops)}$$

Bulk canopy (or surface) conductance (g_c) is the conductance to water vapour diffusion through the vegetation that is subject to stomatal control plus diffusion from wet soil surfaces. Bulk canopy conductance can be calculated using leaf area index and stomatal conductance thus:

$$g_c = (g_s)(LAI) \tag{16}$$

where LAI = leaf area index of sunlit leaves, g_s = leaf stomatal conductance (usually measured on sunlit leaves), and g_c = canopy (or surface) conductance.

For a woodland and forest, LAI can vary between 0.5 for a sparse open woodland of a semi-arid site, to more than 5 for a tall closed rainforest. Defining the proportion of total LAI that contributes substantially to transpiration is difficult, as is assigning an average leaf stomatal conductance. Within the canopy of a single tree stomatal conductance of individual leaves can vary between 100 and 500 mmol m^{-2} s^{-1}, or more.

An alternative way of calculating g_c is to use an inversion of the P-M equation (see Lu et al. 2003 for a discussion of this).

Transpiration calculated for a crop and forest using the P-M equation

For a crop of height = 0.4 m and wind speed = 1.5 m s^{-1}, g_c = 0.02 m s^{-1}, g_a = 0.01992 m s^{-1}, R_n = 15 MJ m^{-2} d^{-1}, daily average VPD = 0.98 kPa and T_{air} = 15°C, and 1 day = 86 400 seconds, then from equation 14:

$$\begin{aligned}\lambda E &= \left[sR_n/\{s+\gamma(1+g_a/g_c)\}\right]+\left[(\rho C_p)(\delta_e)(g_a)\right]/\left[s+\gamma(1+g_a/g_c)\right]\\ &= \left[(0.11)(15)/\{0.11+0.0662(1+.01992/.02)\}\right]+\\ &\quad \left[(1.22)(1.012)(1000)(0.98)(0.01992)(86400)\right]/\left[0.11+0.0662(1+0.01992/0.02)(10^6)\right]\end{aligned}$$

$$\lambda E = 14.89\ \text{MJ m}^{-2}\text{d}^{-1}$$

Therefore:

$$\begin{aligned}E &= 14.89/2.45\\ &= 6.07\ \text{mm d}^{-1}\end{aligned} \tag{17}$$

For a forest with the same climatological values as above, but h = 30 m and wind speed = 3 m s^{-1}, then from equation (15) $g_a = 0.19$ s m^{-1} and a typical $g_s = 0.0067$ s m^{-1}. This is generally much smaller than values seen in crops. Aerodynamic conductance is much larger for forests than crops because they are aerodynamically rougher and so are better coupled to the atmosphere.

From equation (17), E = 4 mm d^{-1}. Despite the larger wind speed over the tops of taller canopies of trees, and the same input of solar radiation, the same leaf temperature and VPD, transpiration is reduced by 35%, mostly because of the much lower stomatal conductance of tree leaves compared to row crops such as *Zea mays*.

Lysimeters

A lysimeter is a piece of equipment that allows quantification of water loss from a patch of soil and associated vegetation. Thus, a **lysimeter** measures the rate of evapotranspiration from a small (typically 0. 25–4 m^2) patch of vegetation (usually low lying grass or crops but occasionally trees) to be measured hourly or daily. A circular steel pipe typically 0.25–2 m in diameter and 1–3 m long is hydraulically rammed into the ground in a patch or representative vegetation (crop, pasture or native vegetation). Clearly this is very difficult in very rocky or stony ground, but relatively easy in sandy soil. Digging up the soil and filling the pipe by hand would disturb the soil structure and soil profile too much and therefore it is best to ram the pipe into the ground hydraulically. A metal bottom is welded in place and access/drainage holes drilled into the exposed side of the pipe or in the bottom plate. These holes should be capped when not in use to exclude light and to trap water that is draining down. These holes can be used to insert soil moisture probes, for example, or to collect drainage water from different depths. Access to these holes is usually via a narrow access tunnel alongside the lower half of the lysimeter. The lysimeter should be placed so that it is surrounded by representative vegetation of the same type as is growing in the lysimeter. Access to the bottom of the pipe is made by digging an access tunnel, or by lifting the pipe and soil by crane.

Two types of lysimeter can be constructed. The first type is a **weighing lysimeter,** which monitors the change in weight of the lysimeter – which includes the soil, vegetation and water – using commercial strain gauges (also called load cells) located beneath the lysimeter. Such gauges can measure several tonnes of weight with a resolution of 0.1 kg and changes in weight can be recorded using data loggers at specified intervals. The other type is the **drainage lysimeter**. In this type, the volume of water that drains past the root zone is measured using drainage ports in the side and bottom of the lysimeter vessel. Concomitant measurements of rainfall are needed if a water balance is to be generated. Lysimeters and associated soil moisture monitoring equipment can be used to address the following questions:

- the hourly, daily and seasonal patterns of evapotranspiration from vegetation;
- the rate of drainage of water past the root zone of vegetation;
- where roots take up water from and how quickly water infiltrates through a soil profile.

Other useful plant ecophysiological techniques to assess plant function

Reductions in water availability (e.g. through groundwater pumping) cause, in the early stages, changes in plant function. Typical sequences of changes are illustrated in Figures 2.9, 4.7 and 6.18. Techniques discussed so far have provided methods to assess changes in soil and atmospheric water content and changes in fluxes of water and carbon from and to leaves and canopies. This section deals with methods for assessing leaf stomatal conductance, the use of stable isotopes to determine the site of water uptake by roots, leaf area index of a site, hydraulic

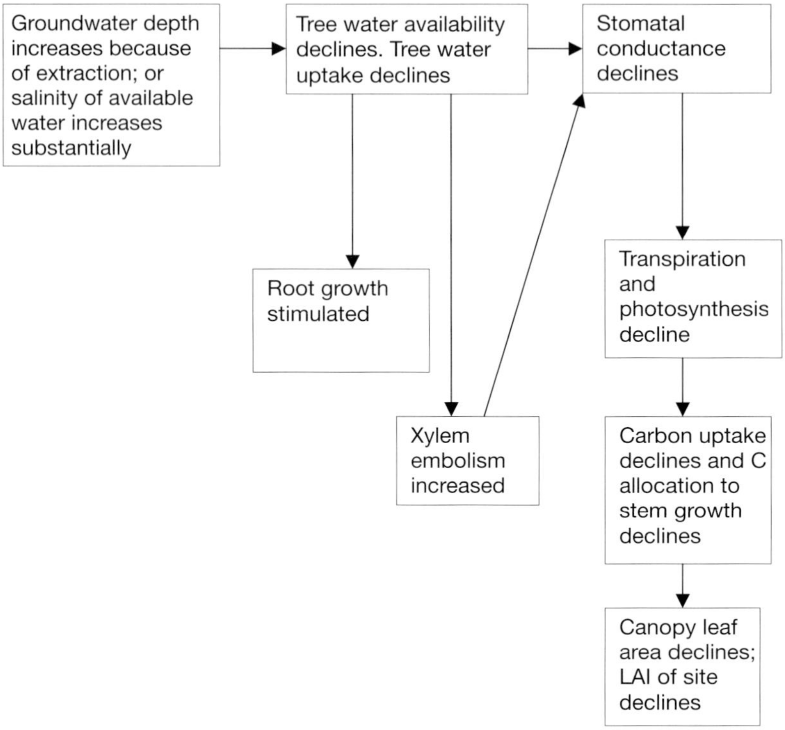

Figure 4.7 Reduced groundwater availability (or increased salinity of available water) causes reduced stomatal conductance, reduced tree water use and reduced growth and leaf area index. The first two may be observed within days, the later two may take months to be measurable in the field. Only after several years can changes in community structure and composition be measured

architecture and xylem embolism, root growth and finally stem growth. If possible, these methods (and those discussed earlier in this chapter) should be applied to control (non-impacted) and experimental plots (impacted by groundwater pumping, for example) that have as little difference in vegetation composition, structure, climate and edaphic conditions as possible. Alternatively, a pairwise BACI design is used. A BACI design allows for measurements **b**efore and **a**fter an event in **c**ontrol and **i**mpacted sites (Underwood 1991), for example before and after pumping of groundwater. Chapter 5 has an example of this design.

Leaf porometers

Leaf porometers measure the degree of stomatal opening of individual leaves. Stomata are the pores in leaves through which water diffuses out, as transpiration, and CO_2 diffuses in, for photosynthesis. Stomatal pores open and close in response to external factors (e.g. atmospheric water content and solar radiation) and internal factors (e.g. leaf water content, xylem sap abscisic acid content – a plant hormone that closes stomata during drought). Leaf porometers measure the stomatal conductance of leaves; very open stomata have a high conductance to water vapour and closed stomata have a very low conductance.

Stomatal conductance (G_s) of leaves is an important variable that influences water use by vegetation. Stomata respond to a large number of variables, including soil and atmospheric water content, leaf water status, temperature and the amount of solar radiation. Different species show different responses to these variables and, even within a species, leaf age and the complex interaction of all these variables makes accurate prediction of G_s in the field extremely difficult. Different species also have differing stomatal sizes and stomatal densities (number per unit area of leaf), making prediction even more difficult.

Two types of leaf porometer are used in plant ecophysiology. The first is a **leaf diffusion porometer** that measures G_s of individual leaves within a period of about 1 minute. The porometer consists of six principal parts. These are:

1 a small hand-held leaf chamber that is lightly clamped onto a leaf;
2 a humidity sensor that measures the humidity of the air space above the leaf within the chamber;
3 an air pump that passes air through a drying agent then over the leaf in the chamber;
4 a timer to determine the time taken for the humidity inside the chamber to increase by 5% as a result of transpiration;
5 an in-built computer to calculate G_s and store the data;
6 a calibration plate.

When a leaf is enclosed in the chamber, humidity in the chamber increases because of transpiration. The operator sets a limit (usually a few percent above ambient humidity) at which the air pump is activated and dry air is pumped into the chamber. When the humidity in the chamber is reduced to 5% below ambient, the pump stops and humidity in the chamber increases. The on-board computer and timer determine the time taken for the humidity to increase by a set amount. A calibration plate of known conductances is used to calibrate the leaf chamber in the field at the ambient temperature. This calibration is vital and must be undertaken several times each day if ambient temperature varies by more than a few degrees. The larger the value of G_s the shorter the time taken for the humidity in the chamber to increase by the required amount. Usually two or three cycles of increasing/decreasing humidity are sufficient to get a stable estimate of G_s.

The second form of porometer (**continuous flow porometer**) has a continuous air flow through a large transparent leaf chamber. The amount of air required to be passed through a desiccant (powder that absorbs water vapour) to keep the humidity constant within the chamber is measured and G_s calculated with an on-board computer. This system requires less calibration but is more bulky than the leaf diffusion porometer.

Sampling of leaves in a canopy with any porometer is a difficult task. Replication of several leaves on each tree and several trees within each species, with replicate plots within a site, is a minimum requirement. However, each leaf in a canopy is in a unique micro-climate; temperature, humidity, wind speed and solar radiation vary in space and time. In addition, variation in height through the canopy is associated not only with variation in micro-climate, but with leaf age and leaf water status. As a compromise, many studies sample only leaves that are fully expanded and fully sunlit. It is generally assumed that the sunlit portion of the canopy contributes the most to water and carbon flux of the canopy.

Studies of G_s using leaf porometers are very useful in addressing the following questions:

- how does G_s vary with position in the canopy;
- how does G_s vary with soil and atmospheric water content, leaf water status, solar radiation input and temperature;
- how do species differ in their G_s response to variation in climate?

Using stable isotopes to determine sources of water

Trees and shrubs are able to access water from the upper unsaturated soil profile, from the capillary zone of a groundwater store or, when growing sufficiently close, from nearby streams and rivers. By comparing the stable isotope composition of water in the xylem with that of the various sources of water available to a plant it is possible to answer the question: from where are plants extracting most of their water? Examples of the application of this technique are given in Chapter 6.

Stable isotopes are isotopes of an element that do not undergo radioactive decay and therefore are stable over time. Three important stable isotopes used in ecohydrology are those of carbon, oxygen and hydrogen. Most carbon occurs as the stable isotope ^{12}C. A very small amount is present as unstable ^{14}C that decays over time and is used for dating (radiocarbon dating) soils, skeletons and other materials. A third isotope of carbon, ^{13}C, is stable and occurs as a tiny fraction of all the carbon on earth.

The most abundant form of oxygen on earth is ^{16}O. This is a stable isotope and accounts for about 99.76% of all oxygen. Another stable isotope is ^{18}O, which accounts for about 0.2% of all oxygen, while ^{17}O accounts for the rest. For hydrogen, ^{1}H accounts for about 99.985% of all hydrogen and is stable, as is deuterium (^{2}H), which accounts for about 0.015%. Tritium (^{3}H) is a radioactive isotope of hydrogen useful in tracer studies of water flow.

The ratio of $^{2}H/^{1}H$ or the ratio $^{18}O/^{16}O$ is variable in different bodies of water. Water, of course, can be composed of a range of combinations of ^{1}H, ^{2}H, ^{16}O and ^{18}O, including the following:

- $^{1}H^{1}H^{16}O$;
- $^{1}H^{1}H^{18}O$;
- $^{1}H^{2}H^{16}O$;
- $^{1}H^{2}H^{18}O$;
- $^{2}H^{2}H^{16}O$;
- $^{2}H^{2}H^{18}O$.

The molecular weight of each of these molecules is slightly different because of the different number of neutrons in each nucleus of each atom. It is the difference in molecular weight, and the fact that different water bodies have different compositions, that allows us to compare the stable isotope composition of water in the xylem of a tree with the composition of water in soil, in rivers and in groundwater.

The ratio of $^{2}H/^{1}H$ or the ratio $^{18}O/^{16}O$ in water varies for each different rainfall event because of fractionation. When water evaporates from oceans, lakes and wet surfaces lighter molecules of water evaporate slightly faster than heavier molecules of water and clouds therefore are slightly depleted in the heavier isotopes than the ocean or lake from which the water evaporated. When water condenses and falls as rain, the rain is slightly enriched in the heavier isotopes. Also, the degree of fractionation at every evaporation and condensation event is slightly different. Consequently, water in soil from a recent rainfall event has a different isotope composition from water that is deeper in the soil and which came from a previous rainfall event. Similarly, water in rivers, streams and groundwater stores have different isotope compositions from each other. Therefore, when the isotope composition of water in the xylem of a branch is compared to the isotope composition of water in nearby streams or in soil or groundwater it is possible in many (but not all) cases to show from which source the water in the xylem was derived. It is assumed (and has been shown experimentally) that when water is absorbed by roots and transported up the xylem, fractionation of the isotopes does not occur. In many cases, water is absorbed from more than one source. While this makes interpretation more difficult, a linear mixing model of water can be applied to determine the relative contribution of the two sources of water used by a tree (Thorburn et al. 1993).

Figure 4.8 shows a typical isotope profile within the soil, showing a decrease from a value of C at the soil surface to A at the water table. (Near the surface, water is usually enriched in the heavy isotopes due to evaporative processes.) Suppose that we measured an isotope ratio value

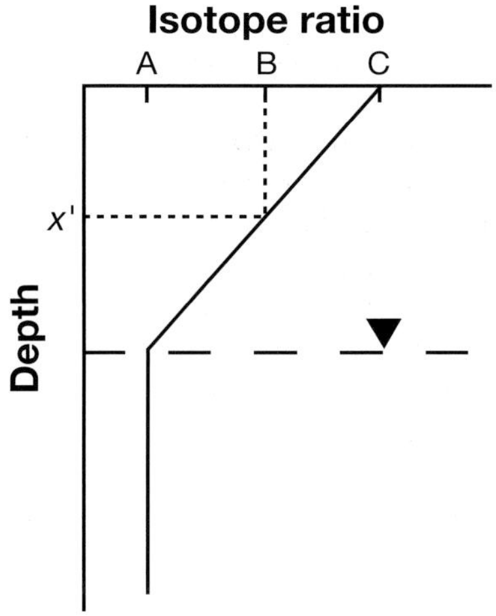

Figure 4.8 Schematic representation of isotope ratios within the soil and groundwater and their use in discovering plant water source. Black triangle indicates depth of groundwater

of C within the plant xylem. Clearly, this would indicate that all (or most) of the water used for recent transpiration was sourced from very near the soil surface. If a ratio value of A was measured, then water must be derived from the watertable or immediately above it. If a ratio value of B was measured in plant xylem, a number of interpretations are possible. Water could either be sourced from the middle of the unsaturated zone (depth x'), or it could be a mixture of water from shallower and deeper depths.

Measuring the stable isotope composition of water requires a mass spectrometer to accurately measure the ratios of $^{2}H/^{1}H$ and $^{18}O/^{16}O$ in the water samples taken from the tree, soil, river and groundwater.

Leaf area index

The theory and a review of methods for determining the leaf area index of a site has recently been published (Jonckheere et al. 2004) and therefore the following is only a brief description of the more commonly applied methods in ecohydrology. Their application is discussed in van Gardingen et al. (1999) and McPherson and Peper (1998).

Leaf area index (LAI), as mentioned in Chapter 1, is the ratio of the total leaf area within a canopy to the ground area covered by the canopy. The most accurate method requires the canopy to be cut down (simple for crops, less easy for trees, especially in forests) and the total leaf area in the canopy to be measured after removing all the leaves from the plants. A leaf area meter is used to measure leaf area. In practical terms, a small sample of trees of known diameter are cut down, all the leaves removed by hand and the total dry weight of all the leaves is measured. The leaf area and dry weight of a small subsample of these leaves are determined to enable calculation of the total canopy leaf area. The LAI of the plot is calculated by scaling from the relationship between leaf area and tree diameter and the total diameter of the trees in the sample plot. This approach is time-consuming, expensive (mostly labour costs) and dangerous (felling large numbers of trees with large numbers of assistants present). It is also a destructive measure that can only be done once at a given plot.

An alternative method is to use a commercial leaf area index meter (such as the Li-Cor 2000) to measure LAI in the field. These meters use a lens with a series of concentric metal rings placed over it. The system measures the amount of light received by a light sensor located beneath the lens. Measurements must be made simultaneously beneath and above the canopy (or in a very large clearing within the forest) and the sun must not be present within the sensor's field of view. Therefore measurements are usually made close to dawn or dusk or in a completely overcast (cloudy) sky. This makes the collection of many measurements within a forest problematic; additionally, the sensor is expensive. The technique assumes that the leaves in the canopy are randomly spaced and oriented randomly – assumptions that are rarely met. Very clumped distributions of the trees, or very sparse canopies, make application of this technique problematic and LAI is usually underestimated. The technique is very useful in uniform tree plantations or crops but can also be used for comparative purposes, such as tracking seasonal changes in LAI, in ecosystems that don't completely conform to the assumptions.

A third method uses a camera with a hemispherical lens. A number of random photographs of the canopy are taken from the ground (i.e. the camera faces upwards) with focal length set at infinity. This method assumes that one-sided canopy leaf area can be calculated from measurements of canopy gap area. Computer software for analysis of digitized photographs is available (Rich 1989, 1990). The camera is usually oriented with a compass so that the top is pointing north; a small pointer is sometimes included in the photograph to indicate north. The camera must be leveled using a bubble level. A uniform sky (dawn or dusk or completely overcast conditions) is required.

Determination of the LAI of canopies is important for answering the following questions:

- whether recent changes in groundwater availability have had an impact on the LAI of canopies, and hence are likely to reflect changes in canopy water use;
- whether vegetation health is declining across patches of landscape.

Measuring hydraulic conductance, xylem embolism and vulnerability curves

Two key components of the hydraulic architecture of trees are measures of the hydraulic conductance of different parts of the conducting pathway (roots, stem, branches and leaf lamina), and the degree of embolism occurring within the conducting tissue. This section describes some commonly applied methods of measuring these components of hydraulic architecture. Examination of hydraulic architecture is useful in ecohydrological studies because, for example, sensitivity of different species to water stress and changes in hydrologic balance is largely determined by differences in sensitivity of xylem to embolism.

Branch hydraulic conductance

Using Ohm's analogy (which states that the voltage (V) driving a current (I) is equal to the product of the current and the resistance (R) to flow of the current; i.e. $V = I \times R$), and knowing that conductance is the inverse of resistance, we can determine that the hydraulic conductance (k) of an excised segment of stem or branch is the rate of flow of water (F; equivalent to current in a wire) divided by the pressure gradient (ΔP) between the ends of the segment (equivalent to the voltage). Thus:

$$k = F/\Delta P \tag{18}$$

One method for a segment of stem or branch uses a partial vacuum to induce flow (see Prior & Eamus 2000; Macinnis-Ng et al. 2004).

Terminal segments of branch (for example) are collected from the field early in the morning, to minimise the chance of xylem embolism already occurring. Branches are cut under water so that, upon excision, no air is sucked into the xylem because of the tensions already existing there. If air is sucked into the xylem the measure of conductance will be lower than that actually occurring because many (most) of the xylem conduits will be blocked by air bubbles.

Branches are placed upright in a bucket of water with the cut end maintained under water, and the bucket covered in black plastic to minimise transpiration from attached leaves.

Back in the laboratory, average minimum xylem conduit length is determined using the method described by Drake and Franks (2003). Water is forced, under low pressure (175 kPa; 1.75 bar) through lengths of branch with both ends cut. The rate of flow of water per unit time (typically measured by collecting the water forced from the cut end in a beaker on a balance) under constant pressure is measured as the branch is repeatedly shortened. Once the average length of the conduit is reached, a sudden increase in the flow rate occurs because the highly flow resistant end walls of the conduit have been removed. An alternative method of measuring average conduit length is to macerate the sapwood lightly with a suitable cellulose stain (toluidine blue) and observe under a microscope. This won't work for many eucalypt species where xylem vessel lengths can be up to 100 cm in length!

Once the average length of xylem vessels has been established, branch lengths used in the determination of conductance should be double this average length.

Branches are removed from the bucket and leaves (including petioles) removed with a razor. If the leaf scars are not sealed (with tape or glue), flow rates measured are likely to be 5–10% larger than if scars are sealed, and the resulting conductance values obtained with unsealed scars will be 5–10% larger than actual values in situ.

Basal and distal ends of the branches are re-cut, under water, and the basal end attached to a suitable diameter hose filled with degassed (water exposed to half vacuum for 10 minutes or until no more bubbles produced), acidified (pH 2.0 with HCl), 0.2 μm filtered water. The hose attaches the branch to a 1 mL graduated pipette (filled with degassed, acidified filtered water). The pipette is held vertically and will be used to measure rates of flow of water through the branch.

As a preliminary study, readers should consider whether pH and ionic composition influence xylem conductance Some studies find an effect, some do not (Zwieniecki et al. 2001 and Macinnis-Ng et al. 2004 respectively).

The majority (90%) of the branch is now inserted into a vacuum chamber, with the remaining 10% protruding out of the chamber. The end with the hose attached to the pipette remains outside the chamber. Four levels of reduced pressure (ΔP = 20, 30, 40, 50 kPa) within the chamber is applied for 3–10 minutes until a steady state flow rate through the branch is achieved, after which time flow rate is measured with a stopwatch. A plot of F (y axis) against ΔP yields a straight line, the slope of which is k (expressed in mg water MPa^{-1} s^{-1}). Hydraulic conductivity (mg water cm MPa^{-1} s^{-1}) can be calculated from this by multiplying k by branch length. Conductance and conductivity per unit sapwood area can also be determined by dividing conductance or conductivity by sapwood area of the basal end of the branch.

An alternative method applies a small positive pressure (5–50 kPa) to the proximal end of an excised branch and measures the flow rate through the branch by measuring the weight of the sap collected from the distal end (Davis et al. 2002).

Whole plant, root, shoot and stem hydraulic conductance

An automated system for measuring root, shoot and stem hydraulic conductance is available for field or laboratory use (Tyree et al. 1995). Shoots are excised from the root system and the root stump (still in the ground) is attached to a multi-sample manifold and hydraulic pressure gradients are applied automatically to the root stump. Care must be taken to ensure that the correct manifold is used for each root/shoot system. Water flow (F) into the root system (measured with the high pressure flow meter) and applied pressure (P) is measured every 3 seconds as the applied pressure is increased at about 3–7 kPa s^{-1}. Root hydraulic conductance (k_{root}) is calculated from the regression of pressure versus flow since $k_{root} = dF/\Delta P$. Following such measurements, the excised shoot is attached to the pressure manifold and a constant pressure (of about 0.5 MPa) applied for 15 minutes or until the flow rate is stable. Applied pressure is then increased and the relationship between pressure and flow established as for the root, thereby measuring total shoot conductance (k_{shoot}). Following determination of whole shoot conductance, leaf and lamina are removed and a new measurement of conductance performed by varying the pressure and establishing the relationship between pressure and flow, to calculate stem conductance (k_{stem}). Whole plant conductance (k_{plant}) can be calculated as the sum of the inverse of k_{shoot} and k_{root}. All conductances thereby calculated can be normalised by leaf area by dividing all values by the total leaf area of the shoot.

An alternative method for calculating root, stem and leaf hydraulic conductance is called the **evaporative flux method**, in which transpiration rate is measured either by the change in weight (every 30–90 minutes) of plants growing in pots, or, for trees in the field, by using sapflow sensors or measurements of stomatal conductance and leaf area and leaf-to-air vapour pressure difference (Berryman et al. 1997). Total tree water flux (Q) is approximated:

$$Q = LA \times G_s \times \theta$$

where LA = total leaf area of the plant, G_s = stomatal conductance and θ is the saturation mole fraction of water vapour (a measure of the amount of water vapour in the air).

At the same time as measuring transpiration rate (as change in weight of pots, or from sapflow or stomatal conductance measurements), gradients of water potential between root and leaf must be made. This requires three separate measures of xylem pressure potential using a Scholander pressure chamber. First, measurements are made using transpiring leaves (Ψ_{leaf}) at different times throughout the day. Second, measurements are made on adjacent leaves that are covered in aluminium foil and plastic (to stop transpiration). Determination of the leaf water potential of leaves covered in foil and plastic provides an estimate of stem xylem water potential (Ψ_{stem}). Finally, for small plants growing in pots, the shoots of replicate plants are covered in black plastic for 6–8 hours to minimise transpiration. Leaf pressure potential is then measured with the pressure chamber and this value is assumed to be equal to the water potential at the soil–root interface (Ψ_{soil}). If measurements are being conducted on whole trees in the field, leaf water potential is measured before dawn and this value assumed to be equal to the water potential of the soil–root interface (O'Grady et al. 2006). See Donovan et al. (2001) for a discussion of when this assumption may not be valid.

From these measurements, K_{plant} (leaf specific whole plant conductance) is calculated from:

$$K_{plant} = E/\Psi_{soil} - \Psi_{keaf} \qquad (19)$$

Furthermore, $K_{root+stem}$ and K_{leaf} can be calculated from $K_{root+stem} = E/\Psi_{soil} - \Psi_{stem}$ and $K_{leaf} = E/\Psi_{stem} - \Psi_{leaf}$ respectively (Tsuda & Tyree 2000).

The high pressure flow meter method has been compared to the **evaporative flux method** for a number of crop species (Tsuda & Tyree 2000) and a regression of the two methods had a slope not significantly different from one (i.e. the two methods gave the same results). Hydraulic conductances (root, shoot, leaf) ranged in values from 0.5 to 10 kg MPa m^{-2} $s^{-1} \times 10^4$ (Tsuda & Tyree 2000) in both methods. However, it is important to note that K is not constant throughout time; diurnal and seasonal variation is apparent. A dependency of K upon E is usually observed, possibly because of significant changes in the hydraulic conductance of the soil–root interface (Williams et al. 1998; Tsuda & Tyree 2000) but also because of the strong temperature dependence of the viscosity of water, which is not accounted for in these analyses.

Leaf hydraulic conductance

Recently, more attention has been given to leaf hydraulic properties determined from **relaxation curves of leaf water potential** (Brodribb & Holbrook 2003, 2004a,b). In summary, a group of leaves that has been allowed to desiccate while attached to a branch is rehydrated, after excision from the branch, for a known period of time. Leaf water potential of leaves before and after rehydration is measured. It is important to note that a leaf cannot be used twice for measurement of water potential because measurement of water potential with a pressure chamber inhibits subsequent rehydration (Brodribb & Holbrook 2003).

Leaf hydraulic conductance is calculated from the following:

$$K_{\text{leaf}} = C\ln[\Psi_o - \Psi_f]/t \tag{20}$$

where C is leaf capacitance, Ψ_o and Ψ_f are leaf water potentials before and after rehydration respectively and t is the duration of the rehydration period.

To find C, the slope of the relationship between leaf relative water content (RWC) and leaf water potential (see pressure volume (PV) curves described previously) is measured. The values of C above and below the turgor loss point are very different because of the effect of the elasticity of cell walls. Below the turgor loss point, this effect is lost.

To calculate K_{leaf} (in mmol m^{-2} s^{-1} MPa^{-1}) the value of C (derived from the slope of leaf RWC and water potential; hence $\Delta RWC/\Delta\Psi$) must be expressed in absolute terms and normalised with respect to leaf area. Thus, C from the PV curve is multiplied by the turgid mass of water in the leaf and divided by leaf area (Koide et al. 1991):

$$C = [\Delta RWC/\Delta\Psi][DW/LA][WW/DW]/M \tag{21}$$

where DW and LA are leaf dry weight and area respectively, WW is saturated leaf weight and M is the molar mass of water (g mol^{-1}) (Brodribb & Holbrook 2003).

Using this method to calculate leaf hydraulic conductance plus measurements of stomatal conductance, Brodribb and Holbrook (2003) showed that the closure of stomata of excised leaves during water stress is strongly coordinated by changes in K, which declines sigmoidally with increasing water stress (i.e. as leaf water potential declines).

Xylem embolism and vulnerability curves

An estimate of in situ rates of xylem embolism can be obtained by comparing the conductance of a branch before and after high pressure perfusion of the branch with acidified, degassed filtered water. The value of conductance obtained prior to infusion represents the in situ conductance. Perfusion with acidified, filtered and degassed water at 175 kPa for 30 minutes removes any emboli present and the value of conductance obtained after should be higher than the initial value because of the removal of present emboli. The difference is expressed as a percentage of the initial value of conductance. Occasionally a decrease in conductance is observed following perfusion, a result attributed to plugging of xylem vessels by bacteria or other material. By sampling branches at different times of day or different seasons, daily and seasonal patterns of xylem embolism can be determined (Prior & Eamus 2000).

An alternative method of measuring xylem embolism is increasingly being applied to establish vulnerability curves for different species and different habitats (Tyree & Sperry 1989). This method relies on centrifugation of excised branches, and is discussed here in relation to vulnerability curves.

Vulnerability curves describe the relationship between xylem water potential and the percentage loss of conductance of the xylem (Fig. 2.14). The most common method for determining a vulnerability curve uses excised branches, with leaves attached, that are allowed to dehydrate slowly in the laboratory. Multiple replicate branches are collected from the field early in the morning so that the in situ degree of embolism is low. Branches are sealed in plastic bags with damp tissue paper to minimise the degree of leaf drying that occurs during transport to the lab (Melcher et al. 2003). Alternatively, branches are collected and the cut end re-cut under water and transported to the lab in a bucket of water in a plastic bag (Prior & Eamus 2000). Once at the lab, leaf water potential for a number of replicate leaves is measured with a pressure bomb and the branches allowed to dehydrate slowly for up to 5 days (Melcher et al. 2003). As the leaves transpire, xylem water potential declines and this is measured periodically with the pressure bomb. Leaf water potential is measured after the entire branch and leaves have been sealed into a plastic bag for an hour to allow all parts of the system (branch and leaves) to equilibrate to a single water potential (Melcher et al. 2003). Paired measurements of xylem water potential and percentage loss of conductance are made over the 2–5 day period using the pressure infusion method described above.

An alternative method of measuring a vulnerability curve uses a centrifuge to induce emboli in excised branches (Alder et al. 1997; Davis et al. 2002). Excised branches, with no leaves attached, are collected, stored in plastic and transported to a lab in sealed plastic bags. The branches are pressure-perfused with acidified, degassed filtered water or dilute (10 mM) KCl solution. Neither Macinnis-Ng et al. (2004) nor Davis et al. (2002) found any difference in conductivity values obtained using water or saline perfusion solutions, but using water alone may cause a decrease in conductivity because of swelling of xylem walls.

Hydraulic conductivity is measured using either the vacuum method (Prior & Eamus 2000; Macinnis-Ng et al. 2004) or the pressure infusion method whereby flow is induced by application of a small positive pressure (Davis et al. 2002). Following this, the branch is placed in a modified centrifuge rotor that holds one or more branches horizontally, with the mid-point of each branch at the centre of the rotor. Spinning the rotor at a known speed induces a known negative pressure within the xylem, and this will cause a certain number of emboli to form. The hydraulic conductance of the branch is measured repeatedly after being subject to increasing speeds of rotation for 5 minutes. As the speed of rotation increases the number of emboli increases and therefore the percentage loss of conductivity increases. A vulnerability curve is generated by plotting the xylem pressure (calculated from the rotation speed and branch length) against the percentage loss of conductivity (Fig. 2.14).

Although vulnerability curves traditionally show the relationship between xylem water potential and loss of conductivity, the concept has also been applied to determine the sensitivity of stomatal conductance to changes in hydraulic conductivity of leaves as leaves are water stressed (Brodribb & Hobbrook 2004).

Measurements of hydraulic conductivity, hydraulic architecture, xylem embolism and vulnerability curves provide important insights into the relationships between water availability (a function of rainfall and soil characteristics), evaporative demand and plant structure and performance. In particular, application of techniques described in this section allow us to address the following questions:

- what is the likely response of vegetation to a change in the availability of water?
- what are the upper limits to vegetation water use, given a known leaf area and water potential gradient?
- how do leaf area, hydraulic conductivity, stomatal conductance, water use and C fixation respond to changes in plant water status?

Root distribution

Roots clearly have major functions, including water and nutrient uptake and anchorage. They are also major sites of carbon storage and carbon turnover, and provide long-term survival strategies (e.g. for lignotubers; see Chapter 1) to cope with fire and drought. They also contrib-

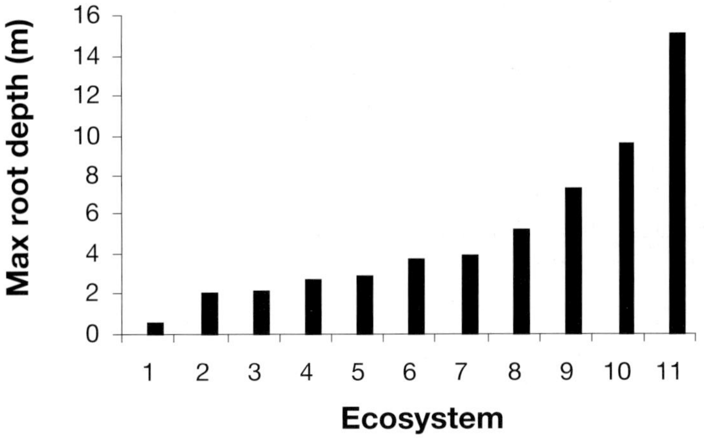

Figure 4.9 Ecosystems differ in the maximum depth to which roots grow. Ecosystems 1–11 are tundra, boreal forest, crops, temperate grassland, temperate deciduous forest, tropical deciduous forest, temperate deciduous forest, sclerophyllous forest, tropical evergreen forest, desert and tropical savanna, respectively

Source: Data from Canadell et al. (1996)

ute significantly to the carbon and nitrogen cycles of soil. Rooting depth (and water relations) have recently been used to explain species distribution patterns on the sandplains of Western Australia (Groom 2004) and there are significant differences, globally, in maximum rooting depth for different vegetation types (Canadell et al. 1996; Niklas 2005). Figure 4.9 shows the distribution of maximum rooting depth for 11 different ecosystems. Clearly, such differences in maximum rooting depth have an influence on the hydrology of ecosystems (Canadell et al. 1996).

Assessing root depth, distribution and patterns of growth has generally been time-consuming, expensive, destructive and dirty work, which is why it remains the least developed major field of study of plant function. While major developments in electronics, data logging and miniaturisation have revolutionised the study of the ecophysiology of plants above ground, it is only recently that similar movement in root studies have occurred. This section provides an introduction to some of the methods of assessing root growth and root function.

Minirhizotrons, root windows, root in-growth bags and root cores

Minirhizotrons consist of an access tube (5–10 cm diameter, 20–100 cm length) with transparent windows. The tube is inserted into the ground and a mini video camera used to view through the windows up and down the tube. Identifiable reference points along each window are used to ensure images are correctly aligned and compared. Tubes can be installed vertically, at angles or horizontally into the exposed face of a trench. Access tubes are often back-filled with soil (Baddeley & Watson 2004). Images captured by digital cameras as they move down the tube can be analysed with software to reveal information about fine root length, root mortality, lifespan, diameter, rooting density and fine root growth rate. The images can also be analysed for mycorrhizal infection and fungal growth, in some cases. Fine roots in most studies are those <2 mm diameter (<1 mm in some studies). A single camera can be used with dozens (hundreds) of access tubes so that spatial as well as temporal changes in root behaviour can be examined. Care should be taken to exclude light from the tube when the tube is not being used as light alters root growth (roots are negatively phototropic). Concern has been expressed that the outside edge of the tube may provide a preferential flow path for water to penetrate into the soil and this may attract roots preferentially to the access tube, thereby skewing the data obtained. Installation of tubes at an angle may help overcome this, and installation horizontally into an exposed face certainly does. However, the trench itself generates an artifact which may influence root growth but presumably this is minimal if data are collected many centimetres away from the side of the trench.

Minirhizotrons are excellent tools for non-destructive assessments of the behaviour of fine roots and, to a lesser extent, coarse roots. The ability to examine the growth of new roots and the behaviour of the same root over time, including decline and death, plus the ability to sample large numbers of locations within a site, provide valuable insights to root ecology. However, they do not produce information about chemical composition nor direct information about biomass and it is difficult to install them in stony soils. There is an extensive literature indicating that shallow roots are under-represented in the data, probably because of the influence of light, soil compaction during installation of the access tubes or altered soil water relations (Hendrick & Pregitzer 1996).

Root windows are structures that allow repeated access to the side wall of a trench. A window (typically 0.5–1.0 m square, usually constructed of perspex not glass) is placed against the side of a trench that has been smoothed and sometimes backfilled to get a reasonably tight fit between window and soil. The trench must be covered to exclude all light at all times. A thin-lined grid is etched/drawn on the window or on a plastic overlay, to allow root growth to be monitored using a algorithmn based on the line intersection method of Newman (1966). Grabarnik et al. (1998) provide further details and developments of this method. Alternatively, digital photographs can be taken at frequent intervals and root length, diameter, growth rate and mortality determined from the images using a digital analysis system (Sandnes et al. 2005).

Root windows are moderately expensive to install but can be used for years and can access roots to depths of several metres. However, it takes many months for roots to become visible in the window and preferential water flow down the inside of the window can be a problem. Changes in soil compaction, light environment and water relations of soil can influence root behaviour in root windows in the same way as in minirhizotrons.

Root in-growth bags are coarse mesh bags filled with sand, vermiculite or sieved soil (Chen et al. 2004; Majdi & Andersson 2005) inserted into the soil profile. Alternatively, they can be bagless sand, vermiculite or sieved soil infills of narrow (5–15 cm diameter) holes. Because the bag (or bagless infill) are without roots at the start of the experiment, when the bag is subsequently extracted (or the hole re-drilled) any roots obtained in the bag must have grown into the bag in the time between inserting and removing the bag. Root biomass, root length density and fine and coarse roots can be separated with this technique. Many replicates are required within a site because the measurements are destructive, and spatial and temporal variability in root growth is large. This method is applied in systems where root growth is rapid. However, if the bags are left too long root turnover occurs and this may not be observed. The use of sieved soil, sand or vermiculite will change the nutrient and water holding properties of the soil and thus influence root behaviour within the sample volume.

Root cores are obtained using augers or similar drilling equipment to extract soil cores, from which roots are obtained either through washing with water or by blowing the soil away with compressed air. Few data have been obtained with this method beyond 1 m depths (Majdi 1996). Larger-scale 'cores' can be obtained by digging by hand or mechanical digger (the **soil monolith** method; Chen et al. 2004). Such coring methods are time-consuming, can be difficult in stony soils and, unless frequently and repeatedly undertaken at a site, yield minimal data about temporal patterns. This method is poor for examining root mortality and decay.

Ground penetrating radar

Ground penetrating radar (GPR) has been used to examine rock structures, burst water pipes and land mines. Most recently it has been applied to the in situ study of roots (Butnor et al. 2001; Barton & Montagu 2004).

Ground penetrating radar produces a series of electromagnetic pulses into the ground from an antenna located above ground. As this pulse moves through the soil it is absorbed, reflected and deflected by water, roots, rocks and other objects. Different classes of objects (soil, rocks, roots) differ in dielectric permittivity and this is the cause of variation in reflection. A receiver measures the reflected signal and records features such as travel time and amplitude. During use, the emitting antenna moves across the ground surface and emits a series of pulses of electromagnetic radiation. High frequency electromagnetic radiation penetrates less into the soil than lower frequencies. For example, 100 MHz will penetrate to 30 m but 2 GHz will penetrate to 20 cm. Soils with a high water content or high clay or salt contents will absorb the pulse quickly and penetration of the signal will be less than 1 m (Barton & Montagu 2004). The benefit of high frequency signals is the finer spatial resolution obtained.

Butnor et al. (2001) showed good correlations between root biomass in the top 30 cm of soil and GPR data, while Barton and Montagu (2004) found reasonable correlations with root diameter, from which biomass could probably be estimated. Root location was also determined in both studies.

This is a new technique and yet to receive widespread application. Data processing and the ability to deal with in situ variation in root depths, orientation, diameters and water content, along with natural temporal and spatial variation in rock, water and salt concentrations, have yet to be resolved. However, if information only about presence and absence of roots is required, as a function of depth, this technique is likely to be of value in the near future.

Knowing the depth, amount and distribution of roots in soil is clearly important when addressing the following questions:

- whether there is groundwater within the rooting depth of vegetation;
- whether there is seasonality in root depth and root growth, and whether this might cause seasonality in the degree of dependency of vegetation on groundwater.

Stem growth

The growth of tree stems is usually assessed by one of two methods. In the first **dendrometers** are attached to stems and changes in stem diameter recorded every 2–4 months (Prior et al. 2004). Dendrometers are narrow (about 1 cm) and thin (about 1 mm) metal bands wrapped around a stem 1.3 m above the ground. A spring holds the two ends of the band together and two nails can be used to stop the band from sliding down the stem. The two ends of the band must overlap by several centimetres. A vernier scale on the metal bands can be used to measure the gradual movement of two marks close to the edge of the two ends. Growth rates of less than 1 mm per month can be measured with this technique, which is simple, but not cheap if sufficient replication (six trees per species) is to be obtained. The dendrometers and can be left on-site for several years.

An alternative method is to measure the DBH of many trees at a site using a tape measure. This method requires far more replication than the dendrometer bands because variation among measurements over time is much larger.

Measurements of stem growth can address the following questions:

- whether groundwater pumping or increased salinity of soil water has reduced growth rate of trees at a site;
- the seasonal patterns of growth of trees.

Hydrological techniques

Hydrological properties crucial to ecosystems were identified in Chapter 3, and include rainfall, surface (or overland) flows, subsurface flows and groundwater recharge. The dynamics and magnitude of each of these fluxes and other properties (depth and quality of groundwater, surface flooding, stream flow) can be measured manually or remotely using automated systems. The intensity of sampling in time and space, the required accuracy, reliability and site accessibility ultimately determine the technology and sampling regime appropriate to a given ecohydrological application. The following survey of technology and approach is not exhaustive, but rather is intended to provide for a range of complexity, pragmatism and accuracy.

The hydrological techniques of most importance to ecohydrology can be divided into four groups:

1 measuring rainfall and its interaction with vegetation;
2 determining the fate of rainfall;
3 measuring stream flow;
4 measuring groundwater flows as recharge and discharge.

Measuring rainfall and its interaction with vegetation

Rainfall is measured most simply with an open container (gauge) placed under a plastic or metal cone of known area. The gauge is calibrated to read depth of rainfall, usually recorded daily. The gauge should be level and placed away from higher objects. No object should be located within an imaginary volume of air above the gauge defined by a 45° angle upward in all directions from the top of the gauge. With careful placement, such simple technology can be as accurate as more sophisticated ones. Typically, local rainfall can be measured within 5–10% of the true local mean. It is worth noting that there is a systematic bias in increasing measured rainfall with increasing height of measurement above ground.

The principal disadvantage with the above technology is that it gives little detail on the intensity of rainfall. This deficiency is overcome with a **pluviograph** (a recording raingauge, e.g. a **tipping bucket** rain gauge), which records the intervals over which fine units of rainfall occur. Mechanical pluviographs yielding analog data have given way to electronic devices with digital output. This latter technology is a standard component of electronic meteorological stations that also record temperature, humidity, solar radiation and other weather and environmental variables. Such automated weather stations are now the widely accepted standard in ecological studies. Automated rain gauges can plot the change in weight of water in the gauge over time and therefore give information about duration and intensity of rainfall.

In ecohydrological systems in which the water balance is influenced by regional processes (see Chapter 2), rainfall can be expected to vary over the domain. This variation will have random and systematic components, and a network of rain gauging is required to give a meaningful picture. Of the systematic components of variation, topographic gradients can be expected to dominate, and it is usual to array gauges along an elevational transect.

One source of error that can significantly affect the accuracy of rain gauges, especially those in arid zones, is evaporation from the gauge after rain has ceased. Many automated gauges have a lid which opens when rain is detected and closes automatically when rain stops. Alternatively, a small amount of light oil can be placed in the gauge at the start of the measurement period to reduce evaporation.

The calculation of average rainfall for a catchment with many rain gauges is an important parameter in many hydrological studies. Three common methods of doing this are the arithmetic mean, the Thiessen Polygon method and the Isohetal method.

Calculating the **arithmetic mean** is the simplest method. If there are numerous gauges that are evenly distributed, and the terrain is flat, the arithmetic mean is acceptable. Note, however, that these conditions are rarely met.

The **Thiessen Polygon** method can increase the accuracy of estimates of average rainfall if gauges are not randomly distributed. Polygons are drawn by constructing the perpendicular bisector of lines joining adjacent gauges. The area within each polygon is calculated and this area is assumed to receive the same amount of rainfall collected by the gauge within it. Each polygon represents a known fraction of the total catchment area and the rainfall in each gauge is multiplied by that fraction. The depth of rainfall for the catchment is then the sum of the rainfall depths (after being multiplied by the fraction of the total catchment area) in each polygon.

The **Isohyetal method** plots the rainfall depth in each gauge and contours (isohyets) are drawn when gauges with equal rainfall are joined on a map. Average rainfall is calculated by determining the fraction of total catchment area bounded between adjacent isohyets and multiplying the rainfall of each isohyet by the fraction of the area bounded by that isohyet, and summing all of these. This is theoretically the most accurate, but also the most time consuming.

Measurements of rainfall across a catchment can be used to address the following questions:

- the spatial and temporal distribution of rainfall;
- the input of water to a catchment, as rainfall.

Interception of rainfall by vegetation

Vegetation intercepts rainfall. Different canopies intercept different amounts of rain, and a larger proportion of rainfall is intercepted during low intensity, short duration rain compared to high intensity, long duration rain. Some of the rainfall intercepted by a canopy is lost as evaporation, but some reaches the soil through stemflow and drip flow. **Stemflow** is the movement of water from the canopy and upper branches down to the soil via the outside of the main stem of a tree, following interception by the canopy and branches. **Drip flow** (or **throughfall**) is the water that drips from a saturated canopy. The evaporation of intercepted water contributes to evapotranspiration from wet canopies and this needs to be accounted for if such measurements (from wet canopies) occur. Estimates of stemflow and drip flow are important for small-scale studies of nutrient recycling because water that has interacted with a canopy has a different chemistry (especially pH and ion content) from rainfall that has not interacted with a canopy. This has been especially significant in studies of the impact of acid rain on vegetation, where highly acidified rain (and mist and fog) leaches nutrients out of foliage. Methods for partitioning rainfall into throughfall, stemflow and interception are discussed extensively by Crockford and Richaradson (2000).

Stemflow can range between zero and 10% of total precipitation. It is influenced by the intensity and duration of rain events and tree characteristics such as branch angle, bark type and other features of the canopy. Stemflow is measured by attaching collecting buckets to individual tree stems and providing a waterproof seal that guides water from the entire circumference of the stem into the collecting bucket. Drip flow is measured by placing rain gauges under the canopy of many trees.

The difference between total rainfall (measured in a large canopy gap where interception is zero) and stemflow plus drip flow, represents interception loss. Total rainfall is measured using a rain gauge located close to the study site but positioned so as not to receive interference from nearby vegetation.

Measurements of interception, stemflow and drip flow are important in answering the following questions:

- the proportion of total rainfall that actually reaches the ground and is available to infiltrate into soil;
- how the chemistry of rainwater entering the catchment changes after being intercepted by vegetation.

Fate of rainfall

Rain that reaches the ground has several potential fates. It can infiltrate into soil, it may run overland into streams and other surface water bodies (lakes, dams) or it can sit on the surface as ponded water. The techniques to assess these fates are now discussed.

Infiltration and saturated hydraulic conductivity

Infiltration is measured in the field using **infiltrometers**. A flooding type infiltrometer is basically a cylinder pushed into the soil with a constant head of water pressure (typically 10 cm height) maintained from a reservoir. The rate of water supply into the cylinder to keep the head of water constant is recorded. Because these infiltrometers measure the saturated rate of infiltration, it is important to minimise the flow of water laterally from the infiltrometer. Consequently a dual ring infiltrometer is used. The outer ring is designed to provide a lateral buffer between the wetting front of the water moving down into the soil from the inner reservoir, and the dry soil. In effect, water moving down through the outer ring ensures the water moving through the inner ring is moving down through the soil and is not moving laterally outwards into dry soil. It is the rate of water flow through the inner ring that is actually measured.

Dual ring infiltrometers are cheap and relatively easy to use. Although the absolute values may be an overestimate, because rainfall generally does not produce a hydraulic head of 10 cm, they are useful for comparing rates of infiltration between different soils, vegetation types and types of land use.

Infiltration rate is extremely variable spatially. Local variation in litter thickness and density, the presence and absence of macro-pores (dead root channels) and compaction (because of grazing by livestock or farm traffic in an agricultural setting) can cause orders-of-magnitude differences in infiltration rate. Consequently, large numbers of replicates are required.

Water repellent soils (**hydrophobic** soils) occur throughout the world and reduce the rate of infiltration of water. Hydrocarbons in the soil coat soil particles and cause soils to repel water at the surface or in deeper layers. Fire can increase the occurrence of hydrophobic soils.

Measurements of infiltration are important to answer questions such as:

- the variability of the rate at which water can enter the soil within an ecosystem;
- how different land use practices influence the ability of water to enter the soil profile;
- the maximum rate at which water can enter the soil at a site and how this compares to the range of rain intensities at the site. Therefore, how much of a problem surface run-off is likely to be.

Hydraulic conductivity of soil at saturation (K_{sat}) is a measure of how quickly water can move through soil that is saturated with water. It is needed for many models of soil water dynamics. A **constant head permeameter** (Amoozegar 1989) can be used to measure K_{sat} in situ at a range of depths.

A constant head permeameter estimates K_{sat} by determining the rate of water flow into a hole and through the soil once the soil is saturated. A constant head (pressure) of water is maintained in the instrument and, once infiltration into the soil is steady state, flow rate is recorded. Flow is converted to K_{sat} using the Glover equation (Amoozegar 1989):

$$K_{sat} = \frac{Q[\sinh^{-1}(H/r) - (r^2/H^2 + 1)^{1/2} + r/H}{2\pi\Delta H^2} \qquad (22)$$

where Q is the steady state rate of flow into the hole, H is the is the head of water driving flow and r is the radius of the hole. The Glover equation is valid if the distance from the bottom of the hole to any impermeable layer or water table (s) is greater than $2H$ and the ratio H/r is greater than 5.

Run-off

When the rate of rainfall exceeds the rate of infiltration, surface **run-off** or **overland flow** occurs. Such flows generally end up in stream flow, although in some cases **sheet flow** moves into wetlands without passing through a defined stream. Alternatively, surface run-off can infiltrate the soil downslope (because surface run-off occurs downslope) and become subsurface lateral flow or groundwater recharge. Run-off is more common on steep slopes (compared to flat terrain) and is lower across sandy soils (which have a high permeability) than on soils with a high clay or silt content (which have a lower permeability).

Measurements of run-off are important to investigate the following:

- the proportion of rainfall lost from a site as run-off;
- the proportion of rainfall driving stream flow during low, medium and high intensity rainfall events;
- how much soil, litter and nutrients are being lost from a site through run-off.

There are several ways to measure the volume of surface run-off and, as with most measurements, the choice of technique depends on the question being addressed. The expected volumes/depths of the surface flows and the required accuracy of measurements influence the choice of technique. Choice also depends on whether the flow to be measured is in a stream or flowing overland.

Surface flows typically have long-flow recession periods which last for hours to days or weeks, depending on catchment properties such as the area above the point of measurement and soil and groundwater stores and the fluxes through them. This means that any measurement has to be done in the context of data on antecedent flows or at least the antecedent rainfall and how rainfall influences flows. River and stream flows can reach very high volumes during floods; robust structures are required to withstand the forces generated by large volumes of water moving at speed. Scouring of the stream bed or stream banks also occurs during floods. Floods can radically reshape a stream's measured cross-section, necessitating re-measurement.

For streams and rivers the critical issue is typically not just how much water is flowing, but how those flows vary over time and at different points along the stream. For example, a system or organism may respond to a particular mean flow frequency distribution (as represented by a flow duration curve) or to a particular flow sequence or flow volume during a particular season. This is not the place to go into these details – they are covered in numerous textbooks and manuals on environmental flow assessment (e.g. King et al. 2000 for South Africa; Arthington et al. 1998 for Australia).

Overland flows can be measured using areas of ground which are enclosed by an impermeable vertical structure such as plastic or metal sheeting or a more permanent structure of bricks or blocks and mortar. The structure is typically rectangular and laid out with its long side parallel to the prevailing slope. A funnel-shaped arrangement of the impermeable wall on the lower side channels the water to a measuring point such as a small weir or a container such as a barrel (partially) sunk into the ground. A rain gauge can be erected on site to measure incoming rainfall. Precipitation that does not infiltrate into the soil will then be directed through or into the measurement structure and the volume running-off can be calculated. Data can be collected on a per rainfall event or per unit of time basis depending on the measuring instruments and site accessibility. The greatest detail is obtained from recording gauges and data loggers for both rainfall and incoming water; the least from daily or weekly measurements that can only give mean values.

Scaling experimental measurements of run-off from small plots to entire catchments or landscapes is difficult because of heterogeneity in soil and topography. The degree of soil saturation or inherent infiltration capacity can vary by orders of magnitude within a small catchment, with little observable systematic or predictable bias in the spatial variation. Similarly, there is temporal variation in surface moisture. There can be spatial patterns in the areas of saturation prior to a rainfall event, usually in low-lying concave parts of the landscape. These areas of land that act as sources of saturation excess run-off (i.e. land that is receiving rain at a rate greater than the ability of the soil to absorb it) change in time and areal extent. Consequently, interpretation of run-off measured at a particular time depends on knowledge of recent rainfall prior to the time of measurement. Finally, scaling local run-off measurements to whole catchments requires assumptions about the patterns and intensity of rainfall at the landscape scale.

Stream and river flows

The main types of techniques to measure stream and river flows can be categorised as follows:

1 instantaneous measurements of current flow, generally with a measured cross-section;
2 measurement of river depth with a hydraulic model to relate flow depths to volumes using a measured cross-section and river bed slope;
3 weirs: structures with pools for stilling the flow and measuring depths of flow over a fixed structure with known profile.

Neither of the first two is able to estimate flow within a river bed unless measurements are taken where the bedrock reaches the surface (e.g. at a sill), since at this point flow through the river bed is zero. In most other circumstances, unless there are deep alluvial deposits or boulder beds below the river, subterranean flows are likely to be very slow and relatively constant.

Current flow measurements

Ideally, measurements of current flow are made using a specifically designed meter, a current flow meter. A cruder method is to take a floating object and use a stopwatch to measure the time it takes to move a known distance. Measurements need to be repeated 5–10 times and at more than two points across the width of the stream, unless it is very narrow, to get a representative value of velocity. However, velocity of flow is, of itself, not terribly useful because a velocity of flow in a small stream is not the same (in terms of water flow) as the same velocity in a large river. Therefore, a conversion of velocity to volume flow is required.

Determination of volume flow from estimates of velocity is difficult. Volume flow (Q) = velocity × CSA, where CSA is the cross-sectional area of the river. Significant errors can be introduced in converting velocity to volume flow. First, velocity varies with stream width and depth and therefore getting an accurate and precise measure of velocity is difficult. Second, estimating the CSA is often difficult, partly because the river is full of water, partly because the CSA varies along the length of the river and partly because of flow within the river bed.

This information can be combined with data from simultaneous measurements of the depth and length of the wetted perimeter (length of the cross-section of the river profile that is in contact with the water), or the cross sectional area, to estimate stream flow volume using an appropriate formula. Two commonly used empirical formulas are the **Manning** equation and the **Chezy** equation.

The **Manning** equation is:

$$V = (1.49/n) \times (R_h^{2/3}) s^{1/2} \quad (23)$$

where V = average stream velocity, R_h is the hydraulic radius (= the wetted perimeter = the length of the outer boundary of the wetting front below the river bed) and s = the energy slope, which can be approximated by the slope of the water surface in the stream. Manning's roughness coefficient (n) typically varies between 0.015 and 0.15.

Alternatively, the **Chezy** equation can be used:

$$V = C\sqrt{(R_h \times S)} \quad (24)$$

where C is the Chezy roughness coefficient and $C = (1.49/n) \times R_h^{1/6}$.

These two equations give similar answers. There appear to be no theoretical reasons to apply one equation in favour of the other.

Current flow measurements are instantaneous and will not give an integrated picture over time. A structure can be rigged to measure and record current flow rates continuously but it is more practical to use the techniques described below, if this is the information required. Instantaneous measurements will also be crude and the accuracy will be low, especially if low flows are to be measured. Instantaneous measurements can be put into context if the relationship between the measured flows and rainfall and the likely flow recession curves are known or can be estimated with reasonable confidence. If this information is not available it will be very difficult to make any meaningful comparisons of measurements at different times.

Hydraulic methods for calculating stream flow

This approach translates depth measurements taken from time to time, or continuously using chart recorders or data loggers, into flow volumes. The volumetric flow is computed using the relationship between the measured depth and the wetted perimeter or the cross-sectional area of the stream or river bed and the slope of the river bed. The Manning or Chezy formulas can be used, as can more sophisticated programs. Periodic measurements of the depth can be translated into total flow over time if there were no significant changes in the mean time or by assuming a particular relationship for interpolating the changes in depth between measured values. Rainfall data will be needed, and relationship between flows and rainfall and flow recession must also be known, before data can be interpolated. Continuous recording removes the need for interpolation as long as the equipment operates properly. The formulas are fairly easy to calculate. The relationship between depth and flux can be complex when high flows result in changes in the cross-sectional profile of the stream at that point due to sediment deposition or river bed or bank erosion. Flow estimates will not be very precise and it is difficult to estimate low flows with much accuracy unless the cross-section is measured very accurately.

Weirs

Weirs are fixed structures with a known profile and a stilling pool that minimises the effect of river turbulence on the flow over the structure, thereby increasing measurement accuracy. The measuring system uses a float in a chamber or well (which is open to the stilling pond) to measure the depth of the water flowing over the structure. The volume or **volumetric flow** of the stream can be calculated from stream depth using a rating curve, or from first principles and the known shape or profile of the weir. Flow velocity profiles are measured through the depth of water in the weir at different stages of river height in order to develop a unique empirical curve for each gauging station. The weir should, wherever possible, be located on bedrock to prevent flow of water beneath it. The erection and maintenance of these structures can be time consuming, and the expense limits replication, but if accurate measurements are needed (particularly of low flows) there is no alternative to a weir.

Various shapes and forms of weirs can be used, ranging from a straight channel (flat bottomed U) section, usually called a flume, to stepped or combination profiles with or without a V-notch. A **V-shaped notch** is the most accurate for measuring low flow rates while a **rectangular notch** can allow more flow through for a given depth. **Flumes** have a known size (width, length) and depth can be measured at their start or end. The rate of flow through the flume is directly proportional to the height difference of water between the start and end of the flume.

Weirs can be portable or permanent structures. Portable structures are relatively inexpensive and easy to install, and can be moved once the study is completed or to gauge another site. Portable weirs should be placed where there is a stream bed structure which indicates that bedrock is close to the surface (e.g. a sill). The river sediments are removed and a plastic sheet is used to line the river bed to conduct the water into a channel or pipe which discharges into the gauging structure. The gauging system for a portable weir typically uses a rectangular box structure to create a minimal stilling pool and a V-notch on the downstream side to measure the flow. There must be sufficient drop such that the water does not overflow where it is captured until the capacity of the weir is exceeded. Weirs can cope with volumes of up 20 m^3 hr^{-1}, making them ideal for measuring low flows, but they typically overflow during storm flows so cannot be used to measure total annual flows or responses to rainstorms. The box structure is made of hot dipped, galvanised 3 mm steel (or fibreglass) and should last 5–10 years depending on the corrosiveness of the water.

Permanent weirs have to be properly designed and bedded in to ensure that they are not washed away and that the stilling pond can cope with high flows. The structure of a weir can be flume or sharp-crested, simple or compound, the latter having a variety of profile shapes. The costs of building this kind of weir range from 5 to 100 times that of a portable weir. The foundation should be on bedrock or other impermeable material, which may require substantial excavation of the river bed and associated environmental damage. The primary advantage of this type of weir is that it can be designed to give accurate measurement of both low flows and high flows up to the design limit. The main maintenance cost is the clearing of the sediment that accumulates in the stilling pool.

Much of the water that runs-off a landscape is captured in streams and rivers. Measurements of stream and river flows are vital to help answer the following questions:

- the water balance of a catchment;
- the water yield of a catchment;
- how the water yield of a catchment changes following fire, logging or other disturbances.

Depth of groundwater and height of flooding

The depth of groundwater relative to the land surface is determined by measuring the water level in a constructed well (a hole in the ground). This sounds simple, but much confusion arises regarding the interpretation of such data, because the values obtained are influenced by the type of well constructed. In an open (or observation) well (i.e. open along its whole length, an unlined well), the measured depth is equivalent to the actual water level of the surrounding aquifer. In many cases, it is more relevant to know the **hydraulic head** of a specific zone or

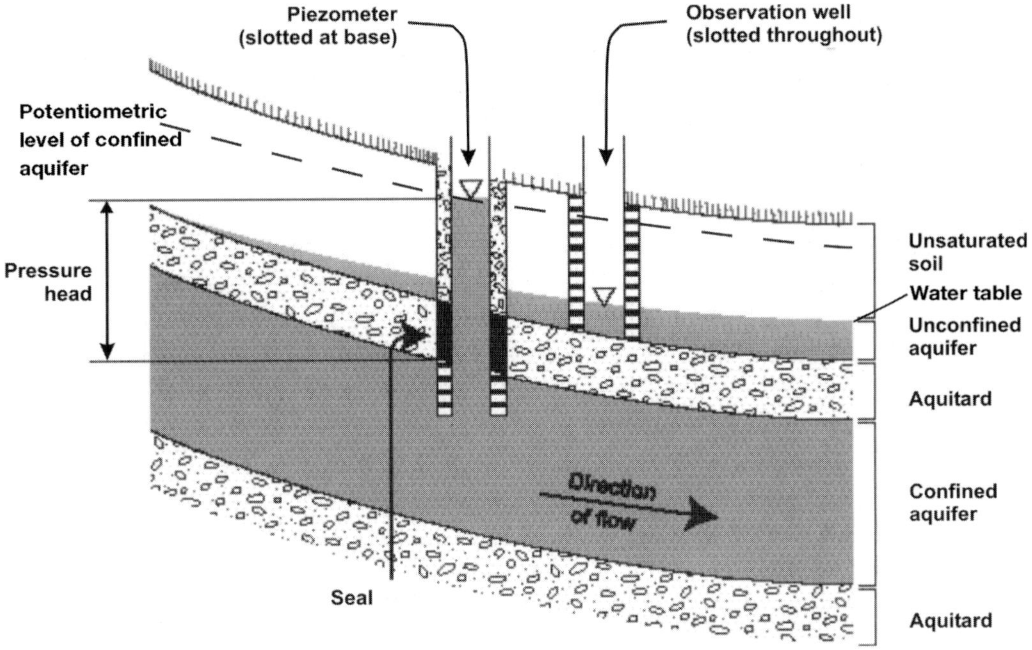

Figure 4.10 Confined and unconfined aquifers within a groundwater system and the methods used to measure water or pressure level within them. Piezometers are used to measure the water pressure or level at a specific depth, and observation wells may cover a broader depth

layer of the regolith. In such cases, the well is lined and open (screened) only over a specific limited length at the depth in question. These wells are termed **piezometers**. Large differences in the actual water table and the hydraulic head associated with any layer beneath that level usually arise only when the lower layer is confined to some degree by an aquiclude (a relatively low permeability strata) or an aquitard (a very low permeability layer).

Groundwater level in a well or piezometer (Fig. 4.10) can be determined by manual 'dipping' with a measuring tape and a sensor that signals contact with water. Increasingly, however, electronic water level recorders are permanently deployed down the hole in association with a single channel data logger, which frequently records the pressure of water above the sensor, from which water levels are calculated. The temporal dynamics assessed through almost continuous recording can give crucial insights into water table dynamics and aquifer processes that infrequent manual sampling cannot provide.

Nested piezometers are frequently used in the field. A nest of piezometers consists of a set of narrow wells closely spaced at a site. Each well has an access screen at a different depth and the set allows determination of the local vertical hydraulic gradient in groundwater systems. Often, the depth of the screen in each piezometer is selected to reflect different aquifers and the intervening aquitards. If the water levels of the deeper piezometers are higher than those of the shallower piezometers, the gradient is upwards and there is the potential for groundwater discharge. If the water levels of the deeper piezometers are lower than those of the shallow ones, the hydraulic gradient is downward and the site is potentially a zone of groundwater recharge.

Similar technology involving height recording of water levels can be used to monitor the dynamics of water levels above ground. In this case, a stilling well is constructed extending above the ground, with openings at the base to permit the flow of water. The height of water

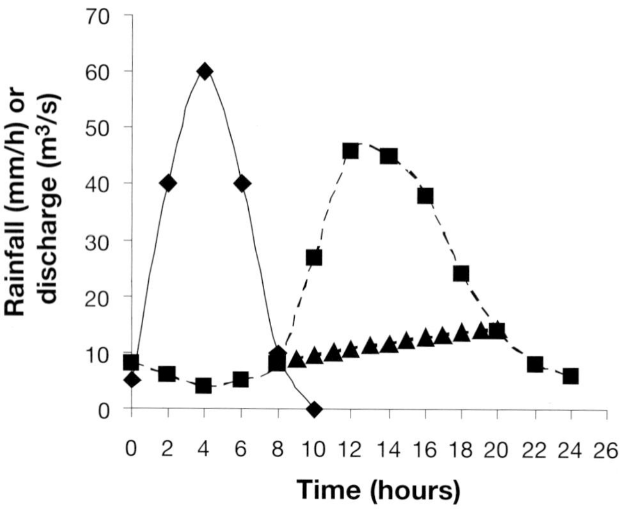

Figure 4.11 A stylised representation of a flood hydrograph, showing rainfall (diamonds, solid line) and river discharge (squares, dotted line). Base flow (triangles) represents flow into the river from subsurface processes, sometimes from groundwater

within the well is dampened from the turbulence or the surrounding flowing water. Water levels can be recorded manually or continuously logged by similar means as for groundwater level measurement.

In many studies of ecohydrology, depth of groundwater is an important parameter. For examination of groundwater dependent ecosystems (Chapter 6), a key question is whether vegetation at a site can access groundwater. Similarly, changes in groundwater depth and flooding are important factors in studies of wetlands and the impact of mining on ecohydrology (Chapter 7) and salinity (Chapter 8).

Flood hydrographs

A **flood hydrograph** plots the rate of discharge (flow; measured in cumecs, or $m^3 s^{-1}$) of a river past a gauging station before, during and after a major rainfall event. Figure 4.11 shows an example of a flood hydrograph and the rainfall event (storm) that caused it within the river catchment. Interpreting such a hydrograph provides useful information to catchment managers concerned about water and vegetation. For example, the total water lost from the catchment as river flow represents water not available to terrestrial vegetation within the catchment or to groundwater recharge.

During the period 0–4 h, the hydrograph shows a slowing rate of river flow as the catchment is drying out since the last rainfall event. This is the **approach segment** of the hydrograph and represents catchment discharge (this might or might not have a groundwater component to the flow). During the period 0–10 h a rainstorm falls on the catchment, with a peak intensity of 60 mm h^{-1} occurring 4 hours into the storm. The period 6–12 hours shows the **rising limb** of the hydrograph, where river flow increases because of rain falling directly into the river and because of surface run-off and subsurface flow into the river. There is an 8 hour delay between the peak rainfall and peak river discharge; the duration of this **time lag** is determined by catchment properties. A catchment comprised of urban and city land use, with large areas of impermeable concrete and roofs, has a lower time lag than an agrarian landscape covered in forest and crop land. Catchments with steep slopes have shorter time lags than catchments with flat terrain. Short lag times are often associated with higher peak rates of discharge and flooding is more likely to occur. After 12 or 13 h, river discharge declines and the slope of this

descending limb tends to be lower than that of the rising limb because subsurface flow into the river is slower than surface flow, which has generally stopped or is much reduced during the descending phase.

Base flow of the river represents water input resulting from subsurface flows, especially groundwater flows. It can vary over time so that when groundwater levels are rising because of recharge, base flow into a river can increase.

Information about groundwater depth and flood depth is important when managing catchments for water yield. These data are specifically important when addressing the following questions:

- how groundwater depth (and hence volume of water available) varies seasonally and inter-annually;
- how quickly the water table responds to rainfall and changes in land use;
- the sustainable yield of an aquifer;
- the threat floods pose to the catchment and to people within the catchment.

Measuring groundwater recharge, flow and discharge

Groundwater flow (either as velocity or volumetric discharge) is rarely measured directly due to the scale and complexity of most natural systems, but there are a number of indirect approaches suitable for most applications. The methods can be classified into methods for estimating rates of flow into the aquifer (groundwater recharge), methods for estimating rates of groundwater flow through the aquifer, and methods for estimating rates of flow out of the aquifer (groundwater discharge).

Estimating groundwater recharge

There are a large number of methods of estimating rates of aquifer recharge. These methods have recently been summarised and reviewed by Scanlon et al. (2002). Most of the methodologies are not universally applicable, rather, different techniques are better suited to different environments. For example, some methods are more suited to measuring high rates of recharge, while others are more accurate at low rates. Some methods are better suited to areas with deep water tables, while others are better suited to shallow water table areas. In the following, we discuss only two of these methods. Readers should consult Scanlon et al. (2002) for a more comprehensive account.

The water table fluctuation method is based on the premise that recharge occurs episodically, and rises in groundwater levels that coincide with rainfall events reflect recharge associated with those events. The method is described in detail in Armstrong and Narayan (1998) and Healy and Cook (2002). For each recharge event, recharge is calculated as:

$$R = S_y \Delta h$$

where S_y is the specific yield, Δh is the rise in the water table, and R is the recharge amount (measured as an equivalent depth; Fig. 4.12). Recharge volumes associated with individual rainfall events are summed to estimate annual recharge rates. The method is best applied in regions with shallow water tables where the water table response to individual rainfall events is apparent. In areas of deeper water table, a slow continuous recharge to the groundwater may occur, which cannot be detected with this method. Other difficulties in applying the method relate to obtaining an accurate value for the specific yield, and distinguishing water table fluctuations due to aquifer recharge from fluctuations due to other processes.

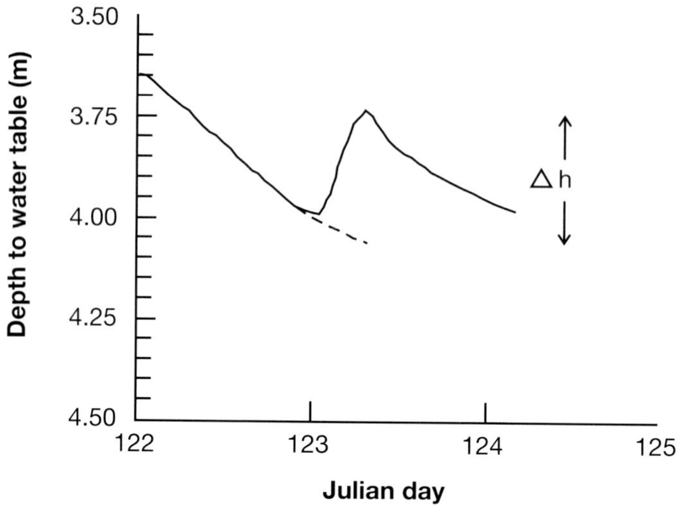

Figure 4.12 Hypothetical variation in water table in response to a rainfall event. The magnitude of the rise in the water table (Δh) is multiplied by the specific yield to determine the magnitude of the recharge

The chloride mass balance method has been used for estimating diffuse aquifer recharge since the 1960s, and has been increasing in popularity since that time. Its main advantages are its simplicity and its ability to estimate very low recharge fluxes with equal or better accuracy than high recharge fluxes (in contrast to most other methods). If the chloride that occurs in soil water and groundwater is deposited by precipitation (and dry fall), with negligible amounts contributed from rock weathering or anthropogenic sources (e.g. fertilisers), then at steady state we can express the chloride mass balance as:

$$C_P P = C_R R + C_Q Q \tag{25}$$

where P is precipitation, R is recharge to groundwater, Q is surface water discharge (run-off), and C_P, C_R and C_Q are chloride concentrations in precipitation (wet and dry fall), recharge and run-off, respectively. In many areas, surface run-off (Q) is not significant and the recharge rate can be estimated simply from:

$$R = (C_p P)/C_R \tag{26}$$

The method then requires only knowledge of the atmospheric chloride fallout ($C_P P$), which is available for a number of stations throughout Australia (e.g. Blackburn & McLeod 1983; Keywood et al. 1997) and can also be approximated based on distance from the coast. Diffuse recharge rates as low as 0.05–0.1 mm y^{-1} have been estimated beneath native mallee vegetation in southern Australia by applying the chloride mass balance to measured chloride concentrations in soil water (Allison & Hughes 1983), and recharge rates up to 200 mm y^{-1} have been estimated from chloride concentrations in groundwater on the Atherton Tablelands, north-east Queensland (Cook et al. 2001).

Estimating groundwater flow

Darcy's law can be used in conjunction with hydrological parameters to determine groundwater flow. Darcy's law states that the groundwater flow rate is proportional to the hydraulic head dh/dl, and is probably the most fundamental equation in hydrogeology. Thus:

$$q = \frac{Q}{A} = -K\frac{dh}{dl} = -k(h_1 - h_2)/l \quad (27)$$

where q is the specific discharge with units of length per time (usually metres per year), Q is the total discharge with units of volume over time, A is the cross-sectional area of the discharging aquifer, K is the hydraulic conductivity (units of length per time), h_1 and h_2 are the hydrostatic heads ($h_2 > h_1$) and l is the horizontal distance between the points at which the two heads are measured. The minus sign indicates flow is in the direction of decreasing head. The procedure is as follows.

1 Measure water levels h_1 and h_2 in boreholes, with h_1 located on the cross-section at which groundwater flow is to be determined and h_2 upslope on a line with h_1 normal to the cross-section.
2 Measure the distance l to find the gradient.
3 Use a value for K based on independent local tests or approximated from available tables based on geologic formation or aquifer texture.
4 Estimate the width and thickness of the aquifer, the product of which is the cross-sectional area A.
5 Solve for flow velocity q or discharge volume Q in Darcy's equation.

The groundwater velocity is calculated as the specific discharge divided by the aquifer porosity.

Applied tracers (conservative dyes, radioactive ions, heat) have also been used to estimate local groundwater velocity. The principle is to inject a quantity of tracer into the aquifer and sample downstream to determine time of arrival. The most obvious problem with these tests is that it is often difficult to predict the flow path of the injected tracer before commencing the test. In practice, this usually means that a large number of observation wells are necessary to ensure that the arrival of the tracer is observed (Fig. 4.13), which limits the utility of this approach. An alternative approach is to inject tracer into a well and measure the rate at which it gets flushed out of that well (rather than the time to arrive in another well). If the water is well mixed, the concentration of the added tracer in the well will decrease exponentially over time. These tests (often referred to as point dilution tests) enable estimation of mean flow rates through the well, which can be related to the flow rate through the aquifer. Point dilution tests carried out in fractured rocks in the Axe Creek catchment, Victoria, are described by Hodgson and Finlayson (1990).

Estimating aquifer discharge

Seepage meters represent the most direct measure of aquifer discharge, and have been used for estimating groundwater discharge to streams, wetlands and the marine environment. In their simplest form, they operate using one of two simple approaches (Kaleris 1998). In one method, a pipe is pushed into the stream bed with the upper part protruding above the water level. The volume of water that must be added to or removed from the pipe to maintain the water level within the pipe at the same height as that outside is recorded – a measure of the rate of recharge or discharge. In the second method, a cylinder vented to a plastic bag is pushed into the stream bed. The bag initially contains a known volume of water, and is allowed to float on or beneath the water surface. After a period, the volume of water in the bag is measured; changes in this volume can be used to calculate the rate of discharge (if the volume increases) or recharge (if the volume decreases).

Where the base flow component of streams represents groundwater discharge, analysis of stream hydrographs can also yield an estimate of groundwater flow. The method relies on

Figure 4.13 Part of the bore network for a tracer test to measure groundwater velocity in a heterogenous aquifer near Oak Ridge, Tennessee. To be certain of measuring the arrival of the injected tracer, tracer concentrations were measured in approximately 50 wells located in a fan-shaped area extending down gradient from the injection well

Source: Photograph courtesy of William Sanford, Colorado State University

interpretation of changes in surface water heights over time, and infers groundwater inflow from the shape of the water height versus time plot. The method is popular because it requires only stream flow (which is commonly available), but it can be somewhat subjective. A simple graphical method is presented in Figure 4.14. For more detail, see Salama (1998).

Hydrochemical methods for determining point groundwater discharge exploit the characteristic differences between the chemistry of rainfall and that of groundwater. As the former infiltrates into the ground, its chemistry changes. As a result, each catchment has its characteristic surface, groundwater and stream water qualities. Contrasting ionic or isotopic compositions of surface water run-off, shallow subsurface and groundwater provide a potential opportunity to quantitatively distinguish the various contributions from these sources. Specifically, tracers such as radon, bromide and chloride can often be used to estimate the groundwater flow component in a stream. In some cases, run-off, shallow soil throughflow and groundwater discharge have quite different concentrations of these tracers, and application and solution of a simple mixing model constrained by the observed values in the gauged stream yield separation of the component fluxes. Contrasts in the isotopic ratios of stable, naturally occurring isotopes of oxygen and hydrogen in waters originating as rainfall, run-off and groundwater can be similarly used to quantify the separate contributions to stream flow of a given isotopic value. Approaches based on water chemistry can also be used to estimate rates of groundwater discharge to wetlands and to the marine environment.

In areas where the watertable is shallow, groundwater discharge may take the form not of localised flow but chiefly of evaporation over a large area. In this case, groundwater must move up through the soil to discharge from a water table below the surface. If we assume that this upward flux is limited to liquid water (as opposed to vapour flux), the rate of flow will depend

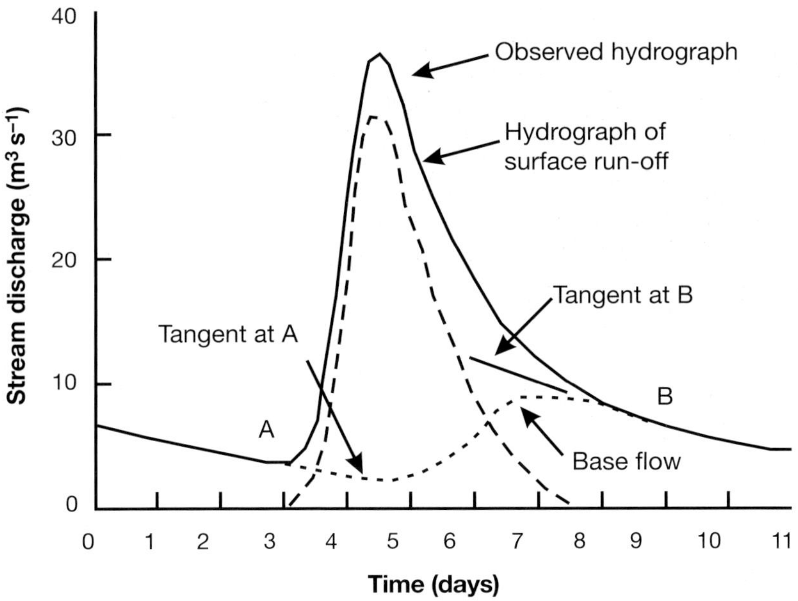

Figure 4.14 Illustration of the base flow separation method for determining groundwater discharge to a stream

Source: Salama (1998)

on the hydraulic properties of the soil, the depth of the water table and the hydraulic gradient created by evaporation (or transpiration). The maximum (i.e. soil-limited) steady upward flux q_m that can be sustained from a water table at depth d is given by:

$$q_m = Aad^{-n}$$

where A, a and n are constants. Values of a and n are obtained from the hydraulic conductivity (K) – soil matric potential (S) relationship, described by:

$$K = a/(S^n + b)$$

where b is also a constant. Typical dependent values for A, n, a and b can be found in Warrick (1988) and Thorburn et al. (1992).

Tracer techniques can also be applied to groundwater flow estimation in diffuse discharge contexts when the principal factor determining the rate of evaporation is not available energy but the resistance of the surface to water movement. Conservative tracers such as chloride develop a long term concentration gradient near the surface due to the balance between the accumulation of salts at the surface due to advection of groundwater, and the diffusion of salts downward due to this same gradient. This concept implies an exponential solute profile with depth, the scale of which is proportional to the effective diffusivity and to the inverse of the discharge (evaporation) rate. Sampling to determine the tracer profile with depth allows the resolution of diffusion from the quantity of interest, the flux of groundwater (Salama 1998).

Conclusion

There are many ecophysiological, chemical and hydrological techniques available to investigate the relationships between vegetation behaviour, performance and hydrology. An appreciation of some of the techniques for determining the key processes in ecohydrology is vital for any further discussion of ecohydrology as a science. This chapter provided a brief introduction to some of these techniques. The next chapter discusses the development and application of models in ecohydrology. Application of models in ecohydrology requires the measurement of many ecophysiological and hydrological variables and thus some understanding of the basic techniques will assist in developing an understanding of these models.

Further reading

Andrade JL, Meinzer FC, Goldstein G, Holbrook NM, Cavelier J, Jackson P & Silvera K (1998). Regulation of water flux through trunks, branches, and leaves in trees of a lowland tropical forest. *Oecologia* **115**: 463–471.

Baldocchi D (1997). Measuring and modelling carbon dioxide and water vapour exchange over a temperate broad-leaved forest during the 1995 summer drought. *Plant, Cell and Environment* **20**: 1108–1122.

Baldocchi D & Meyers T (1998). On using eco-physiological, micrometeorological and biogeochemical theory to evaluate carbon dioxide, water vapour and trace gas fluxes over vegetation: a perspective. *Agricultural and Forest Meteorology* **90**: 1–25.

Baldocchi D, Valentini R, Running S, Oechels W & Dahlman R (1996). Strategies for measuring and modelling carbon dioxide and water vapour fluxes over terrestrial ecosystems. *Global Change Biology* **2**: 159–168.

Brooks KN (1997). *Hydrology and the management of watersheds.* Iowa State Press, Iowa.

Chen XY, Eamus D & Hutley LB (2004). Seasonal patterns of fine-root productivity and turnover in a tropical savanna of northern Australia. *Journal of Tropical Ecology* **20**: 221–224.

Eamus D, Hutley LB & O'Grady AP (2001). Daily and seasonal patterns of carbon and water fluxes above a north Australian savanna. *Tree Physiology* **21**: 977–988.

Eamus D, Chen X & Hutley L (2002). Root biomass and root fractal analyses of an open Eucalyptus forest in a savanna of north Australia. *Australian Journal of Botany* **50**: 31–41.

Goulden ML, Munger JW, Fan SM, Daube BC & Wofsy SC (1996). Measurements of carbon sequestration by long-term eddy covariance: methods and a critical evaluation of accuracy. *Global Change Biology* **2**: 169–182.

Hahn G & Marschner H (1998). Effect of acid irrigation and timing on root growth of Norway spruce. *Plant and Soil* **199**: 11–22.

Healy RW & Cook PG (2002). Using groundwater levels to estimate recharge. *Hydrogeology Journal* **10**: 91–109.

Knight MJ, Minty EJ & Smith RB (1983). *Collected case studies in engineering geology, hydrogeology and environmental geology.* Sydney.

Pearcy RW, Ehleringer J, Mooney HA & Rundel (eds) (1994). *Plant physiological ecology. Field methods and instrumentation.* Chapman & Hall, London.

Phillips N, Oren R, Zimmermann R & Wright SJ (1999). Temporal patterns of water flux in trees and lianas in a Panamanian moist forest. *Trees* **14**: 116–123.

Schulze E-D, Williams RJ, Farquhar GD, Schulze W, Langridge J, Miller JM & Walker BH (1998). Carbon and nitrogen isotope discrimination and nitrogen nutrition of trees along a rainfall gradient in northern Australia. *Australian Journal of Plant Physiology* **25**: 413–425.

Sililo OTN & Tellam JH (2000). Fingering in unsaturated zone flow: a qualitative review with laboratory experiments on heterogeneous systems. *Ground Water* **38**: 864–871.

Zwieniecki MA, Boyce CK & Holbrook NM (2004). Functional design space of single veined leaves: role of tissue hydraulic properties in constraining leaf size and shape. *Annals of Botany* **94**: 507–513.

Plate 1

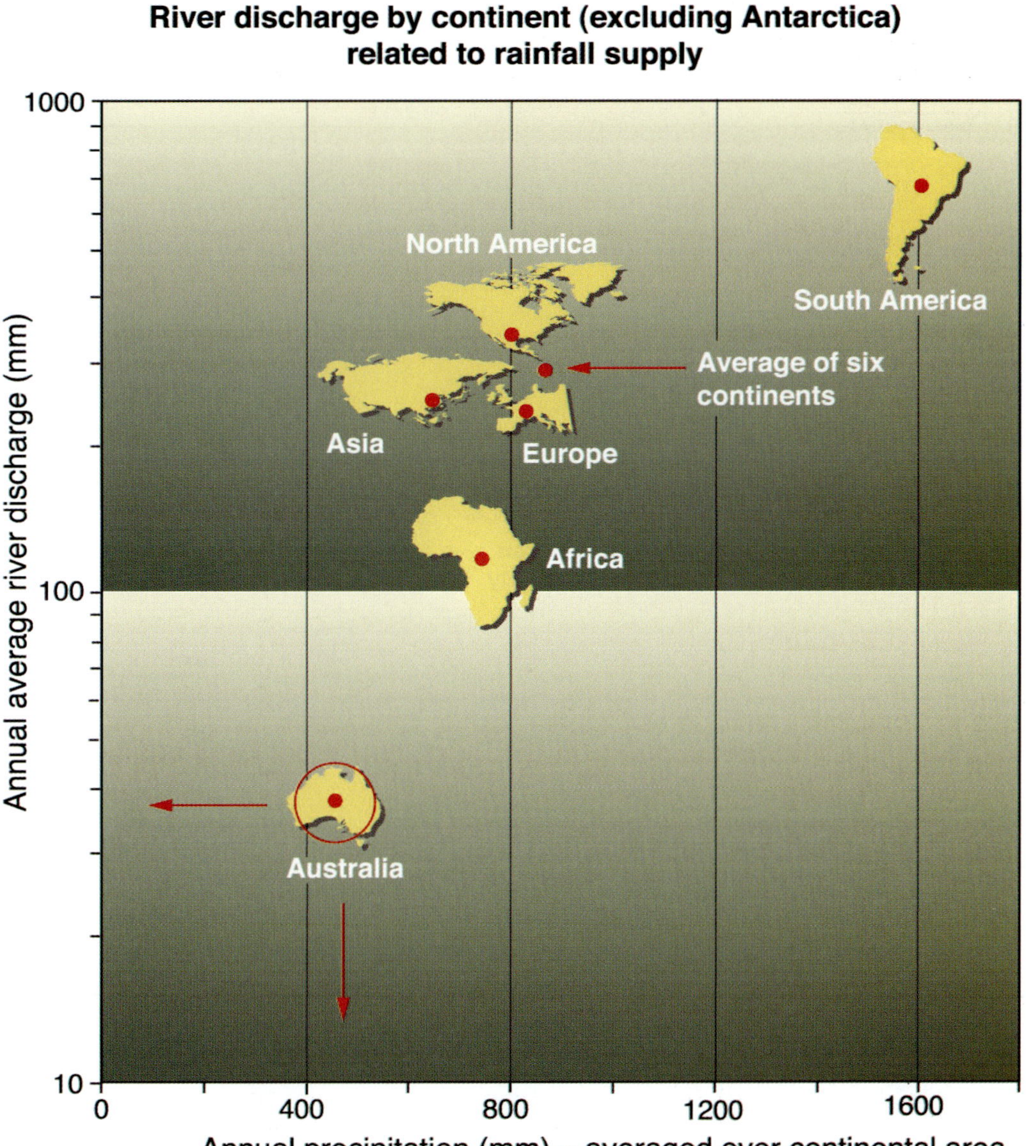

Fig. 1.1 Australia receives less rainfall and discharges less water as river flow than any permanently inhabited continent

Source: © Bureau of Meteorology

Plate 2

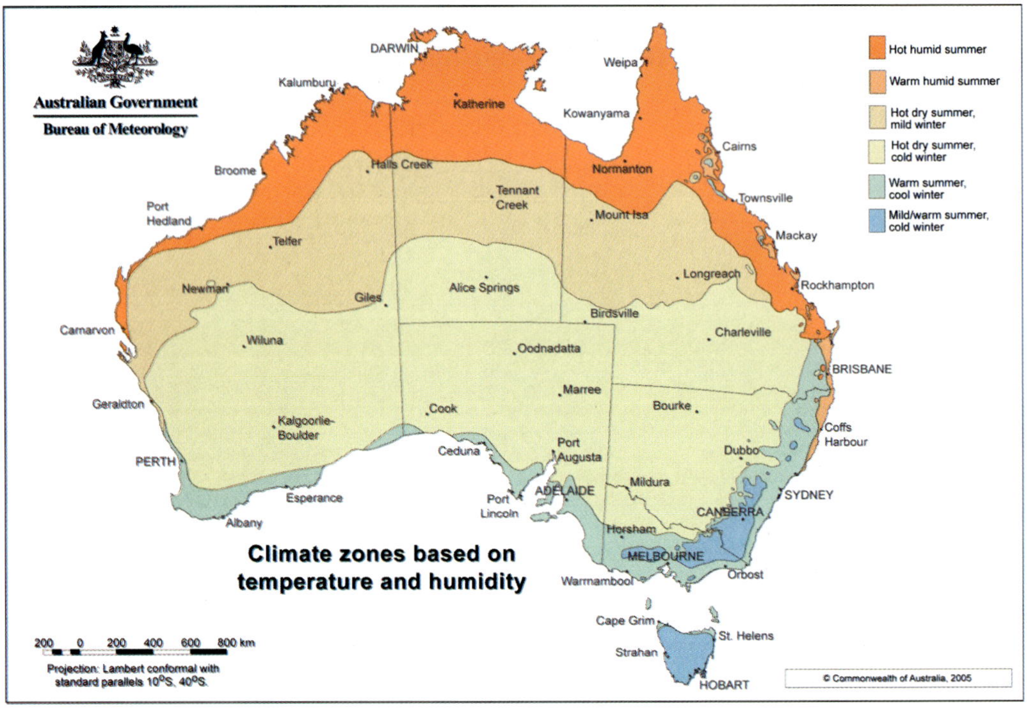

Fig. 1.2 Australian climate zones based on temperature and humidity

Source: © Bureau of Meteorology

Plate 3

(a)

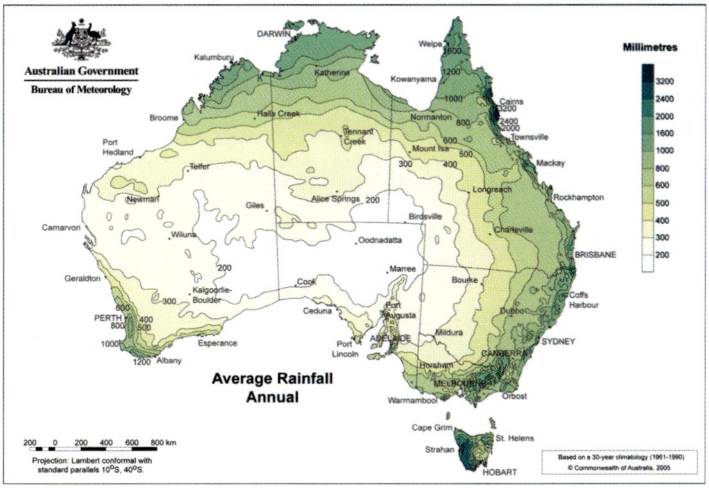

(b)

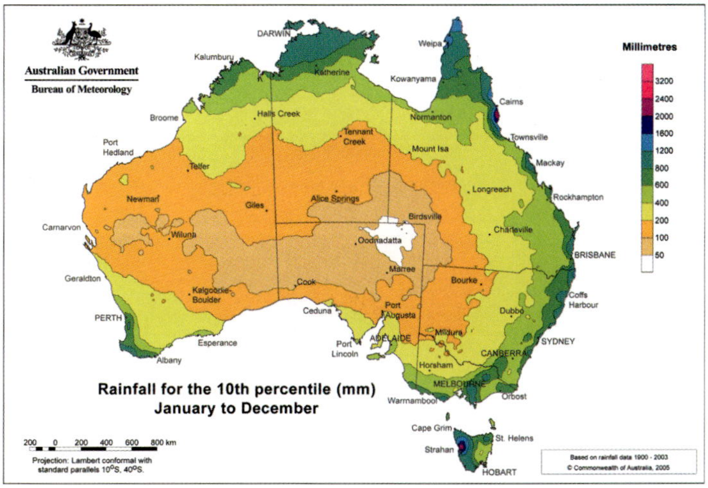

(c)

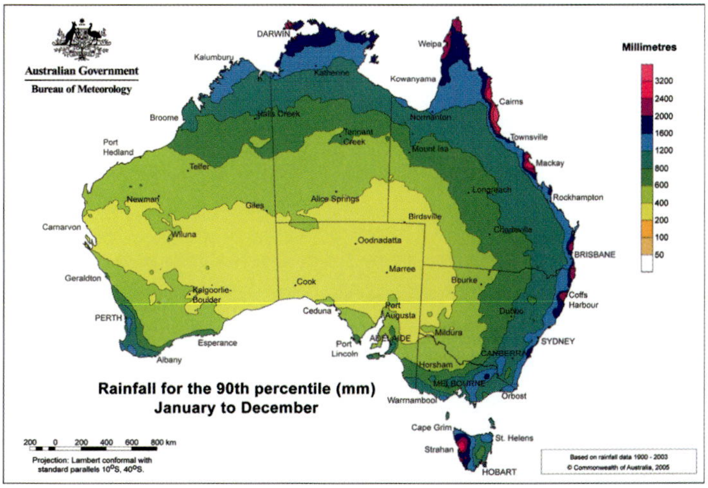

Fig. 1.3 (a) Australian median annual rainfall showing the arid interior and the wetter northern and eastern coasts. Extremes of rainfall, showing (b) 10th and (c) 90th percentile annual totals

Source: Data and graphics supplied by and © Bureau of Meteorology

Plate 4

Fig. 1.6 Fire is a recurrent feature of the Australian landscape. Much of the Top End of Australia burns every 1–2 years

Photograph courtesy of Dr Lindsay Hutley, Charles Darwin University

Fig. 1.8 A medium open eucalypt forest (savanna) of the Northern Territory. This photo was taken in Kakudu National Park, NT, Australia, where annual rainfall is approximately 1500 mm per year. The upper canopy is about 12 m tall. There is minimal grass cover in this photo because it was taken in the late dry season (August)

Photograph courtesy of Dr Lynda Prior, Charles Darwin University

Fig. 1.9 Coastal heath of New South Wales is low-lying (<2m) and densely packed

Plate 5

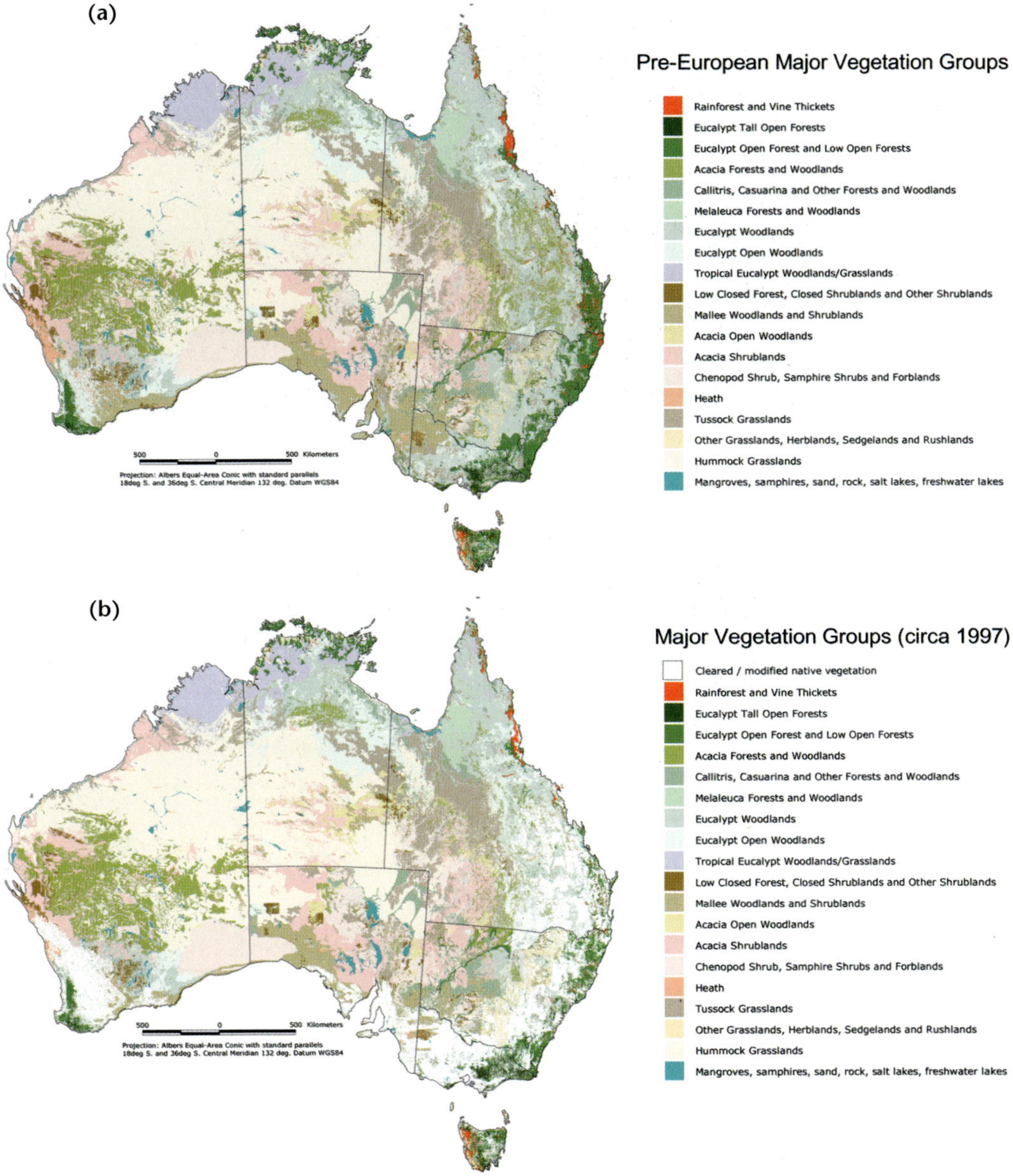

Fig. 1.14 (a) Vegetation types of Australia prior to European settlement were dominated by hummock grasslands, acacia shrublands, acacia forests and woodlands, tussock grasslands, chenopod shrubs and eucalypt woodlands. (b) After approximately 200 years of European settlement, vegetation is dominated by hummock grasslands, cleared and modified vegetation, eucalypt woodland, acacia shrublands and acacia forests and woodlands

Source: Reprinted from the NLWRA (2001)

Plate 6

Fig. 7.2 A savanna after a late season burn in northern Australia. Lower canopies were scorched, taller ones were not

Photograph courtesy of Dr Lindsay Hutley, Charles Darwin University

Plate 7

Fig. 7.7 Sapwood and heartwood can be clearly separated in many species by a distinct colour change

Photograph courtesy of M Zeppel

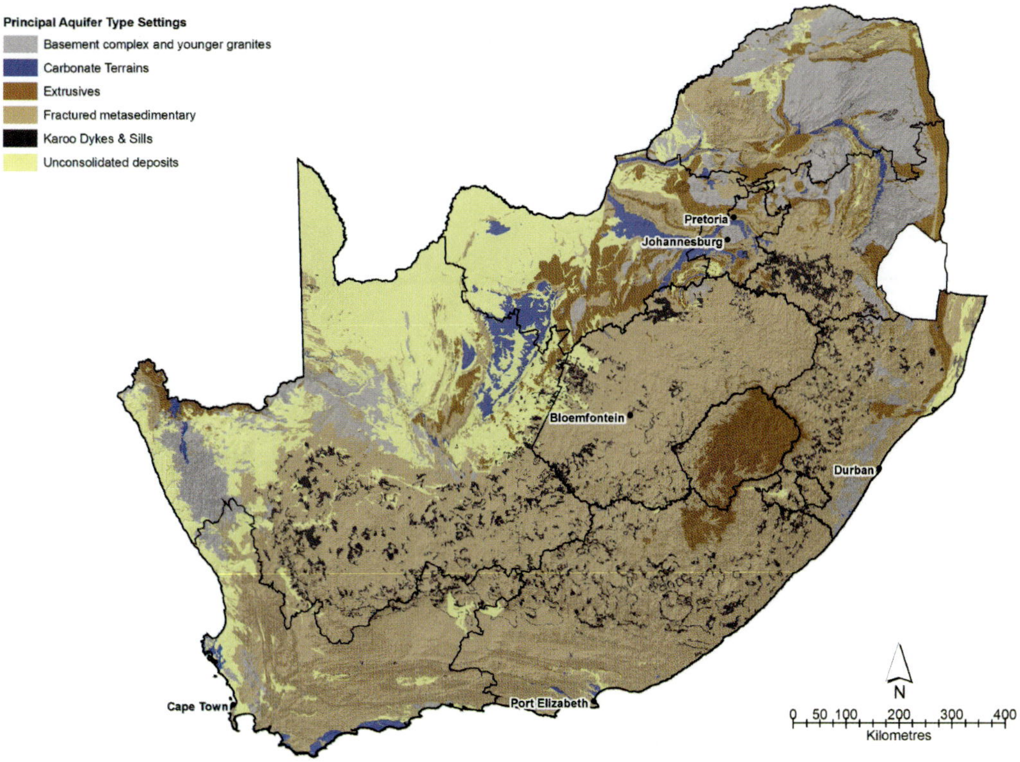

Fig. 10.1 The main aquifer types in South Africa, based on primary lithology

Plate 8

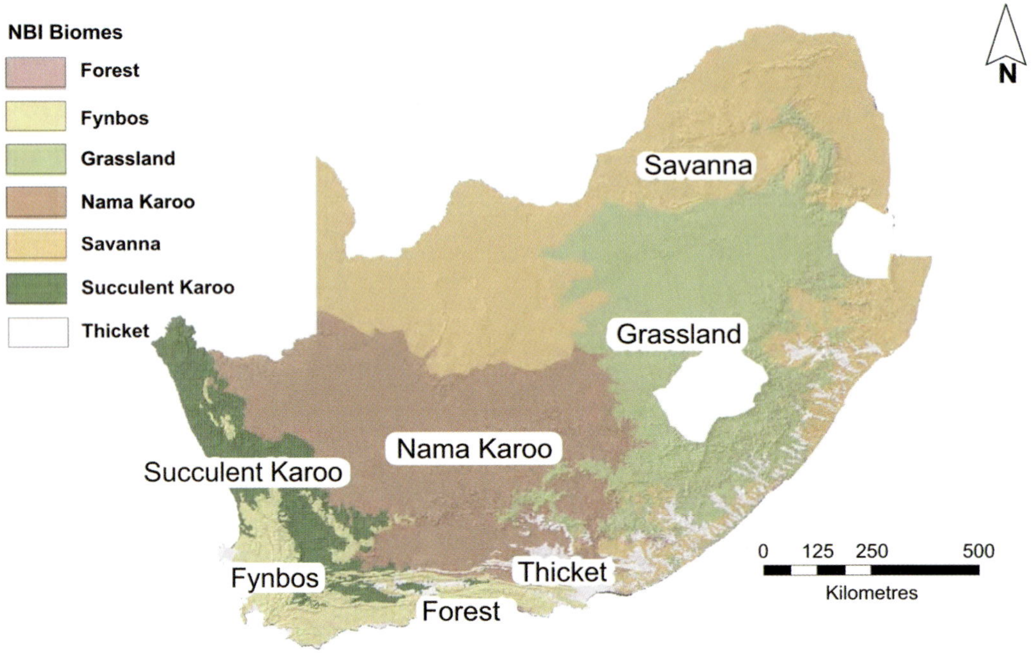

Fig. 10.4 The dominant biomes of South Africa

Fig. 10.9 NDVI remote satellite image of the Limpopo Valley between South Africa and Zimbabwe, showing relatively lush areas in red

Chapter 5

Hydrological models

In Chapters 1–4 the foundations for an understanding of Australian climate, vegetation structure, plant water relations and hydrology were presented. Both theory and techniques were discussed with reference to plant ecophysiology and hydrology. In this chapter we present an important approach that links hydrology and plant ecophysiology, namely, the application of process based models.

This chapter is divided into four sections. In the first, we start with a brief consideration of how to construct a model. We then quickly move on to a more detailed consideration of modelling components of a site's water balance, including stem flow, interception losses, canopy throughfall, surface run-off and soil water storage. The third section of the chapter considers plant growth and how plant growth (especially increase in leaf area and root growth) can be incorporated into ecohydrological models. Finally we discuss a range of system models, including a description and application of the various models to ecohydrology.

After reading this chapter readers should be able to address the following topics:

- how a number of process models incorporate a knowledge of the biophysics of evaporation, interception, transpiration, water uptake and growth of roots and leaves;
- how different ecohydrological models produce different insights into the functioning of terrestrial vegetated landscapes;
- how different ecohydrological models can be used to address different questions relating to landscape function.

Introduction: constructing a model

An ecohydrological model is a simplified mathematical description of a field situation. It describes the relationships between different parameters (e.g. rainfall and soil moisture content), so that one or more parameters (such as tree water use) can be calculated based on the values of others. Models are often used for calculating things that cannot be measured, or cannot easily be measured. In the case of hydrological models, this might be an estimation of a component of the water balance that is difficult to measure, such as the drainage rate beneath the root zone of a forest ecosystem. It might also be prediction of an event that is yet to occur, or analysis of the possible future consequence of a range of different management options. Questions that might be relevant in ecohydrology include:

- What will be the change in run-off from the catchment if a *Eucalyptus* woodland is replaced with a *Pinus radiata* plantation? and
- How will run-off change over time as the plantation grows?

Alternatively, we might ask:

- What will be the change in the water table depth will be if a proportion of a forested catchment is cleared for agriculture?

We might attempt to answer the first two questions by measuring run-off in different catchments under *Eucalyptus* woodland and pine plantations of various ages. We might attempt to answer the third question by measuring water table depth in catchments that have been cleared to different extents. However, this could be an expensive exercise and would not account for other possible differences due to soil type or climate. The results might also be somewhat dependent on the rainfall regime over the period of measurement (e.g. if the period of measurement was one of lower than average rainfall). Alternatively, we could attempt to answer these questions using modelling.

Models designed to answer the above questions for particular field sites might be classified as **predictive models**. At the other end of the spectrum are **generic models**. Generic models are used to analyse flow in hypothetical systems, and are not designed to represent a particular field situation. They are usually designed to assist understanding of processes, and the relationships among different processes, in complex systems. **Interpretive models** fall somewhere between these two. They are used to assist understanding of system dynamics and to gain insight into relationships between different processes and parameters. However, unlike generic models, they are usually designed to represent particular field situations (Anderson & Woessner 1992).

The steps involved in the modelling process are summarised in Figure 5.1. First, the purpose of the modelling exercise needs to be established. We need to ask 'What question is the model intended to answer?' Next, a conceptual model of the system must be developed. A conceptual model is an understanding of the system to be modelled and how it operates. It will require collation and analysis of field data. Clearly, we need to understand a system before we can model it. Once the conceptual model is developed, we should have a good idea of the different components of the hydrological cycle that need to be included in the model, and the required level of detail for each component. At this stage we can make a choice of the mathematical model most suited to our needs.

Model construction and parameterisation involves setting the boundaries for the area to be modelled and the initial selection of values for the different model parameters. We then commence the model calibration process, which involves comparison of model predictions of some field parameters with available measurements, and modification to the model construction based on this comparison. The more data of different types that are available to compare with simulation results the more accurate will be the model predictions. Because they are not based on field data, generic models do not require calibration and in fact cannot be calibrated. Calibration of interpretive models is somewhat less important than is the case with predictive models.

It is also necessary to conduct some type of sensitivity analysis as part of the calibration process. Sensitivity analysis involves asking 'How does the model output change if I change some of the input parameters?'

Sensitivity analysis is important because we never know the values of all the input parameters perfectly. We therefore need to understand how sensitive the output is to these parameters. If the input parameters are poorly known and the output is highly sensitive to them, little trust

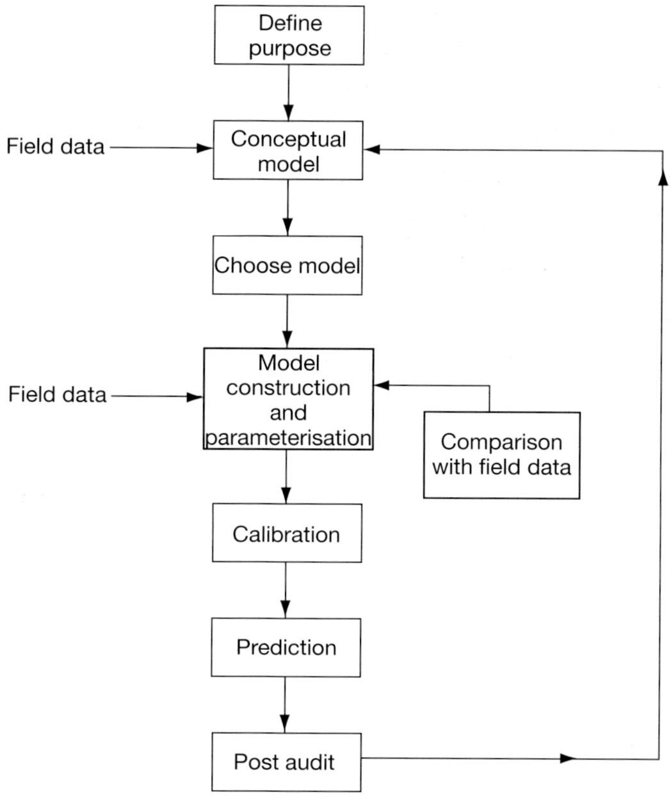

Figure 5.1 Developing a model requires many steps, involving feedback between conception, data, model outputs and calibration

Source: After Anderson & Woessner (1992)

can be placed in the modelling results. Sensitivity analysis can also indicate which parameters should be the focus of subsequent field studies to improve the accuracy of the model predictions.

Once calibration of the model is complete, it can be used for prediction and for answering the question posed at the beginning. The final stage of the modelling process is the post audit, in which the results of the modelling are assessed. This is a feedback loop in which results of the modelling may lead the modeller to conclude that the original conceptual model was a useful and reasonably accurate one. Alternatively, it may result in modification of the conceptual model of the site or even to abandonment of the model, and a repeat of the modelling process with a different choice of model.

One question that must be considered as part of any modelling process is 'Is my model a sufficiently accurate representation of reality?' The operator needs to understand the effect model assumptions will have on model outputs. Unless different models with different assumptions are run, this is very difficult to quantify. A related question is 'Would a simpler or more complex model be more appropriate?' The answer to this question will depend on the sensitivity of the model outputs to the assumptions used by the model, and the availability of the additional data that would be required by a more complex model. Simple models can be as valuable as complex ones.

This chapter describes common modelling approaches for water flow in catchments. We do not discuss approaches to modelling the movement of salts or other dissolved substances, or movement of suspended particles and erosion processes. Readers should be aware, however, that models also exist for these purposes. The first section outlines some of the mathematical equations that form the basis of a number of different hydrological models. The next section briefly describes some simple approaches for included plant growth within hydrological models. The third section presents three different models that describe the soil–water–vegetation system in different levels of detail. This chapter is intended as an overview only, and for a more complete treatment of the subject readers should consult some of the textbooks that cover this subject in more detail. These include Freeze (1978) and Kirkby (1988) for overland flow processes, Konikow and Mercer (1988) and Anderson and Woessner (1992) for the groundwater system, and Hillel (1977), Campbell (1985) and Feddes et al. (1988) for unsaturated zone flow. Other useful reviews include Tiktak and van Grinsven (1995) and Wang and Smith (2004).

Modelling components of the water balance

Figure 5.2 is a simple representation of the **water balance** for a catchment. Precipitation is initially partitioned into **interception, depression storage, infiltration** and **overland flow**. However, interception and depression storage are only temporary reservoirs. Intercepted rainfall will either evaporate or reach the ground surface as throughfall or stemflow to become depression storage, infiltration or overland flow. Depression storage will either evaporate or infiltrate. Soil infiltration will either be extracted by plant roots (transpiration), will drain laterally as interflow or will recharge the groundwater. Ultimately, infiltrated water will either be transpired, will enter the deep groundwater (which is assumed to exit the catchment) or will become stream flow. A number of pathways have been omitted from Figure 5.2 for simplicity. Both recharge and interflow may re-emerge at the surface as springs to become overland flow. Also, overland flow may infiltrate in downslope areas before it reaches the stream.

Different models may describe processes and linkages depicted in Figure 5.2 in different levels of detail, and many models describe only a part of this system. In one dimensional models, surface run-off and lateral subsurface flow are considered only as sinks, and the fate of the flow (i.e. where it goes) is not considered. Similarly, there is no attempt to follow the movement of recharge within the groundwater system.

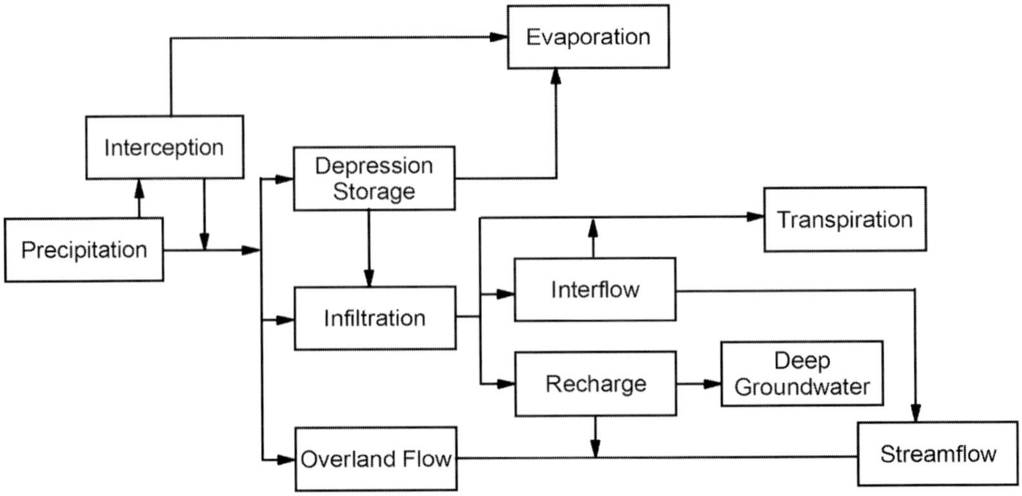

Figure 5.2 Simplistic representation of a catchment water balance

Two and three dimensional models consider horizontal as well as vertical flows. They essentially consider the system as a large number of one dimensional models, where the lateral loss of water from one point becomes a source of water to a neighbouring point.

In terms of representing vegetation systems, because one dimensional models consider water uptake within the soil only as a function of depth they may be most suitable for simulation of cropping monocultures on relatively flat land, and for simulation of forestry plantations and closed canopy forests. However, for simulation of row crops and agroforestry two dimensional models are probably required. For simulating isolated trees and open woodlands, three dimensional models may be needed.

In the following sections, approaches for modelling different components of the water balance are described. First, we will describe some simple approaches to modelling interception, stemflow and throughfall. Moving downwards through the hydrological system, we then briefly describe modelling approaches for depression storage and overland flow. We next describe soil water modelling, and how this is used for determination of interflow and recharge. The next section describes modelling of transpiration and uptake of soil water by plant roots. Finally, approaches for modelling groundwater flow are discussed.

Interception, stemflow and throughfall

Canopy interception loss is that proportion of rainfall that is intercepted, stored and subsequently evaporated from the leaves, branches and stems of vegetation. A number of models of varying complexity have been developed for predicting rainfall interception and stemflow. At the simplest level, we might consider that interception is a constant fraction of rainfall:

$$I = kP \tag{1}$$

where I is the interception loss, P is the precipitation and k is a constant. The value for k would usually be determined empirically.

However, interception is a relatively high fraction of rainfall at the start of a rain event (assuming that the canopy is initially dry), then reduces as the canopy approaches saturation. The above model is clearly a simplification of reality. It is also sometimes useful to distinguish between canopy storage and trunk storage, which the above model does not allow. When canopy storage is exceeded throughfall occurs, and when trunk storage is exceeded we get stemflow (see Chapter 3). One of the simplest and most widely used models for this purpose is that developed by Gash (1979). As with all models, the Gash model contains a number of simplifications and assumptions. One of these is that a vegetation canopy can be described by a single canopy storage capacity, which is constant unless canopy cover changes. It thus assumes that the canopy storage capacity is not a function of the rainfall energy or drop size, for example. The model considers rainfall to occur as a series of discrete events, and that the canopy has sufficient time to dry fully between rainfall events. It is useful to define a parameter, P', which denotes the amount of rainfall required to saturate the canopy. During the canopy wetting stage (P < P'), canopy interception is given by:

$$I = (1 - p - p_t) \text{ when } P < P' \tag{2}$$

where P is the cumulative rainfall, p is the throughfall coefficient (the fraction of rainfall that reaches the forest floor without touching the canopy) and p_t is the proportion of rainfall that goes into wetting up the trunk. I, P and P' have units of depth (usually millimetres) and refer to a particular rainfall event. Equation (2) describes the interception loss for rainfall events which do not saturate the canopy or the trunk. For larger rainfall events which are sufficient to saturate both the canopy and the trunk, interception loss is given by:

$$I = (1 - p - p_t)P' + \frac{\overline{E}}{\overline{P}}(P - P') + S_t \tag{3}$$

where S_t is the storage capacity of the trunk (mm), $\overline{E}$ is the mean evaporation rate during rainfall and $\overline{P}$ is the mean rainfall rate during saturated canopy conditions. $\overline{E}$ and $\overline{P}$ have units of depth per time (usually millimetres per hour). The first term represents evaporation from the canopy during wetting up and after rainfall ceases, the second term represents canopy evaporation during the period of canopy saturation, and the third term is evaporation from the trunk. The total stemflow during the event is given by:

$$S = p_t P - S_t \ \left(P > S_t / p_t\right) \tag{4}$$

and the throughfall is:

$$T = pP + \left(1 - p - p_t - \frac{\overline{E}}{\overline{P}}\right)(P - P') \tag{5}$$

where the first term represents rainfall that does not come in contact with the canopy, and the second term represents throughfall which first contacts the canopy (and may spend some time as canopy storage).

One of the reasons why the Gash model has been so widely used is its relative simplicity. However, simplicity causes problems, and some processes are not well described by this approach. The Gash model is essentially a simplification of the Rutter model (Rutter et al. 1971, 1975). The Rutter model does not assume complete drying of the canopy and trunk between rainfall events and allows throughfall to occur before the canopy is saturated. The rate of throughfall is assumed to be exponentially related to the amount of water stored in the canopy (reaching a maximum value at canopy saturation). There are also more sophisticated models that simulate the pattern of throughfall in three dimensions based on plant architecture (Bussière et al. 2002).

Overland flow

There are a large number of models for estimating **overland flow** from rainfall data. Some of these models have been developed for flood prediction and so concentrate on accurate simulation of run-off from high rainfall events. Models that are designed to predict soil erosion will also focus on high intensity events. For modelling water flows in catchments, however, small rainfall events can be as important as large events. In general terms, surface run-off will occur whenever the infiltration capacity of the soil is exceeded. At the simplest level we might consider that, over a given time period, any rainfall that occurs above a certain threshold value will become overland flow:

$$Ro = P - P_{Ro} \tag{6}$$

where Ro is the overland flow (mm), P is precipitation and P_{Ro} is a constant that represents the infiltration capacity of the soil over the time period under consideration (in mm). It is important to note that the value of P_{Ro} will depend on the time interval being considered. If the analysis takes places using weekly rainfall data, the threshold value will be larger than if daily rainfall is being used.

Of course, the infiltration capacity of the soil is not a constant, and will depend on the antecedent soil moisture. A slightly more complex model might assume that the infiltration capacity is equal to the soil moisture deficit. The maximum amount of water that can be held within the soil at any time (S_{max}) can be expressed as:

$$S_{max} = (\theta_S - \theta_{WP}) \times z_r \tag{7}$$

where θ_S is the water content of the soil at saturation (with units of cm^3 of water per cm^3 of soil), θ_{WP} is the water content at plant wilting point, and z_r is the plant rooting depth. (S_{max} will have units of depth.) According to this model, run-off occurs whenever the amount of precipitation on any day exceeds the available volume within the soil reservoir (Jothityangkoon et al. 2001). Thus:

$$Ro = \begin{cases} 0 & \text{when } P \le (S_{max} - S) \\ P - (S_{max} - S) & \text{when } P > (S_{max} - S) \end{cases} \tag{8}$$

where S is the volume of water within the reservoir on the previous day. The soil moisture storage, S, is updated each day using a bookkeeping approach. On each day, the amount of rain that falls is added to the soil moisture store and the actual evapotranspiration is subtracted from it. The constraint is that the value of S can never be greater than S_{max} or less than zero. The method for calculation of the actual evapotranspiration in this type of model is described below. Models such as this, that record changes in soil moisture storage at fixed intervals over time, are often simply referred to as **water balance models**. Their main limitation is that they do not consider the influence of hydraulic conductivity on infiltration.

A more physically realistic model might consider run-off to occur whenever the rainfall intensity exceeds the soil infiltration rate, rather than simply being based on the rainfall volume and the soil moisture deficit. The soil infiltration rate is dependent upon the moisture distribution in the subsurface, and the soil hydraulic conductivity. It is calculated by solution of the **Richards equation** (equation 12). Also, the above approach assumes that rainfall that does not infiltrate immediately runs off, and so does not permit any water ponding on the surface. In reality, however, the soil surface is not smooth, but possesses a surface roughness which can detain water and prevent it running off. We might assume therefore that the amount of storage is related to the height of these roughness elements. Once the depth of water exceeds the height of the roughness elements, run-off can occur (Fig. 5.3). If we wish to consider only the total amount of run-off over the time scale of a day or a single rainfall event, it might be

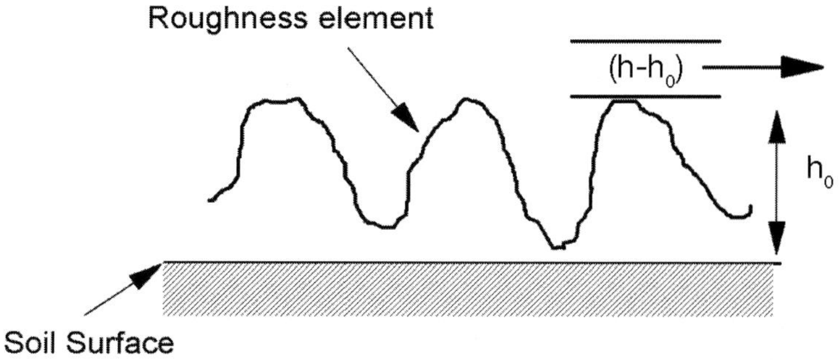

Figure 5.3 Effect of surface roughness on water detention and surface run-off. Run-off occurs only when the height of water on the soil surface exceeds a minimum depth, h_o, which is related to the height of surface roughness elements. The volume of run-off is equal to the height of water minus the h_o

Source: After Verburg et al. (1996)

sufficient to consider that all rainfall above this height runs off. However, if we are interested in run-off processes at smaller time scales and wish to simulate the routing of water within a catchment (using 2D and 3D models), then we also need to consider the run-off velocity. A common approach is to relate the flow velocity to the flow depth using a variant of **Manning's equation**. Thus, run-off might be expressed:

$$Ro = \begin{cases} 0 & \text{when } P \leq (S_{max} - S) \\ (h - h_o)^{5/3}\sqrt{s}/(nL) & \text{when } P > (S_{max} - S) \end{cases} \tag{9}$$

where h is the water height, h_0 is the water height before run-off occurs, s is the slope gradient, L is the slope length, n is **Manning's** n (see Chapter 4), and Ro has units of depth per time (Verburg et al. 1996). The water height is then adjusted based on the calculated volume of run-off (together with any run-on from upgradient). For laminar overland flow, an exponent of 3 rather than 5/3 is recommended, although some models use an exponent between these values to represent flow that is a mixture of turbulent and laminar flow. The height of surface roughness elements, h_0, will be related to the land cover, and models such as SWIM (Verburg et al. 1996) allow it to change due to cultivation practices. SWIM also allows h_0 to decrease during a rainfall event due to the impact of the rain. Manning's n is a constant related to the roughness of the surface. It typically ranges from approximately 0.016–0.03 for unvegetated surfaces, to 0.03–0.06 for short grass (50–150 mm) and 0.04–0.15 for long grass (250–600 mm) (Hudson 1981). Equation (9) thus allows the effect of **vegetation cover** on surface run-off to be simulated.

Infiltration, soil water storage and recharge

We saw above how the amount of run-off could be estimated by comparing the rainfall amount with the storage capacity of the soil reservoir. This is one of the simplest representations for the soil zone. The single value for the storage capacity of the reservoir (or 'bucket') defines how much water the soil can hold. This model does not consider water flow or the distribution of water within the soil, simply the total amount. Water can enter the reservoir from rainfall and can leave either as evapotranspiration (out the top of the reservoir), as drainage (out the bottom) or as lateral flow (out the side). If the reservoir is full, additional water can enter only if an equivalent amount exits. Most usually, this reservoir model approach is used only for one dimensional models, so subsurface lateral flow is not permitted. As briefly described in Chapter 3, in one dimensional flow systems drainage is usually considered to be equal to groundwater recharge.

Of course, not all the water within the soil is available to plants. In single reservoir models, the volume of the reservoir that is available for plants (S_{AV}) is often approximated as:

$$S_{AV} = (\theta_{FC} - \theta_{WP}) \times z_r \tag{10}$$

where θ_{FC} is the water content of the soil at field capacity ($cm^3\ cm^{-3}$). The water content at field capacity is often taken to be the water content at a matric potential of -33 kPa or -10 kPa, and the wilting point is often taken to be the water content at a matric potential of -1.5 MPa. We know that these terms are all somewhat arbitrary. Plants can use water above field capacity, although this water will drain rapidly and so might only be available for a relatively short time after rainfall. The lower limit at which water can be extracted will vary between species, and often be lower than -1.5 MPa (Eamus & Cole 1997; Fordyce et al. 1997). However, in most soils the amount of water that is released below -1.5 MPa is relatively small, so this approximation is often considered adequate. Also, because plants preferentially extract water where root density is highest and can extract only small quantities of water where root density is low, z_r should not

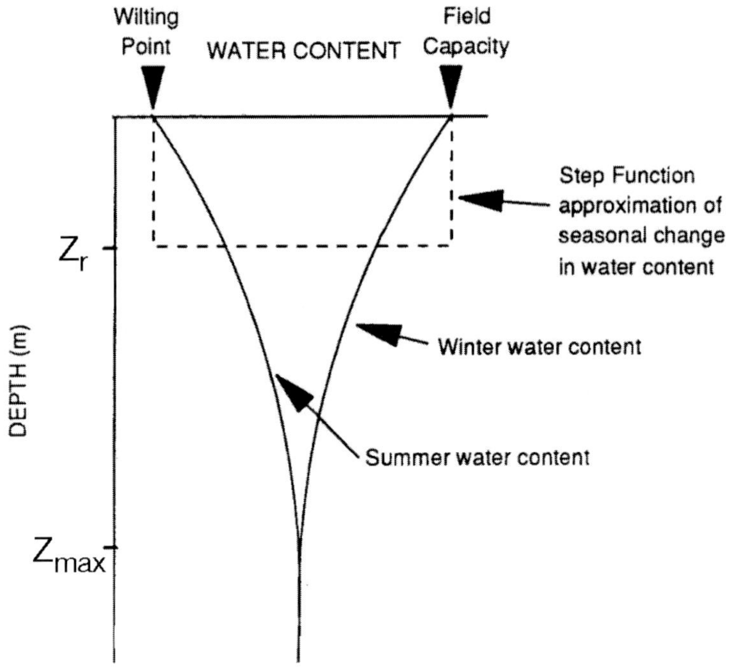

Figure 5.4 Illustration of the step function used to approximate the depth of the plant root zone in single reservoir water balance models

Source: Cook & Walker (1990)

be equated to the maximum depth at which any roots are found. Rather, it represents a mean depth of water extraction, as shown in Figure 5.4.

Multi-reservoir models have also been developed, in which the soil is separated into a number of vertically stacked reservoirs, each representing different depth intervals in the soil profile. In this case, soil water flow is represented by water moving from one reservoir into another one – usually flow to a lower reservoir occurs only when the upper reservoir is full. However, such models often do not permit upward movement of water through the soil, even though this might occur in field situations when the upper layers are depleted due to evapotranspiration but deeper layers remain close to field capacity. These models are also unable to simulate the time delay associated with some of these processes. Movement from one layer to another occurs instantaneously, and drainage from the deepest layer is usually equated with aquifer recharge. The time delay between drainage and recharge is thus ignored, even though it can be significant if the water table is deep.

An important difference between a number of reservoir models is the fate of rainfall in excess of the storage capacity of the uppermost layer. As discussed above, some models assume that this rainfall excess is run-off (Jothityangkoon et al. 2001), whereas other models assume that it drains below the root zone (Kennett-Smith et al. 1994). This represents an important distinction, and the most appropriate method will depend on characteristics of the field site. In particular, Cook and Walker (1990) argued that at their site all this water became drainage because the hydraulic conductivity of the soil was relatively high and the topography was relatively flat. Alternatively, we might consider a system that partitions the rainfall excess into run-off and drainage according to some pre-specified rules. Finch (2001) described a four layer reservoir model that simulates both run-off and drainage. Essentially, the model defines two separate storage capacities for the uppermost layer: one that represents field capacity and a

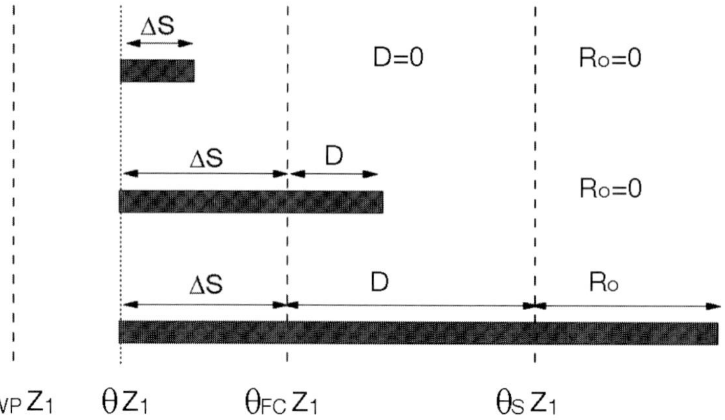

Figure 5.5 Schematic representation of how precipitation (horizontal bars) is partitioned into run-off drainage and change in soil storage according to a simple reservoir model (Finch 2001). $\theta_{WP} Z_1$, $\theta_{FC} Z_1$ and $\theta_S Z_1$ represent the storage of the reservoir at wilting point, field capacity and saturation, respectively; θZ_1 is the storage prior to the rainfall event. If the rainfall event fails to raise the water content to field capacity (top bar), then neither run-off nor drainage occurs and the rainfall simply increases the soil storage (ΔS). If the water content exceeds field capacity but does not exceed saturation (second bar), drainage occurs but there is no run-off. Rainfall is partitioned into that portion required to raise the water content to field capacity (ΔS) and drainage (D). If the rainfall is sufficient to increase the water content above saturation, then both drainage and run-off occur

second that represents saturation. Drainage occurs whenever field capacity is exceeded, but run-off only occurs when saturation is exceeded. Thus:

$$
\begin{aligned}
&Ro = D = 0 && \text{when } P < (\theta_{FC} - \theta)z_1 \\
&\left.\begin{aligned} &Ro = 0 \\ &D = P - (\theta_{FC} - \theta)\times z_1 \end{aligned}\right\} && \text{when } (\theta_S - \theta)z_1 > P > (\theta_{FC} - \theta)z_1 \\
&\left.\begin{aligned} &Ro = P - (\theta_S - \theta)\times z_1 \\ &D = (\theta_S - \theta_{FC})\times z_1 \end{aligned}\right\} && \text{when } P > (\theta_S - \theta)z_1
\end{aligned}
\qquad (11)
$$

where P is rainfall, Ro is run-off, D is drainage (all with units of mm), z_1 is the thickness of the top layer (z_1 must be less than the plant rooting depth z_r) and θ is the antecedent soil moisture, which will always be less than the field capacity. This partitioning of rainfall into run-off, drainage and change in soil storage is shown schematically in Figure 5.5. Drainage from the first layer enters the second layer, where it may still be extracted by plant roots. Once the field capacity of the second layer is exceeded, drainage occurs to the third layer and so on. Recharge occurs when the field capacity of the fourth layer is exceeded.

It is useful at this point to discuss time steps of water balance models a little further. We have suggested that a daily analysis takes place. With these types of models, the output of the model can sometimes depend upon the time step chosen, with longer time steps causing lower predictions of run-off and recharge and higher predicted evapotranspiration (Howard & Lloyd 1979). This is because in many areas rainfall tends to be highly episodic, whereas potential evaporation is less variable. Thus the rainfall excess over evaporation that occurs over short periods of high rainfall is reduced when these are averaged over longer time periods.

The water balance model does not consider that drainage may be limited by the hydraulic conductivity of the soil. As discussed above, it also is unable to reproduce the time lag between rainfall and recharge. A better physical description of water movement in the soil is given by **Richards equation**. The Richards equation in one dimension is:

$$\frac{\partial \theta}{\partial t} = \frac{\partial}{\partial z} K\left(\frac{\partial \psi}{\partial z} + 1\right) - T \qquad (12)$$

where θ is the volumetric water content, K is the hydraulic conductivity, ψ is the soil matric potential, z is depth and t is time. θ and ψ are related by the water retention curve and K is related to θ by the hydraulic conductivity function (Chapter 3). T is a term that represents plant transpiration. The term in the parentheses is the vertical hydraulic gradient, and this equation simply says that the change in water content with time is equal to the difference between the amount of water flowing in and that flowing out (as calculated using Darcy's law) *minus* water lost to evapotranspiration. The main difficulty with this approach relates to the data requirement. The water retention curve dictates the change in soil matric potential and hence hydraulic gradient associated with a change in soil water content. The hydraulic conductivity function determines the amount of water that will drain for a given hydraulic gradient. Both of these functions are difficult to measure, particularly at low values of the volumetric water content (large negative values of matric potential). Equation (12) also requires information on water uptake by vegetation as a function of depth. Methods for estimating this are discussed below.

The Richards equation is easily generalised to two and three dimensions, in which case it can simulate lateral subsurface flow as well as vertical drainage. Lateral subsurface flow (interflow) is calculated from horizontal gradients in matric potential, which are likely to arise whenever spatial variations in soil properties or vegetation characteristics occur. Two dimensional simulations of soil water dynamics are likely to be particularly important for simulation of soil water movement beneath row crops and agroforestry (Schlegel et al. 2004).

Transpiration and soil water uptake

In most water balance models, **actual evapotranspiration** (AET) is assumed to be proportional to **potential evapotranspiration** (PET) and to a coefficient related to the soil dryness and root distribution. The simplest method for estimating potential evapotranspiration is to assume that it is a constant fraction of pan evaporation (usually 0.7–0.9). Alternatively, it can be estimated from other meteorological data. Common approaches to estimating potential evapotranspiration from meteorological data include the Penman-Monteith model and the Priestley-Taylor model. According to one form of the Penman-Monteith model, PET is given by:

$$\lambda E = \frac{sR_n + \rho C_p \delta_e g_a}{s + \gamma} \qquad (13$$

where λ is the latent heat of vaporization of water , E is the evapotranspiration rate, s is the slope of the curve relating saturation water vapour pressure and temperature, R_n is net radiation, ρ is air density, C_p is the specific heat capacity of air at constant temperature, δ_e is the ambient (bulk) air water vapour pressure, g_a is the aerodynamic conductance to water vapour and γ is a psychrometric constant.

At the simplest level, single reservoir models may consider water uptake to decrease linearly with decreasing soil water storage (e.g. Jothityangkoon et al. 2001):

$$AET = \frac{S}{S_{AV}} PET \qquad (14)$$

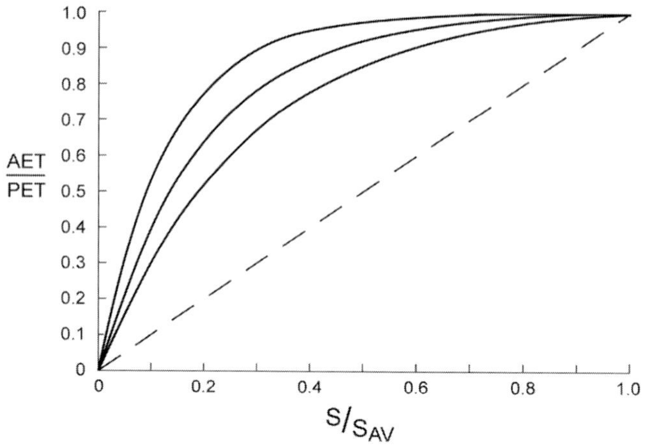

Figure 5.6 Hypothetical relationships between relative evaporation (AET/PET) and relative soil water storage (S/S_{AV}). The broken line depicts a linear relationship (equation 14), while the solid lines denote non-linear relationships used by Cook and Walker (1990)

According to this model, when PET is constant and there is no additional water supply (i.e. no rainfall), the soil water storage will decrease exponentially over time. Cook and Walker (1990) used a non-linear relationship between AET and PET to simulate a more rapid decline in transpiration as the soil approaches wilting point (Figure 5.6), although the principle of the approach is the same. The form of the relationship may vary between different species. For multiple reservoir models, water extraction from the different layers is determined simply from the relative water content of the layer and the proportion of plant roots assigned to each layer.

Models that concentrate on vegetation and atmospheric processes usually estimate actual evapotranspiration using an **energy balance** approach. Using another version of the Penman-Monteith model, actual evapotranspiration is given by:

$$\lambda E = \frac{sR_n + \rho C_p \delta_e g_a}{s + \gamma\left(1 + g_a/g_c\right)} \qquad (15)$$

where g_c is canopy conductance and the other parameters are as defined above. This version of the Penman-Monteith model requires knowledge of canopy conductance, which is a biologically meaningful variable closely allied to stomatal conductance (see Chapter 4). Where several canopy layers are simulated, an energy balance approach is used to partition total net radiation into that reaching each of the canopy layers, based largely on their leaf area indices and surface albedos.

Water uptake from different soil layers is usually related to the **soil matric potential**. van den Honert (1948) noted that under steady state conditions the rate of water flow between two plant parts was directly proportional to the difference in water potential between the parts and inversely proportional to the resistance of the pathway for transport. For plant roots, the equation was written as:

$$Q = \frac{\psi_{rs} - \psi_{rx}}{R_r} \qquad (16)$$

where Q is the rate of water flow, ψ_{rs} is the water potential of the root surface, ψ_{rx} is the water potential of the root xylem, and R_r is the resistance of the root system to water flow. Flow through the plant can be represented by a series of interconnected resistance elements. The demand for water from each layer can thus be determined by the difference between the xylem water potential and the soil matric potential, and by the soil and root resistances to flow. A number of models use this approach to estimating the water extraction from different depths within the soil. In the SWIM model (Verburg et al. 1996), the user is required to enter the minimum xylem water potential that the vegetation can maintain. The maximum rate of water extraction from each layer is then calculated as:

$$T_i = \frac{\psi_{si} - \psi_x}{R_{si} + R_{ri}} \tag{17}$$

where T_i is the maximum extraction rate from layer i, ψ_x is the minimum xylem water potential, ψ_{si} is the soil water potential of layer i, and R_{si} and R_{ri} are the soil and root resistance terms for layer i. The minimum xylem water potential will vary depending upon the vegetation being considered. The maximum extraction rate from the entire soil profile is then calculated by summing the extraction rates from each layer ($T = \sum T_i$) If the maximum extraction rate calculated from this process is greater than the potential evapotranspiration rate, the transpiration is set to be equal to the potential rate and the model calculates the xylem water potential that would be required to produce this transpiration rate. The water extracted from each layer is thus calculated using equation (17) but with ψ_x replaced by the calculated value of the xylem water potential. The soil and root resistances depend on the root length density, the root conductance and the soil hydraulic conductivity.

With all these models, there is a need to define the **root distribution** within the subsurface. With the single reservoir models, the maximum root depth, z_r, is the only parameter needed. With Richards equation approaches and with some multi-reservoir models, the root distribution within the subsurface is required. A number of models use some functional form for the root distribution between the soil surface and depth z_r. At the simplest level, roots can be assumed to be evenly distributed between the soil surface and the maximum rooting depth, although a model which assumes an exponential decrease in root density with depth is probably more realistic. The more sophisticated models allow roots to proliferate within the soil depending on the availability of water and nutrients. The resulting root distribution does not follow a functional form, but the distribution at any time determines the pattern of water uptake. Models that simulate root growth are discussed later in this chapter.

Groundwater flow

In one dimensional models the groundwater system is often not modelled, and the recharge rate to groundwater is the model output. In two and three dimensional models, groundwater flow rates are calculated from hydraulic gradients and aquifer transmissivities. However, even in these cases, it is common practice to model the unsaturated zone using a one dimensional model at a number of spatial locations and to use the recharge and discharge rates calculated from this as input to a separate groundwater model (e.g. Batelaan et al. 2003). The main problem with this approach is that it does not properly simulate the capillary fringe, and the dynamic nature of that zone. It will thus not properly represent processes occurring at the soil–groundwater interface, which may be very important for studies of groundwater use by vegetation. Although models that link soil water flow and groundwater flow do exist (e.g. FEFLOW; Diersch 2004) they are not commonly used.

Within the groundwater system, the variation with time of the water table is modelled by the non-linear **Boussinesq equation**. The Boussinesq equation links storage changes to flow rates and hydraulic gradients in the saturated zone the same way that the Richards equation

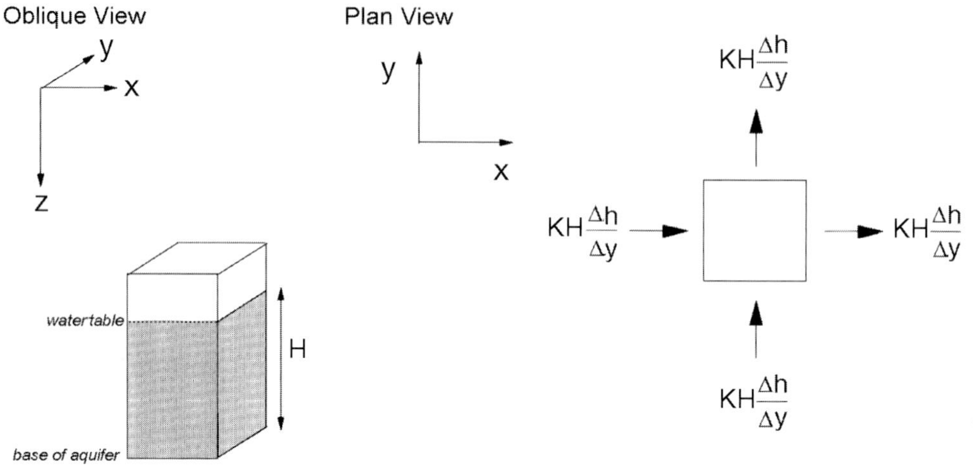

Figure 5.7 Schematic representation of groundwater flows into and out of a unit area of aquifer

does in the unsaturated zone. For an isotropic (i.e. having uniform properties in all directions) unconfined aquifer, this is:

$$S_y \frac{\partial h}{\partial t} = \frac{\partial}{\partial x}\left(KH\frac{\partial h}{\partial x}\right) + \frac{\partial}{\partial y}\left(KH\frac{\partial h}{\partial y}\right) + R \qquad (18)$$

where S_y is the specific yield, h is the height of the water table above some datum point, H is the aquifer thickness (the difference between water table height and elevation of the base of the aquifer), K is the saturated hydraulic conductivity, R is the instantaneous vertical recharge into the saturated zone, t is time, and x and y are horizontal cartesian coordinates. (Groundwater discharge, D, is represented by negative values for recharge, i.e. D = -R.) This equation calculates the change in storage within a unit area of aquifer based on the flows into and out of that unit area. Each of these flows is calculated using Darcy's law. Consider a unit area of aquifer as shown in Figure 5.7. If there is no recharge, the change in storage within the unit area shown will be equal to the groundwater flow into the cell from the left minus that which leaves the cell to the right, together with the flow into the cell from the bottom plus the flow out of the cell to the top of the figure. The first term on the right-hand side of equation (18) gives the difference between the flow into the cell from the left and the flow out of the cell to the right (which we are referring to as the x-direction). The second term gives the difference between the flow into the cell from below and the flow out of the cell to the top. These two terms together with any aquifer recharge represent the change in water within the cell, which is equal to the change in height of the water table multiplied by the specific yield.

Groundwater flow in three dimensions is usually modelled as a series of layers, each layer representing a different aquifer or aquitard, and equation (18) is solved for each layer. In the case of confined aquifers, S_y becomes the storativity rather than the specific yield, h is the pressure head rather than the water table height and R is the net leakage from layers above and below.

As noted above, one problem with this approach is that it does not correctly represent processes at the soil–groundwater interface. In particular, the specific yield is characterised by a constant, which is not strictly correct. The **specific yield** (the relationship between the water added to the aquifer and the rise in the water table) will vary depending on the rate of water table rise or fall. It will also vary with the depth to the water table, and so may change as the

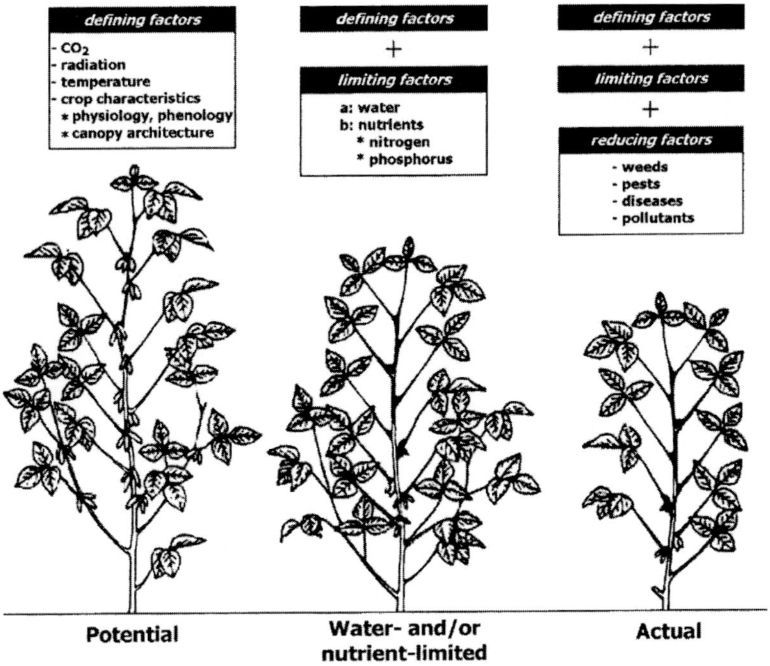

Figure 5.8 Plant growth (or crop production) is the result of a number of interactions between defining factors, limiting factors and reducing factors

Source: van Ittersum et al. (2003)

water table rises and falls (Healy & Cook 2002). Models which link the unsaturated and saturated zones would be required to properly simulate processes that occur within this interface zone.

Modelling plant growth

Accurately modelling the growth of plants is extremely difficult. van Ittersum et al. (2003) note that plant growth (or crop production) can be considered to be the result of a number of **defining factors, limiting factors** and **reducing factors** (Fig. 5.8). Defining factors determine the potential maximum growth rate. They include environmental variables such as CO_2 concentration in the atmosphere, incoming solar radiation and soil and atmospheric temperature, as well as plant physiological characteristics. Growth is limited by limiting factors such as water and nutrients. If these are in sufficient quantities in the soil, growth will occur at the potential rate. Growth may be further reduced by reducing factors, such as weeds, pests, diseases and pollutants. In this section, we very briefly describe the principles of modelling plant growth, particularly as they influence the hydrological balance. The effect of water availability on plant transpiration was discussed above, and plant growth is linked to transpiration. We do not directly discuss the effect of nutrient availability on growth nor the effect of various reducing factors. In the context of hydrological prediction, plant growth is important because of its feedback into hydrological partitioning. Leaf growth increases leaf area index which increases rainfall interception and radiation interception, thereby increasing water use by the upper canopy but reducing evapotranspiration of understorey and soil evaporation. Leaf and stem growth increases the interception storage capacity of the canopy and trunk respectively, which increases interception storage and evaporation of rainfall. Root growth can increase the ability of plants to extract water from the soil, thus directly decreasing drainage but also influencing subsequent infiltration and thus run-off.

Some models allow plant parameters to change over time and this can be used to represent the growth of a plant, although this is generally not considered the same as modelling plant growth. For example, simple models may allow the user to specify variations in leaf area index over time. This could be used to simulate deciduousness of certain species as well as plant growth. Other models allow root densities and maximum rooting depths to be specified as a function of time, which might also be used to represent growth of vegetation. However, a more physically based model would link the growth of roots and leaves to the availability of water, light and nutrients. One of the simplest such representations of plant growth can be found in the RESCAP model (Monteith et al. 1989; Dewar 1997). The model assumes that plant growth is either light limited or water limited. Transpiration under light limited conditions is assumed to be proportional to the daily incident solar radiation, while water limited transpiration is assumed to be proportional to the available soil water storage and the root biomass. Daily transpiration is then the minimum:

$$T = \min\left(\frac{\varepsilon}{q} S_0 (1 - e^{-kL_c}), \sigma R(\theta - \theta_{min}) z_r\right) \tag{19}$$

where T is the transpiration rate (kilograms of water used per day per square metre of land surface area), ε is the light utilisation coefficient, k is the light extinction coefficient, S_0 is the daily incident solar radiation, L_c is the projected leaf area of the canopy, q is the water use efficiency (grams of dry matter produced per kilogram of water used), R is the root biomass (grams of dry matter per square metre of land area) and σ is a constant. The first term inside the parentheses is the light limited transpiration rate; the second term is the water limited transpiration rate. Plant growth is modelled using a linear relationship to transpiration:

$$G = qT \tag{20}$$

where G is net primary production (grams of dry matter per square metre of land surface area per day). It then becomes important to decide how growth is partitioned between above ground and below ground. Above ground growth will generate additional leaf area, which will increase the transpiration rate (and hence growth by intercepting more light and fixing more carbon) if water is not limiting. Below ground growth may increase the plant's ability to extract water from the soil. The simplest approach is simply to proportion G to leaves, roots and stems using pre-defined constants. In the RESCAP model, the proportions of new growth allocated to leaves, roots and stems are denoted n_L, n_R and n_S (with $n_L + n_R + n_S = 1$). The increase in leaf area over time is given by:

$$\partial L_c = GBn_L \tag{21}$$

where L_c is the projected leaf area of the canopy, B is the specific leaf area (m^2 leaf per g leaf), and the increase in root biomass by:

$$\partial R = Gn_R \tag{22}$$

where R is the root biomass (grams of dry matter per square metre of land area). Increasing the root biomass will proportionally increase the transpiration rate when water is limiting. Increasing the leaf area index will increase the transpiration rate under light limited conditions.

The important assumption of this model is the direct linear relationship between growth and transpiration. A number of studies have found this to be a reasonable approximation for crops (Monteith 1986), and it may also be reasonable for young saplings. However, for mature trees transpiration can and does occur without any significant growth. Also q need not be constant and the same daily transpiration rate will not necessarily generate the same daily growth rate.

A more physically realistic approach to plant growth requires modelling of the **carbon balance**. There are a number of models that take this approach. Models such as **TOPOG** and **WAVES** (Zhang & Dawes 1998) take a simple and highly conceptual approach to the carbon balance. The carbon assimilaton rate, A_i, is assumed to be equal to a theoretical maximum rate, modified according to the availability of light, nutrients and water:

$$A_i = A_{\max}\left(\frac{1 + w_W + w_N}{1/m_L\chi_L + w_W/\chi_W + w_N/\chi_N}\right) \tag{23}$$

where A_{max} is the maximum carbon assimilation rate (which must be specified by the user), w_W is the weighting of water relative to light, w_N is the weighting of nutrients relative to light, χ_W, χ_N and χ_L are the relative resource availabilities for water, nutrient and light respectively, and m_L is the modifier of light availability due to temperature. The availablity of light is determined from intercepted radiation and air temperature, and the availability of water is determined from the soil water potential in the root zone (Zhang & Dawes 1998). The carbon assimilation rate is therefore equal to the maximum rate only when all of χ_W, χ_N, χ_L and m_L are equal to 1. When any one of these is less than 1 the carbon assimilation rate is reduced. The rate of carbon assimilation is linked to the rate of transpiration through the canopy conductance term; the canopy conductance being directly proportional to the assimilation rate, and canopy conductance having a role in determining transpiration rate.

Carbon assimilation is not the same as plant growth because a proportion of the assimilated carbon is required for maintenance of plant function. Only carbon assimilation above a critical value leads to growth. One approach is to require the user to specify the partitioning of assimilated carbon to roots, leaves and shoots, and to specify the respiration coefficient, maintenance respiration coefficient and mortality coefficient for each of these pools. The increase in leaf carbon might thus be given by:

$$\Delta c_L = n_L Y_L (A_i - C_L R_L) - C_L M_L \tag{24}$$

where C_L is the carbon content of the leaf biomass, n_L is the proportion of net canopy assimilation allocated to the leaves, Y_L is the respiration coefficient that accounts for the conversion of assimulated carbon to biomass, R_L is the maintenance respiration coefficient, and M_L is a mortality coefficient (Zhang & Dawes 1998). Similar equations are used for stems and roots. More complex models may include a more detailed breakdown of carbon allocation, including differentiation between heartwood, sapwood and tree bark, and between fine roots and coarse roots, and stems and branches. They may also allow the distribution of new carbon between the different pools to vary according to the phenological stage of the plant or the availability of water and nutrients and the need for transport and structural support organs. One approach is to allocate net daily production of biomass in such a way as to minimise any imbalance between carbon acquisition by foliage and water and nutrient uptake by fine roots.

In the case of root growth, the question then arises about where new root growth occurs within the soil profile. This is usually based on soil moisture and nutrient availability, although factors such as soil temperature, soil resistance and mineral toxicity can also be considered.

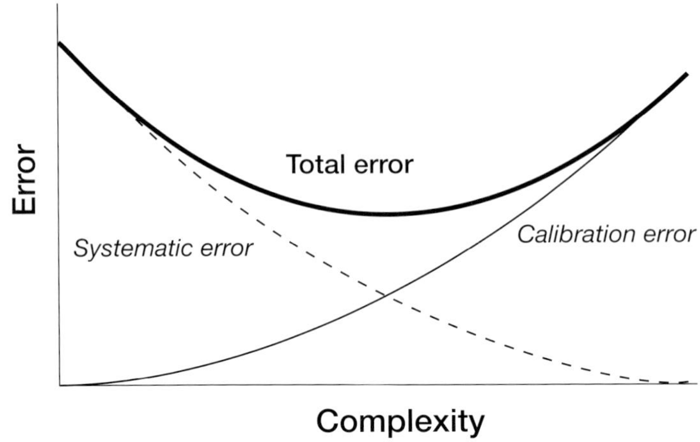

Figure 5.9 Ideal level of complexity for modelling field situations, with minimal total modelling error

Source: Walker et al. (2002)

Carbon allocated to roots is usually converted into root length using a constant parameter for the length:mass ratio.

System models

A very large number of models are available for simulating different components of the hydrological cycle. The models range from very simple ones requiring minimal data input and limited expertise to operate, to very complex models with large numbers of parameters requiring a high level of expertise to operate. Some models simulate only one or two components of the hydrological cycle (e.g. the soil zone or groundwater system), while others seek to simulate the entire soil–vegetation–atmosphere system. Even models that simulate the entire system, however, usually model different components at different levels of complexity, depending on the purpose for which the model was originally designed.

It is important to emphasise that **complex models** are not necessarily more accurate than **simple models**, as they require more data for accurate calibration. In essence, the modeller tries to achieve a balance between the practical need to simplify the complex process that occur in the field situation and the desire to include all the processes that might be important for prediction. The aim is always to reduce error and create greater confidence in the results. Model error can be considered to be made up of **systematic error** and **calibration error**. Systematic error is that resulting from the simplifying assumptions (e.g. if a model neglects macropore flow in a situation where it is important). Calibration error results from incorrect values for the model parameters. Systematic error is reduced by adding more processes to the model (increased model complexity). However, for a given data set, calibration error is reduced by fitting fewer parameters. Conceptually, we might consider an ideal level of complexity for modelling any particular field situation as one where the total modelling error is minimised (Fig. 5.9). Although this ideal level of complexity is a rather abstract notion, conceptually it would depend upon the purpose of the modelling, the particular field situation to be modelled and the available data.

The following section briefly describes three different hydrological models, arranged in order of increasing complexity and dimension. WATBAL is a one dimensional, single reservoir, water balance model. BROOK90 is also a one dimensional model but with a much greater

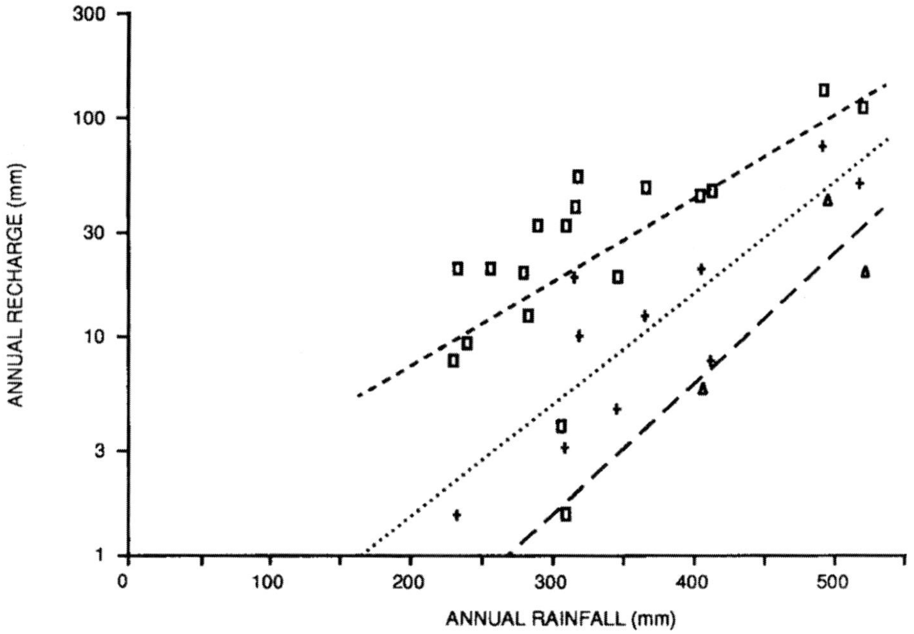

Figure 5.10 Predicted relationship between annual rainfall and annual recharge beneath agricultural land at Wanbi, South Australia (mean annual rainfall 300 mm) for three soils: sand (S_{AV} = 19 mm, squares), loamy sand (S_{AV} = 56 mm, crosses) and sandy loam (S_{AV} = 75 mm, triangles). The model assumes a mean rooting depth of z_r = 0.5 m

Source: Kennett-Smith et al. (1994)

level of detail than WATBAL. TOPOG represents a signicant increase in complexity and is a three dimensional model of transpiration, run-off, soil water flow and groundwater flow that includes simulation of plant growth using a carbon balance approach.

WATBAL

WATBAL (Keig & McAlpine 1974; McAlpine 1970) is one of the simplest models of the soil–vegetation system. It models potential evapotranspiration as a constant fraction of pan evaporation, and models the soil zone as a single storage reservoir. The model requires daily or weekly rainfall and pan evaporation data as input, as well as a value for the pan coefficient (the ratio between PET and pan evaporation). The user is also required to specify the relationship between AET and PET. The volume of the soil reservoir is the only soil parameter required. The reservoir is filled by rainfall, and water is lost from the reservoir by evapotranspiration and drainage. The volume of water extracted by the vegetation depends only on the potential evapotranspiration rate and the volume of water stored within the soil reservoir. The model does not consider interception loss or surface run-off, and assumes that drainage occurs only when the reservoir is full.

Cook and Walker (1990) and Kennett-Smith et al. (1994) used the WATBAL model together with 21 years (1968–88) of daily rainfall and pan evaporation data to simulate drainage beneath agricultural land in the mallee region of South Australia and Victoria. The authors used a pan coefficient of 0.8 to convert pan evaporation data to potential evapotranspiration. The only other parameter required is the maximum available soil moisture storage (S_{AV}), which was estimated using measured values of θ_{FC} and θ_{WP}, and an estimated rooting depth of z_r = 0.5 m. Thus different values of S_{AV} were used to represent the different soil textures found within the region.

The model was used in an **interpretive** sense to examine the sensitivity of annual recharge rate to annual rainfall and to soil type, rather than to provide accurate estimates of recharge rate at any particular field site. No calibration was performed, other than to ensure that the estimated mean annual recharge rates were reasonable and consistent with other estimates that had been obtained within the study region. Figure 5.10 shows the relationship between annual recharge and annual rainfall predicted by the model for three different values of available soil water storage, S_{AV}. The results suggest an approximately exponential relationship between annual rainfall and annual recharge. (It plots as a linear relationship on Figure 5.10 only because the y-axis is logarithmic.) The model was also used to examine the episodic nature of recharge. In particular, the authors found that the wet years of 1973 and 1974 together accounted for between 37% and 83% of the total recharge over the 21 years of record (depending on the value of S_{AV}).

Although no formal model calibration was undertaken, the authors did examine the sensitivity of the estimated recharge rates to the form of the relationship between relative transpiration and relative water storage. Interestingly, they found the estimated recharge rates to be relatively insensitive to this function, which they suggested may have been because most drainage occured only during very heavy rainstorms ($P > S_{AV} + ET$). On these occasions the soil moisture reservoir is filled so actual evapotranspiration is at the potential rate. The authors also noted that the model would probably not provide reliable estimates of recharge for very heavy soils, as it does not consider the limiting effect of soil hydraulic conductivity. Of course, the results are highly sensitive to the value of S_{AV}, which can be difficult to estimate.

One of the attractions of the WATBAL model is its simplicity. The authors used it to gain an understanding of the extent to which recharge rates might vary across a landscape in response to rainfall and soil variability. It was an attempt to understand the extent to which limited field measurements of recharge could be extrapolated across the region. The model was successful to that extent. However, since it does not include a large number of processes, the accuracy of the predictions in any particular field situation would be highly questionable. The model assumes that the recharge process is controlled only by the storage capacity of the soil, and ignores any relationship to the hydraulic conductivity. It makes no allowance for land management or crop nutrient status . It also assumes a constant rooting depth of 0.5 m, even though the region is used for annual cropping and periods of fallow are widely used. Thus the soil will be bare for a period of time and crop rooting depth will increase through the growing season. The adopted rooting depth of 0.5 m is assumed to be an average value. If the model is intended to produce accurate estimates of recharge, the user should compare its results with those of a more complex model that includes some of the above processes. Agreement with the more complex model would provide increased confidence in predictions of the simple model. As it stands, the simplicity of the model might lead users to question the results if it were used in a predictive mode.

BROOK90

The BROOK90 model was designed for **simulating water yield** from small catchments. It focuses on simulation of evapotranspiration processes, vertical soil water flow and generation of storm flow. The model calculates interception, transpiration and evaporation using a modified form of the Penman-Monteith equation (Shuttleworth & Wallace 1985), and uses a one dimensional solution to the Richards equation for flow within the soil zone. There is no provision for spatial variation of parameters within the watershed other than variation of soil properties with depth. Similarly, there is no provision for run-off to infiltrate in downslope areas or for lateral subsurface flow to be used by vegetation in downslope areas. Stream flow is calculated as the sum of run-off, downslope matrix flow, downslope pipe flow and groundwater discharge. However, as the model is essentially one-dimensional it is unable to model the time delay associated with these processes. The model essentially assumes that lateral transport is rapid relative to the time step and so is suitable only for simulating stream flow variations over relatively large time intervals (e.g. monthly) and in relatively small catchments.

Armbruster et al. (2004) used the BROOK90 model to predict the effects of change in tree species composition on water flow in two upland catchments in Germany. The two catchments, Schluchsee (mean annual rainfall 1870 mm) and Rotherdbach (mean annual rainfall 1030 mm), are comparable in terms of land use, relief, soils and geology. Both were originally covered with mixed forest dominated by beech (*Fagus sylvatica*), but these had been replaced with managed spruce (*Picea abies*) stands. Radiation, temperature, vapour pressure, wind speed and precipitation were measured in each of the catchments. Vegetation parameters to represent the current spruce stands were taken from the literature and modified during the calibration process, although root distributions were determined in the field.

Model calibration was achieved using a number of different data sets. This included canopy throughfall that was measured at representative stands, and discharge from the catchment recorded at V-notch weirs. A hydrograph analysis was applied to the stream flow data and used for calibration, by comparing results of the hydrograph separation with flow components calculated by the model. Soil water content data were also measured and used in the calibration.

Once the model was calibrated using vegetation data for spruce, the authors re-ran the model using a set of vegetation parameters chosen to represent the deciduous beech. Comparison of model outputs using spruce and beech data allowed the effect of vegetation type on the water balance of the catchment to be assessed. In both catchments, the model calculates that annual evapotranspiration in the fictional beech stand would be lower and that stream flow would be higher, despite the greater rooting depth of the beech. The differences are largely due to the reduction in rainfall interception during winter, when the beech leaf area index drops to zero (Fig. 5.11). In the low rainfall catchment, mean annual stream flow under beech is predicted to be 30–50% higher than for spruce, while in the high rainfall catchment the increase in discharge is between 7% and 14%. Although the authors calculated hydrological budgets of mixed beech–spruce stands by averaging the results for the pure stands (and weighting according to the proportions of each species), they express caution in using these results to predict changes from spruce to mixed spruce–beech stands because of possible interactions in the mixed communities. In particular, some studies have suggested that root distributions of spruce in mixed stands are likely to be shallower than in pure spruce stands, and root distributions of beech are likely to be deeper (Armbruster et al. 2004). Since the BROOK90 model is one dimensional and has no provision for spatial distribution of parameters it cannot simulate lateral transfer of water between adjacent sites with different vegetation types.

This application of BROOK90 represents an example of a **predictive model** although at this site the vegetation change had already been made, so the prediction relates to a past environment rather than a future one. The authors were able to use a large number of different data sets for model calibration, although this was necessary because a large number of the input parameters were not measured but simply derived from the literature. In particular, because both catchments were currently vegetated with spruce the authors were forced to simply select model parameters for beech (e.g. LAI, seasonal variation of LAI, root distribution) from literature values for comparable sites. The accuracy of the model output would therefore depend somewhat on the appropriateness of the literature values.

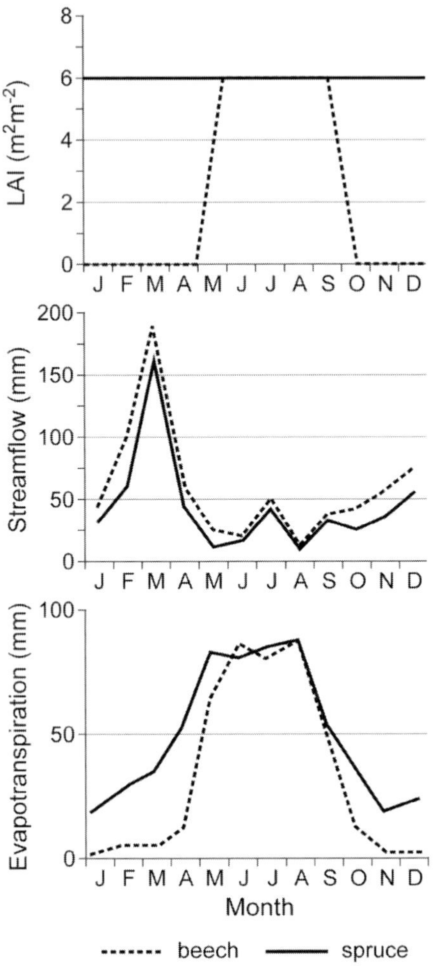

Figure 5.11 Differences in evapotranspiration, stream flow and LAI between two tree species

Source: Armbruster et al. (2004)

TOPOG

TOPOG (http://www.per.clw.csiro.au/topog/) models subsurface hydrological processes using a Richards equation approach, overland flow using Manning's equation and evapotranspiration using the Penman-Monteith approach. Like most models that purport to be three dimensional, it models some aspects of the hydrological cycle in three dimensions but others only in one dimension. Using a daily time step, the model simulates canopy interception of rainfall and the surface energy balance, evaporation and transpiration fluxes, soil infiltration, saturated/unsaturated vertical moisture dynamics, vertical drainage and lateral subsurface flow. Incident solar radiation is varied according to slope and aspect, and almost all the other model parameters can be varied spatially. TOPOG also models leaf, stem and root carbon, so that it can simulate plant growth. However, while the model simulates saturated subsurface flow beneath a shallow water table in three dimensions using Darcy's law, unsaturated flow is modelled only in one dimension (unsaturated lateral flow is not permitted). The model would thus not be able to simulate soil water uptake of row crops or tree belts. Also, the position of the water table is assumed to be parallel to the ground surface and its depth is fixed as a constant. The model thus cannot simulate a dynamic water table that rises and falls in response to recharge events.

The TOPOG model was applied to a small (0.32 km^2) catchment in central Victoria, Australia, to predict the amount of surface run-off from Mountain Ash (*Eucalyptus regnans*) forests (see Chapter 7 for further discussion of the application of TOPOG). Catchment elevation ranged between 590 and 790 m above sea level, with steep predominantly south-facing slopes, mostly between 25% and 35%. A 1:2500 topographic map of the catchment with 10 m contour interval was used to determine surface topography and to divide the catchment into 1117 elements, with an average size of 17 × 17 m. The model required values for 21 different input parameters: 6 climatic parameters, 6 soil parameters and 9 vegetation parameters. Daily values for all the climatic parameters were required for the 12 year simulation period. Two soil layers were used so different values for some of the soil parameters were required for each layer, although soil parameters were not allowed to vary spatially. Vegetation parameters were required both for the Mountain Ash forest, that covers 78% of the catchment, and for the rainforest vegetation that occupies streamside locations and sheltered hollows on south-facing slopes. However, since the plant growth module of TOPOG was not used for this study, vegetation parameters for each vegetation type were fixed throughout the simulation period. Nevertheless, the total number of input parameters that were used was much larger than the 21 parameters required for a system with homogenous soil and a uniform vegetation cover. However, only 5 of the 21 different parameters were directly measured in the field; the others were obtained from the literature or from previous unpublished studies within the region.

The model was run for a period of 12 years (1972–83). Model output includes daily values of rainfall interception, soil evaporation, transpiration, soil moisture storage and stream flow (it assumes that there is no recharge to the deep aquifer). Interestingly, the model predicts **annual evapotranspiration** to be relatively constant, ranging between 511 and 578 mm even though annual rainfall varies between 990 and 1985 mm (Figure 5.12). Soil moisture storage at the end of year varies between 828 and 1107 mm. The large soil moisture storage acts to reduce the year-to-year variation in streamflow. Thus, stream flow (run-off) in 8 of the 12 years varied only between 649 and 879 mm despite rainfall ranging between 1361 and 1985 mm. The lowest predicted stream flow was 474 mm in 1972 (990 mm rainfall) and the maximum was 1102 mm in 1974 (1972 mm rainfall). On an annual basis, predicted stream flows were within 4–13% of observed stream flows, although peak flows are less well simulated than base flows.

An analysis of the sensitivity of **stream flow** to each of the model input parameters was performed as part of this modelling exercise. The most sensitive parameter was the **leaf area index**, which is used for calculation of both the canopy interception and the transpiration rate. A 30% increase in leaf area index produced a 10% decrease in stream flow. The model was also very sensitive to the saturated hydraulic conductivity, with a 50% variation in saturated hydraulic conductivity causing a 14% change in stream flow. Also important were the rainfall interception coefficient and maximum canopy conductance. A 30% change in either of these variables produced a 7% change in stream flow. The rainfall interception coefficient affects the partitioning of rainfall, and the maximum canopy conductance sets the upper limit of plant transpiration (Vertessy et al. 1993). Although not discussed by the authors, an important point in assessing the reliability of the model predictions is the accuracy with which the input parameters can be estimated. Thus if saturated hydraulic conductivity is in error by a factor of 3 (200% error), stream flow will be in error by 56% due to this parameter alone. The inherent spatial variability of many soil properties and the difficulty in measuring them at more than a few discrete locations will contribute to such error. Uncertainties in the different parameters can compound, to give a greater total error in the model output than would be expected from error in any individual parameter. Vertessy et al. (1993) noted that a major limitation on the applica-

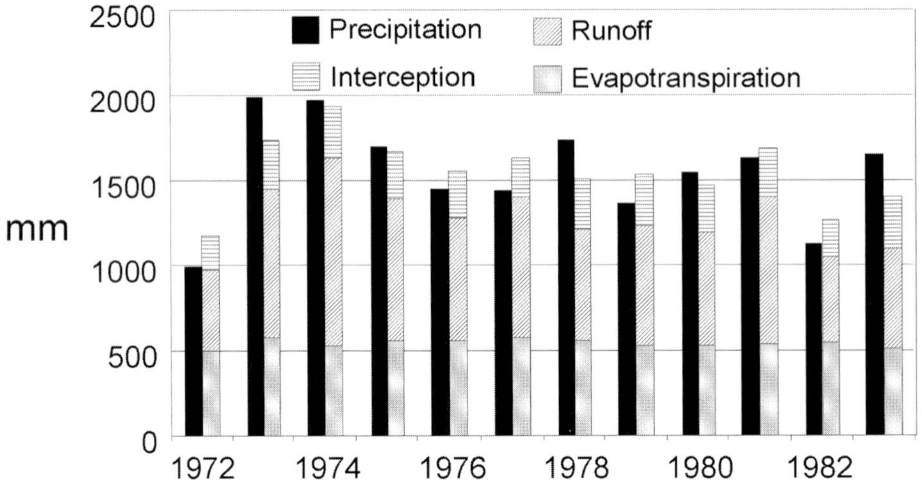

Figure 5.12 Modelled water balance for mountain ash forests. The solid black bar shows the annual volume of precipitation, and the grey bars show the partitioning of precipitation into interception loss, runoff and evapotranspiration (soil evaporation plus transpiration). Where the annual precipitation is greater than the total of interception plus runoff plus evapotranspiration, then an increase in soil moisture storage occurs. Where precipitation is less than the cumulative discharge, then a decrease in soil moisture storage occurs.

tion of models such as this was the lack of soil hydraulic data and the lack of solar radiation measurements. Chapter 7 further discusses the use of TOPOG in generating a water balance for a catchment dominated by Mountain Ash.

Conclusion

Modelling is an important tool for hydrological prediction. A very large number of hydrological models have been developed that describe various components of the catchment water balance in varying levels of detail. This chapter has shown that incorporating soil, climate and vegetation parameters into ecohydrological models allows prediction of fluxes of water through and over soil and through and from vegetation, and movement of water to groundwater as recharge. Estimates of such fluxes are important to land, water, vegetation and river managers as they allow prediction of the effects of land use change and of different management options, on resource management.

Further reading

Baird AJ (1999). Modelling. In *Ecohydrology* (eds AJ Baird & RL Wilby). Routledge, London, pp. 300–345.

Baldocchi D (1997). Measuring and modelling carbon dioxide and water vapour exchange over a temperate broad-leaved forest during the 1995 summer drought. *Plant, Cell and Environment* **20**: 1108–1122.

Coops NC, Waring RH & Landsberg JJ (1998). Assessing forest productivity in Australia and New Zealand using a physiologically based model driven with averaged monthly weather data and satellite driven estimates of canopy photosynthetic capacity. *Forest Ecology and Management* **104**: 113–127.

George RJ, Clarke CJ & Hatton T (2001). Computer-modelled groundwater response to recharge management for dryland salinity control in Western Australia. *Advances in Environmental Monitoring and Modelling* **2**: 3–35.

Lane PNJ, Croke JC & Dignan P (2004). Runoff generation from logged and burnt convergent hillslopes: rainfall simulation and modelling. *Hydrological Processes* **18**: 879–892.

Le Maitre DC & Versfeld DB (1997). Forest evaporation models: relationships between stand growth and evaporation. *Journal of Hydrology* **193**: 240–257.

Walker GR, Zhang L, Ellis TW, Hatton TJ & Petheram C (2002). Estimating impacts of changed land use on recharge: review of modelling and other approaches appropriate for management of dryland salinity. *Hydrogeology Journal* **10**: 68–90.

Chapter 6

Groundwater dependent ecosystems in Australia

Groundwater is a valuable resource in all arid and semi-arid countries. Most of Australia is arid or semi-arid and a large proportion of the agricultural industry, rural towns and some cities (such as Perth) are dependent on groundwater. In addition to human needs for groundwater for drinking and irrigation purposes, many ecosystems of Australia are also dependent on groundwater and this has only recently been documented and investigated in a serious manner.

This chapter provides a detailed description of the nature, extent and issues surrounding **groundwater dependent ecosystems** (GDEs). The chapter starts with definitions and examples of GDEs and discusses difficult issues such as:

- how we determine if an ecosystem is groundwater dependent;
- what measurements can be made to assess groundwater dependency;
- what methods can be employed to monitor the health of these systems if groundwater resources are being used for drinking, irrigation and other uses.

Following discussion of the principles surrounding the issue of GDEs in Australia, four case studies are presented. These provide examples of the issues confronting the management of groundwater dependent ecosystems, highlight the main methods used in the study of these ecosystems and show how the results can be used to aid in formulating a better understanding of the functioning and management options for these ecosystems.

After reading this chapter readers should have an understanding of the following issues:

- defining groundwater dependent ecosystems, and their importance;
- methods to determine groundwater dependency and methods to assess ecosystem structure and function as indicators of ecosystem health;
- the meaning of 'degree of dependency of a GDE';
- measuring tree water use and inferring groundwater dependency in savanna and riparian forests of the Northern Territory;
- differential responses of different vegetation types on the Gnangara Mound to groundwater extraction;
- defining the Great Artesian Basin and associated mound springs.

Groundwater dependent ecosystems: an overview

An ecosystem can be defined as a set of living organisms (plants, animals, microorganisms) that interact among themselves and with their environment (soil, climate, atmosphere) at a

Table 6.1 Rates of groundwater use state by state, 1983–97

	Total use groundwater 1983/84 (GL)	Total use groundwater 1996/97 (GL)	% change in groundwater use 1983/84–1996/97
NSW	318	1008	217
Vic	206	622	202
Qld	1121	1622	45
WA	373	1138	205
SA	542	419	–22
Tas	9	20	122
NT	65	128	97
ACT	n/a	5	–
Total	**2634**	**4962**	**88**

Data from the Australian Water Resources Assessment 2000

given site. Often an ecosystem does not have clearly defined boundaries. Consequently, although we shall routinely talk about groundwater dependent 'ecosystems', in reality it is not possible to examine the dependency of the ecosystem itself because of the physical size, complexity and large numbers of species present within most real ecosystems. What is generally measured or inferred is the groundwater dependency of some component (usually an easily identified, large plant species) of the ecosystem, from which it is inferred that if the current ecosystem structure contains that species and that species is groundwater dependent, the ecosystem can be said to be groundwater dependent. Throughout this chapter, we apply the commonly used terminology of groundwater dependent ecosystems while acknowledging that generally only one or two components (species) of an ecosystem are examined.

Groundwater resources are extensively used worldwide for human consumption, irrigation and industrial use. Within Australia, the utilisation of groundwater has been increasing almost exponentially for the past 150 years. However, since the mid 1990s it has become increasingly apparent that some ecosystems are dependent, to a greater or lesser extent, on the availability of groundwater. Consequently a reduction in the supply, quality, volume, timing or location of available groundwater can have adverse effects on the health, growth or maintenance of these systems within the landscape.

Groundwater resources are utilised throughout Australia to support increased demands for irrigation, mining and human consumption. Table 6.1 shows how groundwater extraction increased between 1983 and 1997. It is clear that many aquifers are being extracted at rates that exceed the rate of recharge and therefore extraction is unsustainable.

The allocation of groundwater is a contentious issue in Australia (see Chapter 9). Arguably, the main concern is how much water should be allocated to the environment, as an increased allocation to the environment is almost invariably linked to a decreased allocation to human consumptive use. Debate on this issue has increased since 1994 when the Council of Australian Governments endorsed the Strategic Framework for Water Reform, which requires the provision of water for the environment (see Chapter 9 for a full discussion).

Effective management of the landscape within the Strategic Framework for Water Reform requires allocations of water (surface and groundwater) to maintain the health and integrity of terrestrial, riparian, wetland and stygian ecosystems. With only a few notable exceptions, however, the groundwater requirements of these ecosystems are extremely poorly understood and even our understanding of the surface water needs of most river and wetland systems is fragmentary and rudimentary.

In addition to increased use of groundwater, it is now realised that there are many ecosystems that are groundwater dependent. There are three simple classes of GDEs (Eamus et al. 2006). These are:

1 aquifer and cave ecosystems, where stygofauna (organisms living in groundwater) reside within the groundwater resource. These ecosystems include karstic, fractured rock and alluvial aquifers. The **hyporheic** zone is in this category because these ecotones often support stygobites (obligate groundwater inhabitants);
2 all ecosystems dependent on the surface expression of groundwater. This class includes the following:
   - base flow rivers and streams;
   - many wetlands, including, for example, the permanent wetlands and damplands of the Swan Coastal Plain and wetlands of the basalt plains of Victoria;
   - estuarine seagrass beds;
   - some floodplains;
   - mound springs of the Great Artesian Basin and Pilbara;
   - riparian vegetation along rivers of central Australia;
   - saline discharge lakes of the western Murray Basin;
   - many low-lying forests (e.g. paperbark swamps) of northern Australia.

   Thus examples of this class of GDE are found all across the Australian continent.
3 all ecosystems dependent on the subsurface presence of groundwater, often accessed via the capillary fringe (non-saturated zone above the saturated zone of the water table) when roots penetrate this zone. This class includes terrestrial ecosystems such as:
   - River Red Gum (*E. camuldulensis*) forests of the Murray River basin;
   - *Banksia* woodland on the Gnangara mound of Western Australia (see below);
   - riparian vegetation of the wet-dry tropics of Australia (see below).

Implicit within classes 2 and 3 are the associated faunal, microbial and fungal populations of the ecosystems. However, within this text we focus primarily on consideration of the woody vegetation components of the ecosystem.

It is clear that GDEs contribute significantly to the biological diversity of Australia and have cultural, social, economic, ecological and conservation value. It is unfortunate that for the past 10 years allocation of groundwater to ecosystem health has generally been confined to maintaining river flows and river health, with little consideration given to terrestrial GDEs (Murray et al. 2003).

Recognition of groundwater dependent ecosystems

It was not until the 1990s in Australia that GDEs were recognised by the hydrological and ecological communities as a class of ecosystems. Even today, these ecosystems are rarely recognised in texts covering hydrology or ecology. However, through extensive lobbying of local and state governments, and finally through their lobbying of the federal government, the provision of groundwater allocations to the maintenance of ecosystem health (including but not limited to GDEs) was incorporated in the mission statements of various government departments charged with managing natural resources, and into various Acts of Parliament and management plans covering most of Australia (see Chapter 9). The provision of water (allocating a known volume or a proportion of available water) for environmental health is generically known as an **environmental flow**, a key outcome of the realisation that ecosystems are themselves valid and valued users of water. Unfortunately the term environmental flow has somewhat narrowly become especially associated with allocating flows of water down rivers to maintain river health (Murray et al. 2003). This is discussed extensively in Chapter 9.

Although there are many types of GDEs and many locations of known GDEs, the question of how to establish dependency of an ecosystem (or at least some components of an ecosystem) is particularly difficult. We now discusss it in relation to above ground ecosystems. We omit discussion of aquifer and cave ecosystems (class 1 above) since this text is only concerned with above ground ecosystems, and principally with woody vegetation. Le Maitre et al. (1999) review much of the information available (to 1999) on interactions between vegetation and groundwater, while Eamus et al. (2006) and Murray et al. (2006) review the more recent literature on groundwater dependent vegetation.

Determining groundwater dependency of ecosystems

Ecosystems reliant on surface expressions of groundwater

Groundwater use may be inferred from positive answers to one or more of the following:

1. Does a stream/river continue to flow all year, despite long periods of low or zero rainfall and hence zero surface flows?
2. For estuarine systems, do salinity levels fall below that of seawater in the absence of surface water inputs (e.g. tributaries or stormwater)?
3. Does the total flux (volume flow per unit time) in a river increase downstream in the absence of inflow from a tributary?
4. Are water levels in a wetland/swamp maintained during extended dry periods?
5. Is groundwater discharged to the surface for significant periods of time each year or at critical times during the lifetime of the dominant vegetation type? If such a resource is present, evolution will have ensured that some species will be using it.
6. Is the vegetation associated with the surface discharge of groundwater different (in terms of species composition, phenological pattern, leaf area index or vegetation structure) from vegetation close by but not associated with (i.e. accessing) this groundwater?
7. Is the annual rate of water use by the vegetation significantly larger than annual rainfall at the site and the site is not a run-on site (either subsurface or surface run-on)?
8. Are plant water relations (especially pre-dawn and midday water potentials and transpiration rates) indicative of less water stress (potentials closer to zero; transpiration rate larger) than vegetation located nearby but not accessing the groundwater discharged at the surface? The best time to measure this is during rainless periods.
9. Is occasional (or habitual) groundwater release at the surface associated with key developmental stages of the vegetation (such as flowering, germination, seedling establishment)?

Affirmative answers to one or more questions leads to the inference that the system is a GDE. However, this inference does not provide any information about the nature of the dependency (obligate or facultative – see below) nor about the groundwater regime (timing of groundwater availability, volume of availability, location of surface expression, pressure of the groundwater aquifer required to support the surface discharge of groundwater) needed to support the ecosystem. These issues are discussed later.

More direct evidence that a vegetation is using groundwater can be obtained by comparing the stable isotope composition of groundwater, soil water, surface water (where relevant) and vegetation xylem water (Thorburn et al. 1993; Zencich et al. 2002). Where there is sufficient variation in isotopic composition among these sources it is possible to identify the single or the most dominant source of water being used by different species at different times of year (Zencich et al. 2002). Thus the use of stable isotopes can provide information about spatial and temporal variation in groundwater dependency within and between species and ecosystems.

For base flow systems (i.e. rivers and streams showing significant flows during periods of zero surface or lateral flows), measurements of the chemical composition of river and groundwater supply can identify and quantify the amount and timing of groundwater inflows into the river (Cook et al. 2003; see the Daly River case study below).

Ecosystems reliant on subsurface presence of groundwater

Groundwater dependency of vegetation reliant on subsurface presence of groundwater can be inferred from positive answers to one or more of the following questions.

1. Is groundwater or the capillary fringe above the water table present within the rooting depth of any of the vegetation?
2. Does a proportion of the vegetation remain green and physiologically active (principally, transpiring and fixing carbon, although stem diameter growth or leaf growth are also good indicators) during extended dry periods of the year?
3. Within a small region (and thus an area having the same annual rainfall and same temporal pattern of rainfall across its entirety) and in an area not having access to run-on or stream or river water, do some ecosystems show large seasonal changes in leaf area index while others do not?
4. See questions 5, 6 and 7 above in the previous section.
5. Are daily changes in groundwater depth observed and are they larger than can be accounted for by the sum of lateral flows, percolation to depth and changes in atmospheric pressure (i.e. is vegetation a significant discharge path for groundwater)?

Stable isotopes (such as deuterium (^2H) and ^{18}O) can be used for identification of these systems too, as can artificial labelling with tracers such as lithium. When tracers are added to the groundwater, subsequent uptake into vegetation is usually conclusive proof that access by that vegetation is occurring. However, the presence of a tracer in a shallow-rooted species can occur if neighbouring deep rooted species exhibit hydraulic lift and the shallow-rooted plants then 'harvest' this water (Caldwell et al. 1998). When a close match between groundwater isotope composition and xylem isotope composition is made, we can conclude that the vegetation is using groundwater. This is discussed in more detail in the first two case studies below.

Groundwater dependent ecosystems require the input of groundwater to maintain their current composition and functioning. Removal of groundwater from these ecosystems or a change in the timing, quantity, quality or distribution of groundwater will influence the composition, structure and function of these GDEs. For example, changing the availability of water for transpiration (and hence carbon fixation by photosynthesis) and/or the recruitment of seedlings into the adult population (through reduced seedling survival) influences species composition. Changes in vegetation composition and structure generally result in changes in associated fauna. Thus, when old trees with large holes are removed from the landscape, birds that rely on them for nesting are forced out; when eucalypts are removed from the landscape, koalas are lost.

The fact that vegetation is using groundwater at some time in the year or at some stage in the life cycle of the vegetation, can mean either the vegetation is always dependent (obligate use) or only sometimes dependent (facultative use) on the availability of groundwater. This issue of obligate or facultative dependency, and the nature of the response of vegetation to declines in groundwater availability, are now explored.

Complexity in groundwater dependency

Plant species within ecosystems may exhibit differing degrees of dependency on groundwater (Hatton & Evans 1998) and can span the spectrum from obligate to facultative dependency. In species with **obligate dependency** the absence of groundwater results in loss of the entire population within an ecosystem and therefore the pre-existing ecosystem structure is altered and that ecosystem could be said to have disappeared too. Stygofauna (fauna of cave and aquifer ecosystems) and artesian mound springs are unique ecosystems in that all species show obligate dependency and therefore the loss of the groundwater resource causes the entire ecosystem to be lost as a unit. Usually, however, not all species of an ecosystem show obligate dependency and therefore some species may remain and some may be lost when groundwater resources are lost. It is a difficult philosophical point to determine whether the loss of a single species is sufficient to justify the view that the ecosystem formerly present has been lost by the loss of that single species.

Facultative dependency is seen in species (and ecosystems) when groundwater is used if it is available but its absence can be accommodated for many years. Paperbark (*Melaleuca* spp.) swamps in the Northern Territory and River Red Gums in South Australia probably show facultative dependency. Showing that a species is obligately dependent can be extremely difficult if access to groundwater is required only intermittently (e.g. for sporadic seedling germination or to recharge soil water stores occasionally).

Some ecosystems can also alternate between high and low dependency at different times of year. Rivers flowing in the dry season in seasonally wet-dry climates show high dependence in the dry season, because all the flow in the dry season is maintained by groundwater input (base flow), but show little dependence in the wet season when most of the flow is from recent rainfall (see below for a discussion of riparian ecosystems as GDEs). A key issue for managers, therefore, is both the degree of dependency and the timing of this dependency, as these influence the management options. For example, extracting groundwater in the wet season might have minimal impact on river health and ecology but extraction in the dry season could have major detrimental impacts.

In addition to information about the timing and degree of dependency of an ecosystem, information is required about the way ecosystems respond to changes in groundwater availability (**response functions**). There has been little research in this area, but two possible forms of response can be proposed. We may expect some systems, such as alpine bogs and glacial lakes, to show a linear response. Other ecosystems, such as salt marshes and other wetlands, may show more of a step function, with minimal change until a threshold of water availability is reached. Figure 6.1 shows two different sensitivities of each type of system; in these hypothetical examples, we have nominated relative ecosystem health and relative groundwater availability as the dependent and independent variables. However, it is important to note that variables spanning a matrix of scales can be adopted to represent ecosystem health, from cellular measures of health to measures of whole organisms, populations and ecosystem characteristics. In addition, groundwater availability per se is not the only attribute of importance for groundwater dependent ecosystems. Additional attributes of importance include:

- the timing of groundwater availability (e.g. wet-dry season, extreme drought years);
- the location of availability (e.g. break of slope, springs);
- the quality of groundwater (e.g. salinity, heavy metal contamination);
- the quantity (e.g. an increase in groundwater depth may take the water out of the root zone).

Responses to changes in any groundwater attributes (including availability) vary with organism age, species and ecosystem type. Consequently the formulation of generically applicable rules about how ecosystems respond to changes in groundwater availability remains elusive (see case studies below for examples of rules).

Because of the problems listed above, establishing the degree of groundwater dependency is difficult and time consuming. Three approaches have been taken in the recent past.

1 Measuring the proportion of annual water use that is derived from groundwater and assuming that this is a measure of the degree of dependency. When there is a consistent

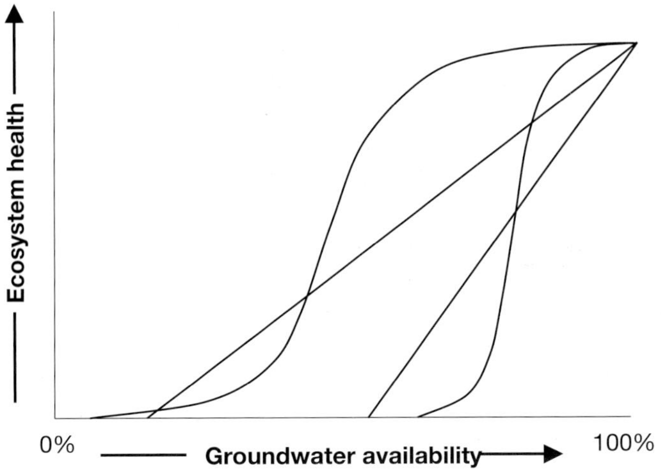

Figure 6.1 A hypothetical relationship between the availability of groundwater to a GDE and the health of that ecosystem. We can envisage linear or curvilinear (approximating a step function) response functions, and the slopes of these response functions can differ among sites

use of groundwater each year this approach may be applied, but it is costly and takes a minimum of 12 months of field work. It does not differentiate between an obligate use of groundwater (where its absence will have a severe negative impact) and a facultative use (where its absence won't have a severe effect). Both uses could give the same numeric value, but the management implications of loss of groundwater availability differ. Furthermore, when groundwater use becomes significant only occasionally (e.g. for the last two years of a 10 year cycle of low rainfall) this approach will not accurately reflect the degree of dependency unless the period of study is longer than the cycle length.

2 Scott et al. (1999) and Shafroth et al. (2002) quantified relationships among patterns of change in groundwater availability (depth, rate of decline in depth and duration of excessive depths of the water table) and vegetation responses. Froend and coworkers have applied this approach to *Banksia* woodland and wetland vegetation of the Gnangara and Jandakot mounds in Western Australia (Froend et al. 2004) to ascertain the response of phreatophytic vegetation to separation from the groundwater source. This was achieved through either long-term monitoring of vegetation vigour and composition relative to groundwater regime, or via shorter-term drawdown experiments where the water table was manipulated to induce seasonal or inter-annual change in plant water source partitioning. The former technique provides a more accurate determination of the degree of dependency at the plant community or population level, but it requires long periods of study. The latter approach is particularly useful for identifying individual dependency on groundwater.

3 Temporal patterns in soil moisture availability, rainfall and vegetation attributes known to be influenced by soil moisture content (e.g. leaf area index and vegetation water use) can be used to estimate the degree of groundwater dependency. From these patterns, rules about the likely temporal dependency (rather than the degree of dependency assessed through quantitative analyses of groundwater use) are deduced. This approach underpins the work of Cook et al. (1998).

Measuring vegetation to assess health of GDEs

A key issue for water and landscape resource managers is the assessment of vegetation response to changes in groundwater availability. Vegetation responses tend to be the focus of many studies because it is vegetation that shows the proximal and immediate utilisation of groundwater. The fauna associated with the flora are only distally and secondarily dependent on groundwater through their dependence on the vegetation. If it is inferred or shown that an ecosystem is dependent on groundwater, the first question asked by managers is whether they can monitor ecosystem health to show that groundwater allocations are adequate or groundwater extractions are not being detrimental. This section provides a brief discussion of what can be measured in the field to assist in this.

Table 6.2 details some of the parameters and processes that can be assessed across a range of spatial and temporal scales when examining vegetation function. The choice of measurements is determined by the resources (labour, equipment, time and money) and staff expertise available. In addition, the parameter measured should have a defined relationship with groundwater levels, have early warning capabilities and consider the 'lag' effects between changed groundwater levels and environmental condition and/or health (Eamus et al. 2005). Short-term simple measures, such as leaf water potential, are generally only of value if comparative values can be obtained from similar sites that are not thought to be affected by declining groundwater availability. More complex and detailed small-scale measures (e.g. chlorophyll fluorescence) require expert knowledge to interpret because so many factors (light levels, drought stress unrelated to changes in groundwater availability, temperature) can influence fluorescence. Chlorophyll fluorescence is the release of light energy from chloroplasts and the amount of fluorescence can be used to assess the photosynthetic potential of leaves. It has been used to quantify the stress response of leaves in the field (Bolhar-Nordenkampf & Oquist 1993).

Long-term studies of several species and populations provide the best information and have been undertaken in research labs but rarely within commercial domains. Ideally, a

Table 6.2 Indicative parameters/processes across various temporal and spatial scales that help assess impact of changes in groundwater availability

	Spatial scale		
	Individual	**Population**	**Ecosystem**
Short-term studies	Chlorophyll fluorescence Stomatal behaviour Tree water use Rates of xylem embolism Soil-to-leaf hydraulic conductance declines	Changes in water relations Abundance and density Age and size distribution	Seedling population composition, density and abundance
Medium-term studies	Growth rate Root depth profiles Identification of sources of water used by plants Canopy fullness	Stand density Rates of flowering and seed-set Canopy water use Xylem vulnerability to embolism Huber value and xylem conductivity	Leaf area index Root depth profiles Water and carbon fluxes between landscape and atmosphere
Long-term studies	Growth rate Canopy fullness Life span	Rates of recruitment and mortality Biomass accumulation Age and size distribution Distribution in the landscape	Species composition Flora/fauna interactions Ecosystem distribution in the landscape Ecosystem water use

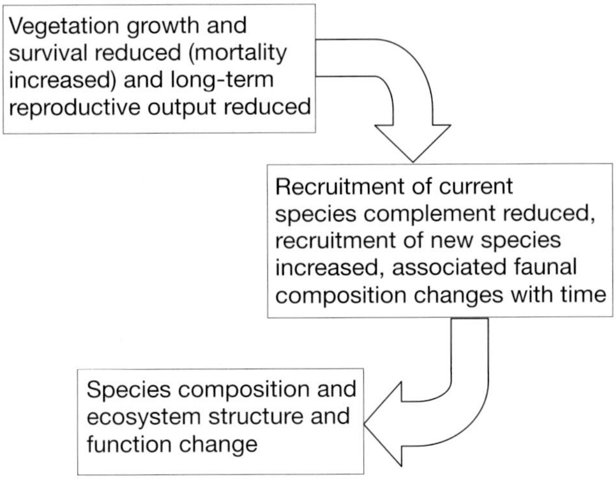

Figure 6.2 Changes in groundwater availability are associated with changes in growth, survival, reproduction, mortality, recruitment and community structure and function

pairwise **BACI design** is used. A BACI design allows for measurements **b**efore and **a**fter an event in **c**ontrol and **i**mpacted sites (Underwood 1991), for example before and after pumping of groundwater. The third case study in this chapter provides an example using a BACI approach.

For large-scale assessments of vegetation responses to changes in groundwater availability, or for small but fast assessments, the **health, vigour** and **mortality** of various components of the vegetation (grasses, shrubs, trees) are generally visually assessed. Crown dieback, canopy cover and, when sufficient time is available, tree growth rate (stem increment) are most frequently determined. However, some comparative data are required from control sites to ensure that it is not a widespread climatic event (drought) or other factor (disease) causing the changes in plant health, vigour and mortality. Recruitment to the population can also usefully be measured as this provides information about possible future changes in population composition.

Resilience is a measure of the ability of a population or community to withstand stress. It remains a useful yet problematic concept for field measurements. Most of the measures in Table 6.2 are correlative measures used at sites with known hydrologic regimes. A comparison of vegetation attributes at sites with different hydrologic regimes can then be used to infer responses to changes in hydrologic regime. However, the resilience of a given vegetation type to given stress is essentially unknown, which means that a time lag between perception by vegetation of a change in hydrologic regime (e.g. flooding or increased depth to groundwater) is likely. This lag could be several years, which poses significant problems for resource managers. If it takes two years for vegetation to respond to increased depth to groundwater but only six months for the vegetation to show substantial mortality, how often should vegetation health be assessed and can the depth to groundwater be manipulated fast enough to halt the decline in vegetation health once it has been measured? These are very important questions for resource managers. Table 6.2 provides some measures over different spatial and temporal scales that can assist.

Broadly speaking, changes in groundwater availability influence five key **ecological processes.** These are growth, reproduction, recruitment, mortality and, if changes in hydrologic regime are severe enough, changes in ecosystem structure and function (Fig. 6.2). Thus we can consider the impacts of a change in groundwater availability at the individual, the population and the ecosystem or community level. Generally, landscape management is focused at a community or ecosystem scale. However, it is generally easier to measure responses of individuals of a species or range of species to changes in hydrologic regime, than to measure community or ecosystem responses. The problem with measuring individuals of a single or multiple species is deciding how to choose which individual(s) of which species to study. To determine which individuals and which species to measure we must answer the following questions:

- Should the individuals chosen for study be seedlings, saplings or mature trees? Which is more important to, or more representative of, the ecology of the site?
- Should trees, vines, herbs or evergreen or deciduous trees be chosen as representative of the community as a whole?
- Is the species or the life form selected going to reliably indicate the response of other species to changes in groundwater availability?

It is frequently difficult to know the correct answers to these questions. Some ideas why are discussed below.

Seedlings have a smaller root system than mature trees, and might therefore be expected to show an earlier response to changes in local hydrology if groundwater that was formerly close to the surface moves deeper into the soil profile. However, if groundwater is often deeper than the roots of seedlings, trees may be expected to show a response sooner than seedlings, which tend to rely on soil moisture stores and exhibit increased deciduous habit (drop their leaves more than mature specimens of the same species) because trees use more water and access water from deeper in the profile. Within an ecosystem, it is rarely apparent which species, for example, might be genetically better at avoiding drought or tolerating drought, or which species is deeper rooted. Consequently, determining which species to examine on a limited budget with limited time is problematic.

This chapter has so far examined the issues of defining GDEs, identifying GDEs and defining the dependency of GDEs. Some of the problems and underlying concepts have been explored. The remainder of the chapter provides a series of case studies in which issues of ecohydrological importance in relation to GDEs are addressed.

Case studies in groundwater dependent ecosystems

The first case study concerns a study of **tropical savannas** undertaken in the Northern Territory (NT) of Australia. This study was undertaken because of the need to increase groundwater extraction to support the increased population and horticultural industry in the NT. The issue addressed was whether savannas will suffer from groundwater extraction.

The second study concerns a very different and more complex ecosystem – **riparian vegetation** along the Daly River in the NT. The question addressed was whether groundwater and river water extraction have a detrimental impact on the riparian vegetation of the Daly River.

The third case study investigates the relationship between vegetation health and groundwater depth in the **Gnangara Mound** around Perth in Western Australia, where groundwater pumping has occurred for decades. There has been a 30 year period of **below average rainfall** and many sites are showing declines in vegetation health. This case study discusses management implications of providing water for the maintenance of ecosystem health.

Finally, we discuss the special case of the **Great Artesian Basin**, a vast aquifer underlying about 20% of the Australian continent. A set of GDEs is associated with **artesian springs** that naturally occur from this aquifer; the nature and values of these unique systems are explored.

Case study 1: Howard River East – a proposed new groundwater bore field

Background and issues

The population of Darwin and close environs in the Northern Territory of Australia has increased since the mid 20th century from tens of thousands to approximately 100 000. The horticulture and grazing industries have also increased substantially. As a consequence of the increase in human population numbers, beef production and horticulture (which requires irrigation in the dry season) demand for potable water is rapidly increasing. Because of the pristine nature of groundwater in the NT and the favourable hydrological environment (large amounts of rainfall each year), groundwater abstraction will be increased in the NT.

In the early 1990s a large new groundwater bore field was proposed for the Howard River East catchment just south-east of Darwin. The site is overlain predominantly by open eucalypt woodland and forest (better known as savanna; Fig. 6.3), but patches of paperbark (*Melaleuca*) swamp and rainforest occur in low-lying areas and riparian forest lines the rivers (see case study 2). These three ecosystems have significant ecological, conservation and tourism value and this, together with the six month dry season, led to concerns that removing large volumes of groundwater would have a significant negative impact on the ecosystems. Furthermore, in 1996, state and territory governments adopted the National Principles for Provision of Water for Ecosystems which were formulated to ensure the maintenance of ecosystem health and, where relevant, to maintain and restore ecological processes and biodiversity of groundwater dependent ecosystems (see Chapter 9). As a result of these policy developments, the degree to which vegetation in the catchment might be impacted had to be assessed before bore fields could be developed in the NT. Consequently a multi-disciplinary team investigated the impact of groundwater pumping on the catchment.

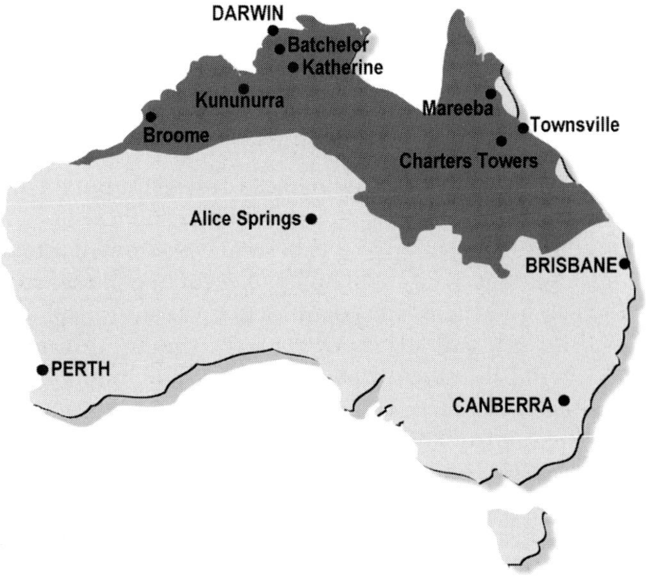

Figure 6.3 Distribution of savannas of northern Australia. The Howard River East site is located 30 km south-east of Darwin

Source: © CRC for Tropical Savannas

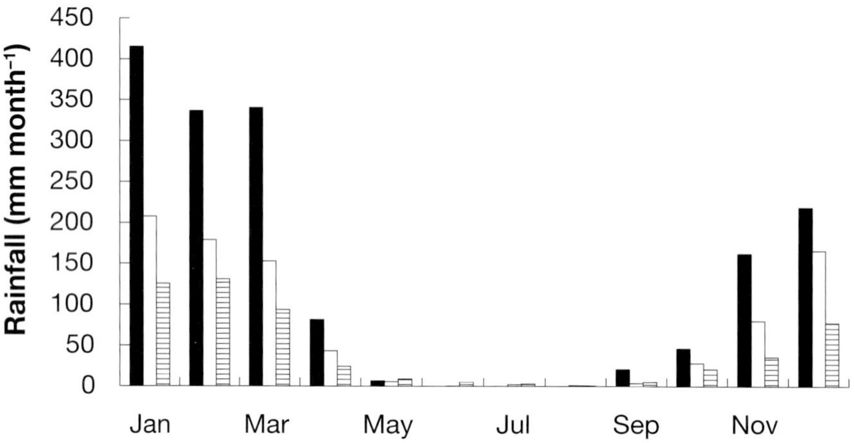

Figure 6.4 Seasonal patterns of rainfall occur across the Top End of the NT. Data shown are for Darwin (solid bar), Katherine (open bar; 300 km south of Darwin) and Newcastle Waters (striped bar; 720 km south of Darwin). This illustrates the gradient in rainfall between coastal (Darwin) and inland (Newcastle Waters) sites in monsoonal Australia

Source: Redrawn from Eamus et al. (2000)

Aims of the study

The study had the following aims:

1 to develop an understanding of how climate influences vegetation water use and the hydrological balance, and vice versa;
2 to provide an estimation of the rate of water use by the entire catchment, from small-scale sampling within the catchment;
3 to identify the sources of water used by vegetation (soil, river or groundwater);
4 partitioning of water flow at different times of year to lateral flow, vegetation water use and groundwater recharge and thereby generating a water balance for the catchment.

Understanding how climate influences vegetation water use

Temperatures at the site are uniformly high, with mean annual minimum and maximum temperatures being 23°C and 32°C respectively. Mean daily maximum temperatures show a small range between seasons (30.4°C in winter, 33.1°C in summer). Rainfall is highly seasonal, with 95% of annual rainfall (typically 1700–1800 mm y^{-1}) occurring in the wet season (October–March inclusive) (Fig. 6.4) and there is no significant (in terms of soil recharge) rainfall in May–August inclusive. Because rainfall is seasonal, the timing of maximum groundwater extraction (dry season) will be different from the time of maximum recharge (wet season). Mean annual pan and potential evaporation at the site are high (2700 mm and 2260 mm respectively). Evaporation is only slightly seasonal (6–7 mm d^{-1} in the dry season; 5–6 mm d^{-1} in the wet season). The lower vapour pressure deficit of wet season air and increased cloud cover in the wet season cause a reduction in pan evaporation, despite warmer temperatures. Vapour pressure deficit influences evaporation from soil and vegetation. In the wet season, vapour pressure deficit is typically 1.5 kPa but in the dry season this doubles to 3.1 kPa. **Leaf-to-air vapour pressure difference** is a better measure of evaporative demand as this is the actual gradient in water vapour pressure between leaf and air. This can increase to 5 kPa or more in the dry season.

The number of hours of sunshine is a major factor determining evapotranspiration from soil and vegetation (see Chapter 3) and therefore has a major impact on the hydrological cycle of a site. The Howard River site has considerably more sunshine hours than all other major cities in Australia and consequently a large potential rate of evaporation.

Soil moisture content of the top 1 m of soil varies seasonally at the study site, with peak soil water content occurring in the late wet season (January–March) and minimum water content occurring in the late dry season (August–October). Throughout the region, soil moisture declines during the dry season because of evaporation from the surface, percolation to deeper layers and transpirational use by trees. The top 50 cm of soil may have only 2–7% volumetric soil moisture at the end of the dry season, but at 100 cm depth soil moisture ranges between 10% and 15% in the dry season (Duff et al. 1997).

To summarise, during the dry season evaporative demand greatly exceeds rainfall and the hydrological balance is characterised by discharge (as transpiration from vegetation and groundwater discharge to the Howard River) rather than recharge. Seasonal changes in soil water content reflect the seasonal patterns of rainfall, with peak soil water content occurring in the wet season and declining throughout the dry season. The decline in dry season soil water content is the result of transpiration by vegetation (see below), lateral flow to the Howard River, surface evaporation early in the dry season and deep percolation to the water table.

Plant water relations and physiological activity are maintained in the dry season

We saw above that if vegetation remained physiologically active in prolonged dry periods, or leaf water potentials were higher than expected from the lack of rainfall in the dry season, then groundwater dependency might be occurring. In the NT study, daily water potentials of the dominant evergreen tree species (Fig. 6.5) were often below the classically defined **wilting point** of crop plants of –1.5 MPa (Myers et al. 1997). However, because these species are sclerophyllous, a water potential of –2.0 MPa does not represent a seriously stressed plant and the trees remained very active in the dry season, as shown by the maintenance of open stomata for much of the day and high rates of photosynthesis and transpiration (Prior et al. 1997a, b). Therefore, it was concluded that trees were dependent on deep stores of water, possibly

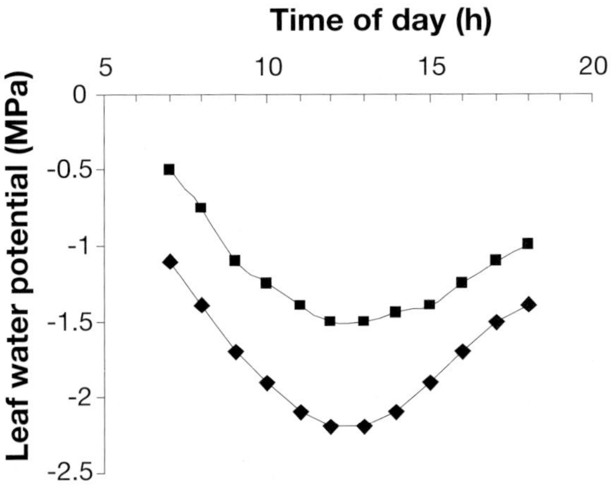

Figure 6.5 Stylised diurnal changes in leaf water potential during the wet season (squares) and the dry season (diamonds) of an evergreen tree in the NT. Water potentials are lower in the dry season because the upper soil profile (top 1 m) is very dry, there is no rainfall for several months and solar radiation input, temperature and vapour pressure deficits are high

including groundwater. A study of soil physics was undertaken to establish whether the water was soil water or groundwater. Previous work had shown that stable isotope analyses of water in soil, branches and groundwater could not resolve this issue because of the lack of differentiation in isotope composition between soil water, groundwater and xylem water.

Physics of soil water storage

How much water was available in the soil to 5–6 m depth was a key question, as soil cores revealed tree roots to those depths. If there was sufficient water within this depth of soil to support the trees throughout the dry season, it was unlikely that groundwater was being used. Soil cores to a depth of 10 m were taken and immediately stored in sealed jars to reduce evaporation. These samples were used to determine soil water content as a function of depth.

Water content of the upper soil profile (to 6 m depth) was uniformly high (10–15% or 0.1–0.15 g g^{-1}) in the wet season, and declined in the dry season due to transpiration by vegetation, lateral flow to the Howard River (see below) and percolation to deeper levels. Groundwater hydrographs using nested piezometers (Chapter 4) showed water table depth to fluctuate between 0.5 m and 14 m below ground level, presumably because of percolation to depth, lateral flow and possible uptake by vegetation.

The large differences in gravimetric water contents at saturation and field capacity did not agree with observed changes in water table depth. The amount of water required to raise the water table from 9 m to 1 m below ground level, assuming the soil in the water table is saturated and soil in the upper soil profile is at field capacity, is equal to $\Delta h\theta(\rho_b/\rho_w) = 8000 \times 0.2 \times 1.4/1 = 2240$ mm, where Δh is the change in height of the water table (in mm), θ is the difference in water content between saturated and field capacity, ρ_b is the soil bulk density and ρ_w is the density of water (Cook et al. 1998). This is far larger than average annual total rainfall (and hence cannot be correct). Therefore the soil water content data are incompatible with observed changes in groundwater depth unless the soils are a **swelling clay soil**, as discussed below.

When swelling clays are removed from the ground a significant overburden pressure is removed and soil moisture release curves measured in the lab are not representative of the behaviour of the soil in situ. Thus, the hydraulic potential (Φ) of a swelling clay soil is given by:

$$\Phi = \Psi_u + \Omega$$

where Φ is matric potential of the loaded soil, Ψ_u is the matric potential of the unloaded soil and Ω is the overburden potential at depth z.

For a uniform soil:

$$\Omega = \beta g \rho_{wb} z$$

where β is the compressibility factor, g is acceleration due to gravity, ρ_{wb} is the wet bulk density and z is depth. Consequently, for a swelling clay soil at depth 5 m, with a compressibility of 0.5, a wet bulk density of 1.4 g cm^{-3} yields an overburden potential of 34 kPa. Therefore this soil will be saturated ($\Phi = 0.0$) at an unburdened matric potential of -34 kPa. The water content of the saturated soil in situ (with overburden) will be the same as the water content of the unburdened soil at -34 kPa. As depth below ground surface increases, the in situ water content of soil (at saturation, field capacity or any other matric potential value) will decrease because of the effect of the overburden pressure.

From knowledge of the soil moisture release curve and assuming a constant dry bulk soil density of 1.4 g cm^{-3}, a total of 1655 mm of water was calculated to be stored in the soil to a depth of 9.1 m (an average of 182 mm per metre depth).

If it is assumed that at the end of the dry season soil has an unburdened matric potential of –1500 kPa (the so-called wilting point for plants), the water holding capacity of soil (the difference between the water content of soil at saturation and the water content at wilting point) at the soil surface is 0.2 to 0.3; at depth, this declines to 0.05 to 0.08. If average observed water table fluctuations are 7 m, a soil water deficit of 400–500 mm is apparent (i.e. 400–500 mm is needed to raise the water content from wilting point to saturation). Most of this water (250–400 mm) is required to wet the soil up to equilibrium (using a water table depth of 9 m) and only a small additional amount (100–200 mm) is required to cause the water table to rise the 7–10 m observed annually (Fig. 6.6). Can we establish that there is sufficient recharge of groundwater, each year, to account for the observed rise in water depth?

How old is the groundwater and is recharge rapid?

Studies of **groundwater chemistry** allow estimations of the age of groundwater and the rate of groundwater **recharge.** Groundwater age was estimated by measuring **chlorofluorocarbon** (CFC) contents of groundwater. CFC-12 is relatively stable, soluble in water and entirely man-made. The concentration of CFC-12 in the atmosphere and in groundwater recharge has been increasing steadily since the 1950s because of increased industrial use. Groundwater age is determined by comparing the concentration of CFC-12 in groundwater with the known concentration in the atmosphere for the past 50 years. In unconfined aquifers estimates of groundwater age can be used to determine the rate of recharge of groundwater if aquifer porosity is known.

CFC-12 concentration declined with depth (Fig. 6.7). Such a trend is expected if groundwater age increases with depth and the groundwater equilibrated with the atmospheric CFC-12 concentration at the time the rain fell and started moving down through the soil. Apparent CFC-12 age was calculated by converting the measured concentration to an equivalent atmospheric concentration based upon the known solubility of CFC-12 in water. Using this fact, it is possible to estimate the maximum CFC-12 age as a function of depth (Fig. 6.8).

From these data, the age gradient of water was calculated as approximately 0.6–1.3 y m^{-1} and this corresponds to a rate of vertical movement of groundwater of about 0.8 m y^{-1}. Assuming a constant porosity value with depth of 0.25 and assuming one dimensional flow, this gives a recharge rate of 200 mm y^{-1} and a groundwater age of about 30–40 years. Therefore

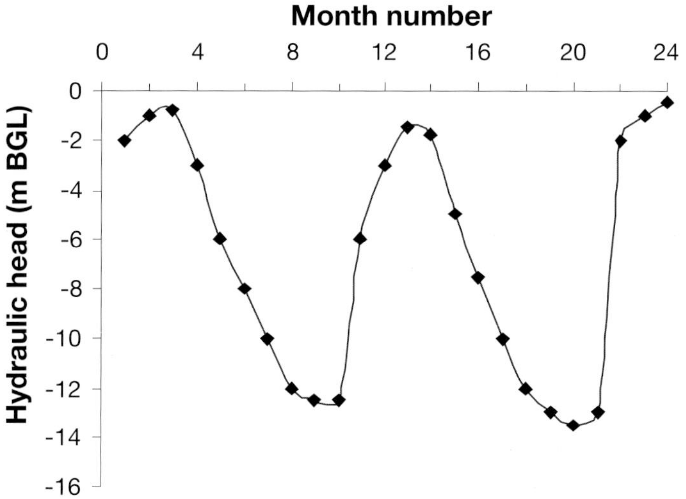

Figure 6.6 Changes in hydraulic head over a 24 month period below a eucalypt savanna in the NT. The lowest hydraulic head occurs in the dry season (July–September; months 7–9, 19–21) and the highest occurs in the wet season (December–March; months 1–4, 13–16)

Source: Redrawn from Cook et al. (1998)

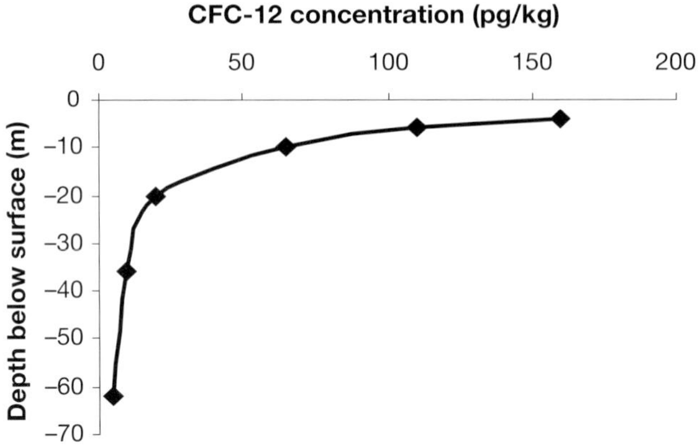

Figure 6.7 Schematic of the change in CFC-12 concentration with depth for water taken from piezometers below a eucalypt savanna in the NT

Source: Redrawn from Cook et al. (1998)

it was concluded that the rate of recharge calculated (200 mm y^{-1}) is sufficient to cause the average observed water table fluctuations of 7 m (Cook et al. 1998).

Vegetation water use

Three steps are required in the estimation of **vegetation water use** at a catchment scale. First, determine the species composition and size distribution of the trees and understorey. Second, establish the relationships among species, tree size and water use. Finally, scale up from measurements at the tree scale to canopy and catchment scale.

Vegetation structure within the savannas is moderately simple, with an open eucalypt woodland/forest (foliage projected cover 30–70%; tree height 14–16 m; LAI 1–1.2) dominated by two tree evergreen species (*Eucalyptus tetrodonta* and *E. miniata*). The understorey is dominated by a continuous cover of C4 annual grasses in the wet season (to 2 m) which senesce in the dry season. Semi-deciduous and deciduous small trees are present all year in the understorey. A vegetation survey of representative plots was undertaken to establish species composition and size class distribution to allow scaling up from estimates of individual tree water use to whole catchment water use.

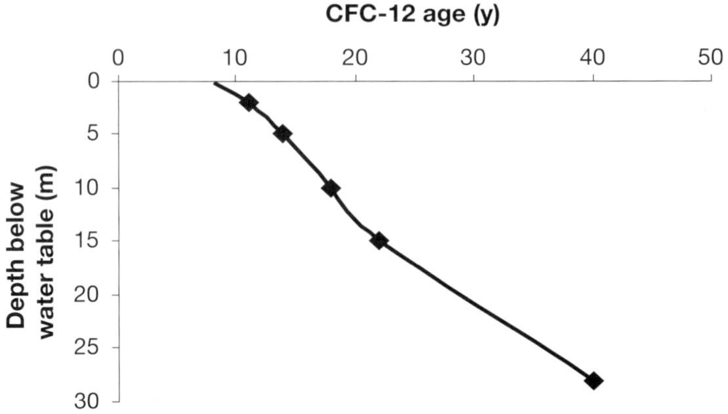

Figure 6.8 Schematic of the maximum measured CFC-12 age as a function of depth below a eucalypt savanna of the NT

Source: Redrawn from Cook et al. (1998)

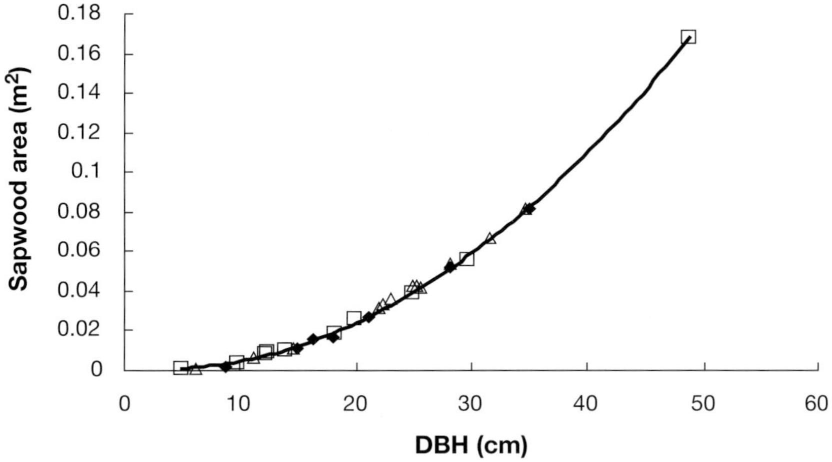

Figure 6.9 Five different tree species growing in savannas of northern Australia shared a single relationship between DBH and sapwood area, allowing a simple scaling from DBH to sapwood area and hence stand water use

Source: Redrawn from O'Grady et al. (2000)

Establishing tree water use for the dominant species was undertaken using sapflow sensors. Daily rates of tree water use for all dominant species, using a complete range of tree sizes (small, medium and large) and for wet and dry seasons, were measured. In addition the relationship between diameter at breast height (DBH), sapwood area and leaf area were established within replicate plots using methods outlined in Chapter 3. Remarkably, and fortuitously, the same regression described the relationship between DBH and sapwood area for the five dominant species at the site (Fig. 6.9). Consequently, scaling water use from DBH required only a single regression, making the task much simpler.

Once knowledge of tree size distribution and the relationship between tree size and water use was established, an estimate of canopy water use was calculated (O'Grady et al. 2000). Understorey water use was measured using open-top chambers at four key times during the year to determine transpiration rates of grass, shrubs and bare soil. Finally, using eddy covariance, a point-scale estimate of whole ecosystem water use was determined for comparison with the scaled estimates from the sapflow and open-top chamber studies (Hutley et al. 2000). These data are discussed below.

Catchment water balance

Two key questions drove the Howard River study. First, just how much water was being used by the savanna vegetation? This is the amount of water that must be preserved from rainfall each year, to maintain vegetation health. Second, if groundwater was extracted, would the savannas be affected? To address these questions, a catchment water balance was developed, with the major fluxes of water quantified through the work described above.

A **catchment water balance** requires information about seasonal and annual fluxes of rainfall, stream flow, evapotranspiration and groundwater recharge. Table 6.3 and Figure 6.10 summarise the wet season, dry season and annual fluxes of water within the catchment. Rainfall is clearly concentrated in the wet season, with dry season fluxes having little impact on soil moisture stores. Evapotranspiration is similarly largest in the wet season, for two main reasons. First, the leaf area index of the ecosystem is largest in the wet season when grasses are present and the deciduous trees are in leaf. In the dry season the grasses die off and the deciduous and semi-deciduous trees lose all or many of their leaves. Consequently vegetation water

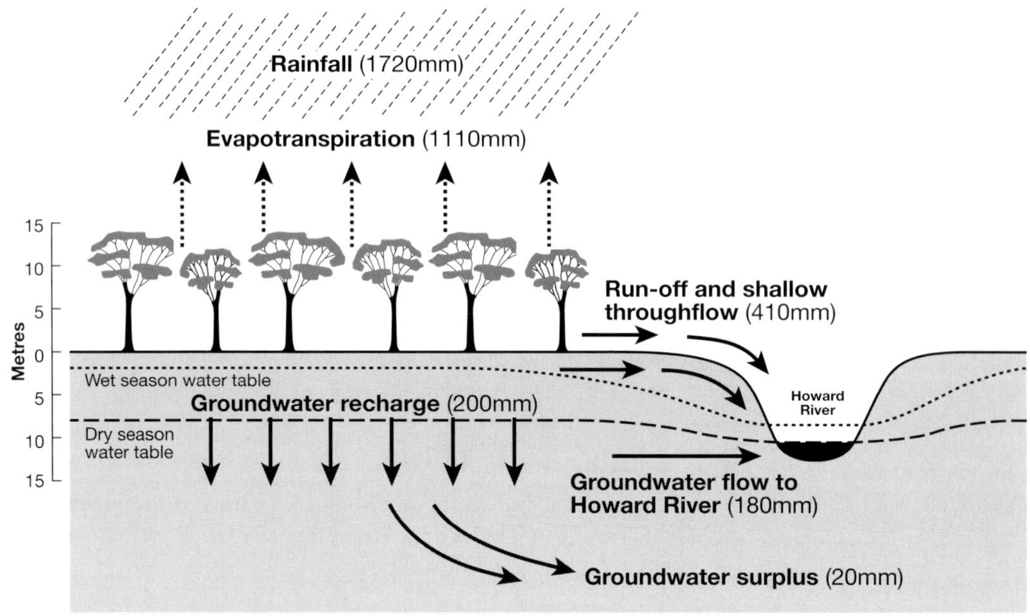

Figure 6.10 Schematic outline of the catchment water balance for a north Australian savanna

Source: Cook et al. (1998)

use is largest in the wet season. Second, in the dry season, the soil surface is extremely dry and therefore evaporation from the soil is much reduced. Significant run-off occurs during the wet season when large storms result in surface run-off either because the surface soil is saturated and therefore further inputs of rain result in run-off or because the rate of rainfall exceeds the rate of infiltration. River flow through the Howard River is therefore much higher in the wet season than the dry season. Dry season river flow is driven by groundwater flow (base flow) and represents groundwater discharge. Consequently the Howard River, like many rivers in seasonally dry climates, is groundwater dependent for at least part of the year.

Table 6.3 Seasonal and annual water balance for a savanna site in northern Australia

	Wet season flux (mm)	Dry season flux (mm)	Annual flux (mm)	Method
Rainfall (A)	1585	135	1720	Rainfall gauging
Evapotranspiration (B)	810	300	1110	Heat pulse and eddy covariance
Run-off (Howard River flow) (C)	570	20	590	Stream gauging
Change in soil/GW storage (D)	205	–185	20	A–B–C
GW recharge (E)	–	–	200	Groundwater dating
Base flow from GW (F)	160	+20	180	E–D
Total change in soil and GW storage (G)	365	–165	200	D+F

GW = groundwater

Source: Data from Cook et al. (1998)

Groundwater recharge was determined from a knowledge of groundwater age and estimated at 200 mm y^{-1}. The difference between the sum of the total (annual) amount of recharge, the total evapotranspiration and the total river flow (200 + 1110 + 590 = 1900 mm), and the mean annual rainfall (1720 mm) is the amount of recharge that eventually enters the Howard River as base flow (180 mm y^{-1}; row F in Table 6.3).

Conclusion

At the start of this study, the following questions were asked.

1 What are the daily and seasonal patterns of water use by the vegetation?
2 Is the vegetation dependent on groundwater at any time?
3 What is the rate of recharge of the groundwater resource?
4 What is the likely impact on these ecosystems of removing groundwater?
5 What is the sustainable yield of the aquifer?

The following answers were developed.

Water use by trees did not vary much between the wet and dry seasons, despite large variations in rainfall and soil water content. Water use by trees was typically 1.2–1.8 mm d^{-1}. In contrast, understorey water use was highly seasonal, with about 2 mm d^{-1} transpired from the understorey in the wet season when grasses were growing and physiologically active, falling to about 0.5–0.8 mm d^{-1} in the dry season when the grasses senesced and only the understorey shrubs were physiologically active. Water flux on a daily basis followed the daily time course of solar radiation input.

Savanna vegetation was not groundwater dependent, because the volume of water stored in the root zone was sufficient to support the transpiration rate of the vegetation throughout the dry season. However, the possibility that trees were accessing the capillary fringe above the water table could not be discounted and therefore if the depth of the water table increased because of groundwater extraction, this source of water could become inaccessible.

Patches of other types of vegetation (paperbark swamps, rainforests) in low-lying positions within the landscape could be more dependent on groundwater than the savannas and are dependent on the lateral flows of water moving from the savannas.

Groundwater recharge was estimated to be about 200 mm per year. However, a large fraction of this (180 mm per year) moves out into the Howard River as base flow and therefore if river health (i.e. river flow) is to be maintained at this level, groundwater extraction could not exceed 20 mm per year (equivalent to 2.52 GL).

River flow in the dry season and vegetation in low points of the catchment are likely to be the most sensitive to overextraction of groundwater. It is likely that riparian vegetation, paperbark swamps and rainforest patches would show declines in growth, vigour and recruitment if groundwater extraction was too high for too long.

Case study 2: riparian vegetation of the Daly River

Background and issues

The **Daly River** (Fig. 6.11) is one of the largest perennial rivers of the Northern Territory, with a catchment of almost 53 000 km^2 (Fig. 6.12). Dry season flows are dominated by the input of groundwater (base flows) from three limestone aquifers underlying the river. The catchment contains a number of very important rivers, including the Katherine, Flora, Edith and Douglas Rivers, which have tourism and conservation value.

Figure 6.11 The Daly River at its best

Source: Photo courtesy of Dr A O'Grady

With an expanding agricultural industry plus rising demand because of a larger population, demand for water is increasing. The management of water resources in this highly seasonal climate is a significant issue. Furthermore, in the Daly River district, 200 000 ha of land have been cleared for development. Large-scale clearing of land for improved pasture and irrigated horticulture will influence the regional water balance through changes in recharge, discharge, evapotranspiration and river flows. To ensure sustainable management of the whole catchment (including management of water, the ecology and the regional socio-economy – the triple bottom line), a project was initiated to establish the environmental water requirements of the vegetation of the Daly River. Riparian vegetation (vegetation in and alongside streams and rivers) was of particular interest as it has an important role in maintaining key **ecosystem services** (Costanza et al. 1997) such as:

- maintaining stability of river banks (erosion control);
- minimising soil and water surface flows to the river (turbidity control);
- maintaining biodiversity through enhanced species richness;
- providing corridors for the movement of fauna across the landscape;
- providing habitat for associated fauna with high conservation value.

Aims of the study

The volume of water extracted from aquifers and rivers of the region will increase with time. Because of the importance of riparian vegetation and its potential for dependence on river and groundwater, the aims of this study were to:

1 describe the climate and topography of the Daly River and its riparian vegetation;
2 characterise daily, seasonal and spatial patterns of soil water availability, leaf water potential and water use of dominant tree species of the Daly River;
3 determine the relationships among tree water use, species and tree size;

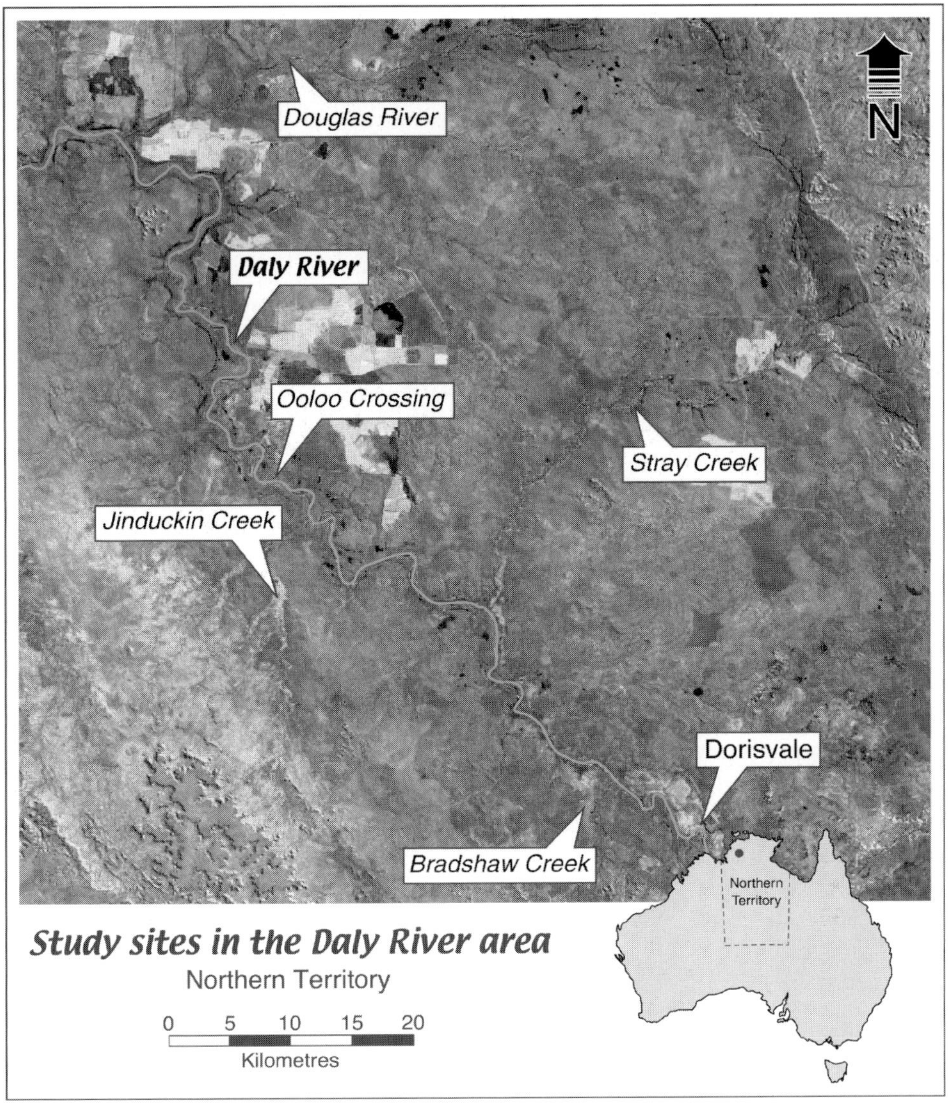

Figure 6.12 Location of the Daly River study, NT

4 determine the sources of water used by riparian vegetation;
5 assess the groundwater dependence and environmental water requirements of Daly River riparian vegetation.

Climate, topography and riparian vegetation of the Daly River

Climate determines vegetation structure, composition and water use. Therefore a knowledge of climate is the first place to examine relationships among vegetation types and water use. The **climate** of the Daly River region is similar to that of the savanna sites of the Howard River case study. Rainfall is highly seasonal, but is less than that of the savanna sites (approximately 1200 mm compared to about 1800 mm). Average daily maximum temperatures are above 40°C but annual daily minimum is less than that of the savanna sites (20°C compared to 23°C). Average daily pan evaporation rates are about 6 mm.

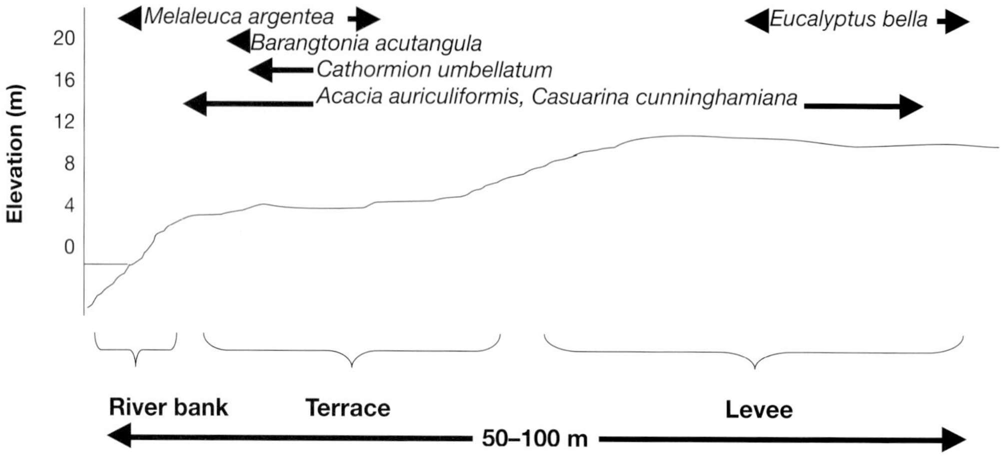

Figure 6.13 Schematic of the topography and approximate species distribution of the Daly River riparian vegetation

Riparian vegetation structure of the Daly River is complex (Figs 6.13, 6.16). *Melaleuca argentea* and *M. leucadendra* (paperbark trees) tend to be restricted to the steep river banks and lower terraces and are generally flooded each year. Behind this paperbark zone lies a closed monsoon forest, with species such as *Barangtonia acutangula, Nauclea orientalis, Catharmion umbellatum* and *Strichnos lucida* along the terraces. Eucalypt species, especially *Eucalyptus bella,* occur only on the levees away from all but the most exceptional floods. Other species on the levee include *Erythphloem chlorostachys* and *Planchonia careya.* A few species have a wide distribution across this elevation gradient, including *Acacia auriculiformis* and *Casuarina cunninghamiana.* As distance from the river increases past approximately 100 m, riparian vegetation is replaced by eucalypt woodland and open forest. Some of the species found within the riparian zone are typical of monsoon closed forest (including *Barringtonia acutangula*).

The **topography** of the Daly River and its associated riparian vegetation is also moderately complex. The river can rise and fall by 10 m between wet season maxima and dry season minima. The river bank is steep and seasonally flooded as the river rises. Behind the bank is a relatively flat corridor 20–40 m wide (the terrace) with a slight incline. Behind the terrace lies a levee bank which has a steeper pitch and can be 20–80 m wide. This levee grades into the surrounding savanna found throughout. The vertical rise from river to top of the levee is 10–15 m and access to groundwater from the top of the levee therefore requires roots to penetrate approximately that depth of soil (see below). Trees therefore have three potential sources of water for transpiration, namely recent rainfall (in the upper soil profile), groundwater and streamwater. To establish the **possible impacts of groundwater extraction** it is important to establish the degree and timing of any groundwater dependence of the vegetation.

The three major limestone aquifers in the region are **highly connected** to surface waters, with run-off recharging them in the wet season. Discharge of groundwater to the Daly, plus evapotranspiration, causes a dry season decline in groundwater levels. The duration and magnitude of the wet season also cause inter-annual fluctuations in groundwater level.

Soil water availability, leaf water potential and water use by riparian vegetation

A knowledge of soil water content and how it varies spatially and temporally provides valuable insight to the water relations and behaviour of vegetation at a site. Given the spatial heterogeneity of the river bank and temporal variation in rainfall, it is important to understand how

Table 6.4 Mean water use (± SE) expressed on a sapwood area basis during each of the sampling periods

Season	Water use ($m^3\ m^{-2}\ d^{-1}$)	
	E. bella	*M. argentea*
August 2001	2.2 ±0.29	1.9 ±0.41
September 2001	3.0 ±0.43	2.8 ±0.36
May 2002	3.2 ±0.38	2.3 ±0.38

Source: Data from O'Grady et al. (2002)

soil moisture content varies over time and space. Similarly, leaf water potential provides important information about the water status of plants and provides insight to the relative availability of water to vegetation. Consequently both of these parameters were measured in this study, in addition to measures of tree water use.

Soils in the top 1 m on the levee and on the river bank had low matric potentials at the end of the wet season and the end of the dry season. This is because the soils were either very sandy and shallow, close to the river (and hence stored very little water), or were deeper and with slightly more clay on the levee but with large depth to groundwater and rapid and significant drainage to depth past 1 m. Most importantly, the matric potential of these soils was in the majority of cases considerably lower than the pre-dawn leaf water potential at any location. The matric potential of the upper 1 m of soil was typically –1 to –5 MPa while pre-dawn water potentials were usually between –0.1 and –0.5 MPa. Therefore, trees were accessing water from depths greater than 1 m.

Midday leaf water potentials ranged from –1.5 to –2.5 MPa with little variation between wet and dry seasons. Because neither pre-dawn leaf potential nor midday leaf water potential showed strong seasonality, it was concluded that **water availability does not become limiting in the dry season** compared to the wet season. In addition, the range of midday water potentials are not indicative of water stress, further indicating water availability is not limiting water use in the wet or dry seasons (see below).

Tree water use increased with increasing tree size for all species. The relationship between DBH and water use ($m^3\ d^{-1}$) for paperbarks (*M. argentea)* located on the river bank and *E. bella* (located on the levee) were significantly different (Fig. 6.14), with Melaleuca trees using less water at a given size than the eucalypts.

Despite highly seasonal rainfall, vapour pressure deficit and daily mean temperatures, tree water use was aseasonal (Table 6.4). This is a first indication that these riparian trees, from the river bank to the levee, are not entirely reliant on soil moisture and that they are therefore groundwater dependent. Such aseasonal water use was observed in the savanna study previously discussed.

Using the distribution of tree size for the riparian vegetation plus the relationship between tree size and water use, it was possible to estimate canopy water use. At the Oolloo site water use was relatively aseasonal but at the Douglas/Daly site water use increased in the wet season compared to the dry season (Table 6.5). Combining all sites and all seasons allowed the relationship between tree basal area and water use to be established (Fig. 6.15), from which riparian water use was calculated.

Sources of water used by riparian vegetation

The stable isotope deuterium was used to determine the sources of water used by the riparian vegetation. The **deuterium** concentration of soil water, groundwater, river water and xylem sap were compared in *E. bella* (on the levee) and *M. argentea* (on the river bank) at different

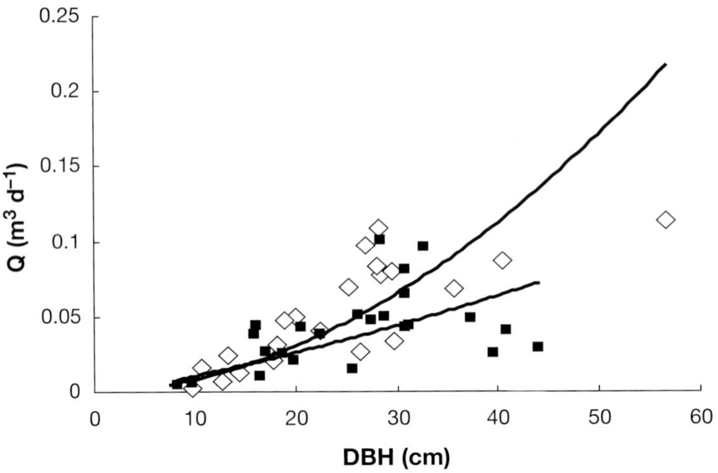

Figure 6.14 Daily tree water use (Q) in *E. bella* (◇) and *M. argentea* (■) as a function of tree size, along the Daly River. Data from sites and seasons have been pooled as there were no significant differences in water use attributable to site or season

Source: Data from O'Grady et al. (2002)

Table 6.5 Stand water use (mm d^{-1}) by riparian vegetation at three sites along the Daly River. T1, T2, transects one and two respectively

Plot	Basal area (m^2 ha^{-1})	Stand water use – August	Stand water use – October	Stand water use – December
T1 (Oolloo)	98.17	4.22	4.89	4.28
T2 (Oolloo)	25.76	1.98	1.88	1.46
Douglas/Daly	52.88	2.32	2.54	4.32

Source: Data from O'Grady et al. (2002)

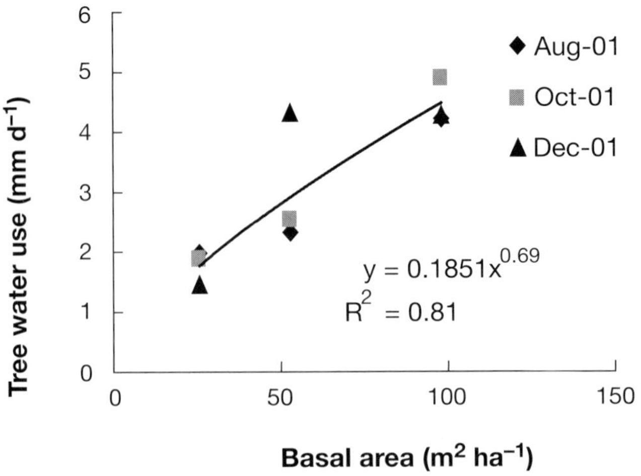

Figure 6.15 The relationship between stand water use (mm d^{-1}) and stand basal area (m^2 ha^{-1}) at three sites along the Daly River

Source: Data from O'Grady et al. (2002)

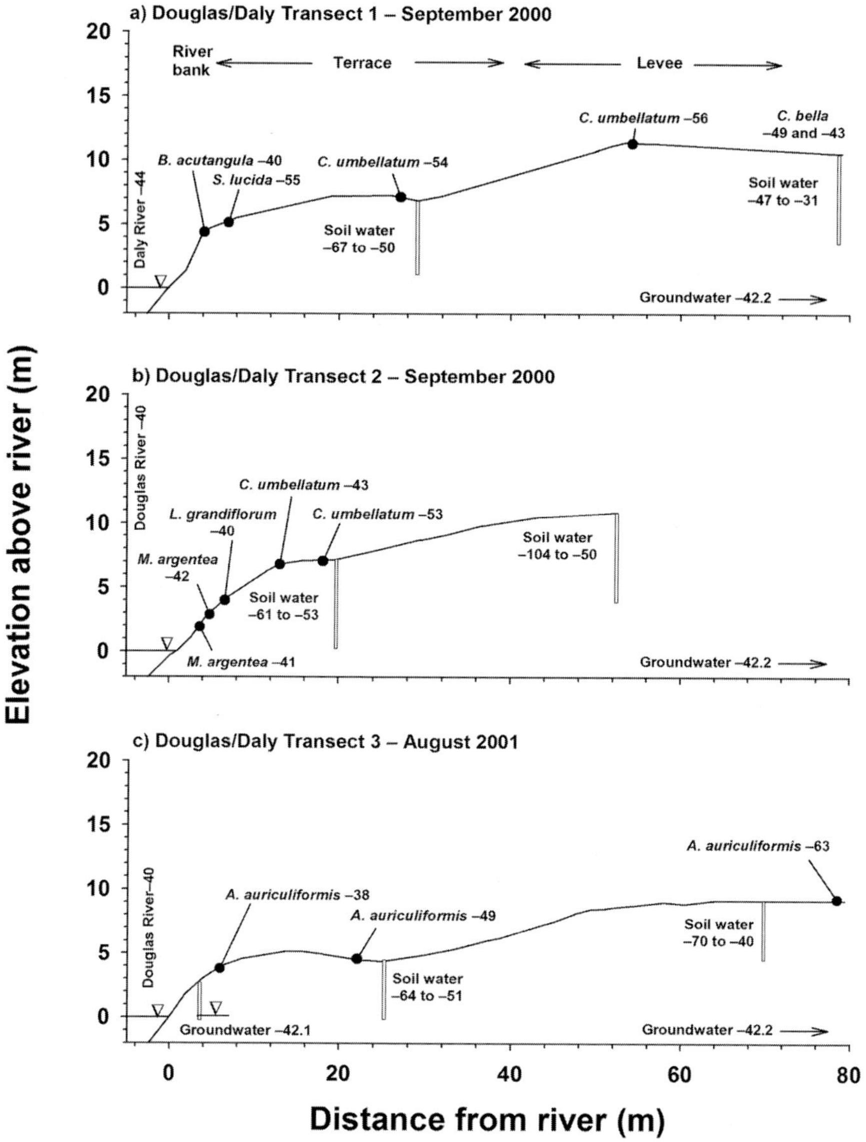

Figure 6.16 Cross-section of the riparian zone for three transects sampled at the junction of the Douglas and the Daly Rivers in 2000–01. Distances and elevations are all relative to river level at the time of sampling (base flow conditions). Deuterium values (‰) for selected trees, river and groundwater also shown. When present, the height of the water table in test holes is shown with an inverted triangle

Source: Data from O'Grady et al. (2002)

times of year (Fig. 6.16; Table 6.6). These data, together with leaf water potential data and transpiration data, were then used to assess the **dependency** of the riparian vegetation on groundwater.

The deuterium signature of xylem sap varied according to position within the landscape. Trees located on levees and the upper terraces generally had depleted signatures (-71 to -43‰) but trees closer to the river had more enriched signatures (–42 to –38‰) (‰ means per thousand).

Table 6.6 Deuterium content of xylem sap of *M. argentea* and *E. bella*

Site	Date	*M. argentea* (‰)	*E. bella* (‰)	Daly River (‰)
Dorisvale	May 2000	–55.9	–100.	–52.5
	Sept 2000	–56.4	–85.3	–45.4
Oolloo	May 2000	–45.2	–70.0	–46.4
	Sept 2000	–45.6	–63.3	–44.2
	Aug 2001	–47.5		–45.9
	Oct 2001	–45.1	–58.3	

Source: Data from O'Grady et al. (2002)

Groundwater had a moderately consistent signature (–45‰) and during the dry season Daly River water was essentially the same (–44‰) while Douglas River water (a nearby river that joins the Daly River) was slightly enriched (–40‰). Clearly, base flow of both rivers is dependent on groundwater influx, with small differences between rivers reflecting the influence of surface evaporation, the relative volume of groundwater influx (relative to bank discharge early in the dry season) and the rate of flow of the rivers (which influences the extent to which evaporative enrichment occurs).

Individuals of *Acacia auriculiformis* and *Casuarina cunninghamiana* were found at several points throughout the riparian zone, including locations close to the river and locations far from the river, up on the levee. Xylem sap of both the Acacia and the Casuarina became more depleted with increasing distance from the river. Such results clearly indicate that these species show **facultative dependence** on groundwater – if it is available they use it, if it isn't they can survive well without it.

The isotope signature of xylem sap of *M. argentea* (located on the river banks) showed little variation throughout the dry season and was close to the value of the river and groundwater signatures, indicating consistent access and utilisation of groundwater. *M. argentea* and *Barangtonia acutangula* appear to be **obligate users** of groundwater because not only is their xylem sap signature consistently similar to that of groundwater, but their distribution appears to be restricted to areas where groundwater depths are shallow. In contrast, the signature of xylem sap of *E. bella* (located on the levee) and most other species became more enriched throughout the dry season, reflecting the increased reliance on deeper and deeper stores of water. These species are facultative users of groundwater. The proportion of soil water utilised by facultative species increased with distance from the river and hence with increased depth to groundwater, but the contribution of groundwater probably increased towards the end of the dry season as the upper soil profile dried out.

That the riparian vegetation was using groundwater could be inferred from two additional lines of evidence. First, pre-dawn water potential data showed that none of the trees were water stressed during the dry season. Therefore, either the volume of soil water available to the roots was sufficiently large (as observed in the Howard River savanna study) or groundwater was being accessed. That groundwater was being accessed can be deduced from the fact that the annual rate of tree water use (1300 mm; O'Grady et al. 2002) exceeds annual rainfall (1157 mm). Note that in addition to water use by trees, understorey transpiration and surface evaporation should be included in estimating ecosystem water use.

The variation in sources of water used by riparian vegetation in the Daly River study confirm the results observed for eucalypts on the River Murray floodplain (Mensforth et al. 1994; Jolly & Walker 1995). *E. largiflorens* and *E. camaldulensis* use stream water, soil water, recent floodwater and groundwater depending on the depth to groundwater, time since flooding, salinity of groundwater and location on the landscape.

Determining the inflow of groundwater to the Daly River

In addition to establishing the groundwater requirements of the riparian vegetation along the Daly River, it was possible to establish the **contribution of groundwater** to Daly River flows by analysing the chlorofluorocarbon (CFC), radon and magnesium content of groundwater and river water (Cook et al. 2003).

In the Daly River study, a 117 km length of river was examined. At several points, warm springs (water temperature approximately 35°C) occur where groundwater is discharged into the river, possibly from depths of 200–300 m. In addition to these warm spring inflows, shallow (cool) groundwater inflows occur. Finally, some water stored in the river banks (recharged in the wet season) discharge into the river during the dry season. Therefore, three inputs to the Daly River support dry season flows.

Daly River water was collected at the end of two dry seasons, when river flow was least, from depths of 0.5 m. The concentrations of CFC, magnesium and radon in the river were determined and groundwater concentrations were also measured. In addition, river temperature was recorded over several days, along with spring water temperature.

Using a simple mass balance equation:

$$(I(x_2 - x_1))/Q_2 = (c_2 - c_1)/(c_i - c_1)$$

where I = groundwater rate of inflow per unit stream length, x_2 is the point of groundwater inflow, x_1 is the point upstream, Q_2 is the rate of stream flow downstream of the groundwater inflow at x_2, c_2 is the concentration of a tracer at point x_2 and c_1 is the concentration at point x_1, it is possible to use temperature differences and differences in magnesium, radon and CFC above and below the point of groundwater inflow, to estimate the volume of groundwater inflow.

From this approach, Cook et al. (2003) estimated that groundwater inflow was between 20% and 48% of total river flow. This approach underestimates total groundwater inflows because it ignores several additional processes. However, it gives an estimate of the minimum contribution of groundwater to river flow.

Determining ecosystem water requirements of riparian vegetation

Estimating the proportion of groundwater in total tree water use is difficult because of the poor resolution of the isotope data and significant variation in riparian forest structure along the river. Total tree basal area and riparian forest width was highly variable and therefore simplifying assumptions were made to estimate the proportion of total vegetation water use derived from groundwater.

Two scenarios were modelled (O'Grady et al. 2002). In the first, all trees within 20 m of the river bank use 100% groundwater; in the second, all trees within 40 m of the river bank use 100% groundwater. Trees outside these widths were assumed not to use groundwater. The isotopic composition of trees along the transect was used to guide the choice of 20 m and 40 m.

The Daly River is 80 km long between the Dorisvale Crossing and confluence of the Douglas and Daly Rivers. Average riparian forest width is 96.5 m, average basal area is 72 m^2 and average tree water use is 3.2 mm d^{-1}. From these data it can be calculated that trees within the first 20 m of the river used 1.9 mm d^{-1} of groundwater (0.08 ML km^{-1} d^{-1}). In the second scenario (trees within 40 m of the river that source 100% of water use requirements from groundwater), groundwater use was calculated as 2.4 mm d^{-1} (0.24 ML km^{-1} d^{-1}).

Large errors are associated with these estimates of groundwater use, including errors associated with measurements of sapflux velocity (such as wound width), basal area distribution and the fact that in the years of measurement groundwater depth was small because of a series of wet years and therefore it was readily accessed by vegetation.

Conclusion

From this study, the following conclusions were drawn.

The climate of the region is tropical monsoonal with a distinct wet and dry season. The topography of the river bank and its associated levees is complex and this results in a complex vegetation structure. Soils in the upper 1 m exhibited low matric potentials and these were lower than leaf pre-dawn water potentials, indicating that trees were accessing water from much deeper layers. However, the isotope signature data showed that the contribution of groundwater to water transpired by trees was highly variable in space and time. Some species appeared to be obligate groundwater users, and these species were located closer to the river than facultative groundwater users, which tended to be located further from the river (and further from the groundwater).

Stand water use showed little seasonal variation for two of the three transects along the river, in agreement with data for the Howard River savannas study. A robust relationship between tree size and tree water use was observed, again similar to that for the savanna case study.

Using two different scenarios, it was possible to estimate the contribution of groundwater to riparian vegetation water use. It was concluded that, despite large associated errors, the estimate of groundwater use of about 1.9–2.4 mm d^{-1} was valuable for resource planning.

Case study 3: *Banksia* woodland of the Gnangara Mound

Background and issues: a hydrological and ecological perspective

The most easily accessed groundwater stores in and around Perth, Western Australia, are unconfined (water table) aquifers. The **Gnangara Mound** is one of two unconfined aquifers which are recharged principally through rainfall (typically about 900 mm y^{-1}, but showing a long-term reduction in the past 30 years) and infiltration from water bodies. Rainfall recharge to these two aquifers is about 343 000 ML y^{-1}, with total storage being about 26 000 GL. The licensed quota for groundwater extraction is 4–15 million m^3 y^{-1} (4000–15 000 ML y^{-1}).

The Gnangara Mound covers about 2200 km^2 from the Swan River to the Moore River and Gingin Brook. Its highest point is about 75 m above sea level and water flows away from that point towards the Indian Ocean, the Swan River and several streams throughout the area.

The Gnangara Mound supports a range of diverse and significant ecosystems, including *Banksia* woodlands, wetlands and several springs and riparian vegetation. These have been the site of extensive investigation to establish the relationship between groundwater extraction and ecosystem responses (Froend and co-workers). It has become apparent that groundwater levels have been steadily declining since the mid 1970s, despite the existence of a regional management program for the Gnangara Mound for the past 20 years and the application of **ecological water requirements** (EWR) for managing the mound. This decline in water level is the result of:

- increased public and private extraction;
- reduced recharge because of reduced rainfall over the past 20 years;
- reduced aquifer recharge because of increased water use by forestry (pine) plantations.

In addition to the decline in groundwater availability (indicated by increased groundwater depth; Fig. 6.17), significant mortality of overstorey trees has occurred at several sites adjacent to active bore fields (see below). Regular monitoring of permanent transects (see below) since

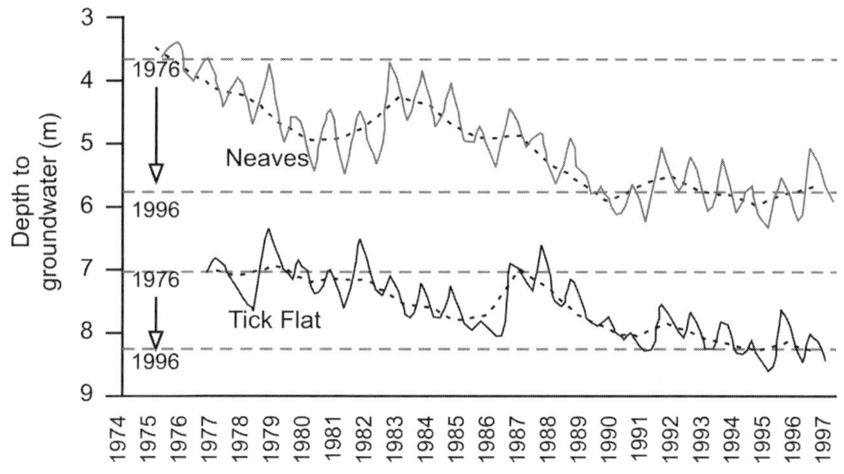

Figure 6.17 Increasing depth to groundwater at Gnangara Mound, 1976–96

Source: Data from R. Froend, pers. comm

1976 has revealed significant changes in vegetation structure and composition on those transects (Groom et al. 2000b, 2001).

Changes in overstorey vegetation were confined to plots on the lower and middle elevation sites (i.e. sites with the smallest depth to groundwater), which typically had a 2 m increase in groundwater depth. A significant increase in the number of species in the understorey was recorded on the transect associated with the bore field. Generally, on the Gnangara Mound, a change in terrestrial vegetation from a **mesic** to a more **xeric** vegetation is under way, indicative of a change in water availability. Loss of macro-invertebrate biodiversity may also be apparent.

Aims of the study

The aims of the study were:

1 to quantify the extent of vegetation changes on the Gnangara Mound;
2 to identify the sources of water being used by vegetation;
3 to produce management guidelines for protecting vegetation on the Gnangara Mound.

Vegetation–groundwater studies on the Gnangara Mound

In 1988 a vegetation survey transect was established next to a major abstraction bore. Importantly, vegetation surveys preceded the start of groundwater abstraction. By having 'before' data on vegetation structure it was far easier to monitor changes in vegetation structure during and after (BACI) the interference caused by groundwater extraction. The transect was located within the expected **cone of depression** of the bore (estimated to be up to 1.5 km wide). A control transect was also established outside the sphere of influence of any bores (Groom et al. 2000b). Vegetation surveys have been undertaken at regular intervals since 1988 and became incorporated into a major research program on vegetation–groundwater interactions on the Gnangara Mound which commenced in 1997.

Prior to pumping from the bore, groundwater levels had been declining for several years because of reduced recharge. With the onset of pumping, groundwater levels fell dramatically and rapidly and continued to decline because of a series of drier than average years.

In 1988, the vast majority of adult trees of the three *Banksia* tree species were found to be healthy on both the control and bore transects. In contrast, by 1991 between 26% and 80% of two species had died on the bore field transect. There was little change in the percentage of mortality between 1988 and 1991 on the control transect. Similar enhanced rates of tree mortality occurred on other bore sites (Groom et al. 2000a) on the Gnangara Mound. Reduced water availability has a direct effect on transpiration and photosynthetic carbon gain through reduced stomatal conductance and leaf area, but also has a secondary effect through increases in leaf temperature to supra-optimal levels because of reduced transpirational cooling – an important consideration in the hot and dry summer of the Mediterranean climate of this region.

Within the understorey, there is a mixture of shallow-rooted, medium-rooted and deep-rooted species. By 1993, more than half of the medium- and deep-rooted species had lost more than half of their adult population. In contrast, shallow-rooted species showed only one-third mortality, and only a 19% reduction in total numbers of plants. For the control transect, fewer species and a smaller percentage of absolute plant numbers showed a decline over the same period. It must be concluded that the medium- and deep-rooted species were suffering from the reduced availability of groundwater, while the shallow-rooted species were better adapted to the drying conditions and were not dependent on the availability of groundwater. Consequently the reduced availability of groundwater had less of an impact on the shallow rooted species (Groom et al. 2000a).

Banksia woodland species on the Gnangara Mound show an **annual cycle of groundwater dependency**. In the wet season (winter), rain is relatively abundant; the soil profile gradually refills (with water) and recharge of the unconfined aquifer occurs. Trees located such that their deep tap roots can access the saturated (groundwater) zone or capillary fringe will use groundwater, which can account for about 20–30% of the total water transpired in the wet season. As the dry season progresses and the upper soil profile dries, less water is absorbed from this zone and the contribution of groundwater to the total water transpired increases to as much as 70–80%. Stomatal control of transpiration increases and a decline in leaf area per tree can occur, to further limit tree water use. Trees located at ever-increasing distances (heights) from the groundwater due to topographic elevation (on hills) become less and less dependent upon groundwater because the roots cannot reach the capillary fringe. Also, the depth of the soil profile is larger and hence it can store larger volumes of water.

Clearly there are temporal and spatial variations within a single community and within a single species, in the degree of groundwater dependence. It is also likely that inter-annual variation occurs, depending on climate and hence rainfall for recharge. Wetter than normal wet seasons can be characterised as causing less dependency on groundwater; similarly, as depth to groundwater increases, dependency on groundwater tends to decline (but see the River Red Gum case study for exceptions).

The responses of *Banksia* woodland to reduced groundwater availability are similar to those observed in most trees in response to drought. This sequence is summarised in Figure 6.18. However, it should be noted that the rate of decline in water availability is an important factor in determining the rate and degree of progression through the various responses of plants to drought. There are short-term and readily reversible changes in plant attributes (changes in stomatal conductance, tree water use) and longer-term, less reversible changes (changes in leaf area, Huber value and xylem vulnerability to embolism) as water availability declines (Table 6.2).

Sources of water

Inferential evidence for groundwater dependency can be made when the water potential of vegetation is shown to be much higher (closer to zero) than expected during periods of low rainfall or drought. Alternatively, stable isotopes can be used to compare the composition of water in groundwater, soil water and vegetation. This is now discussed.

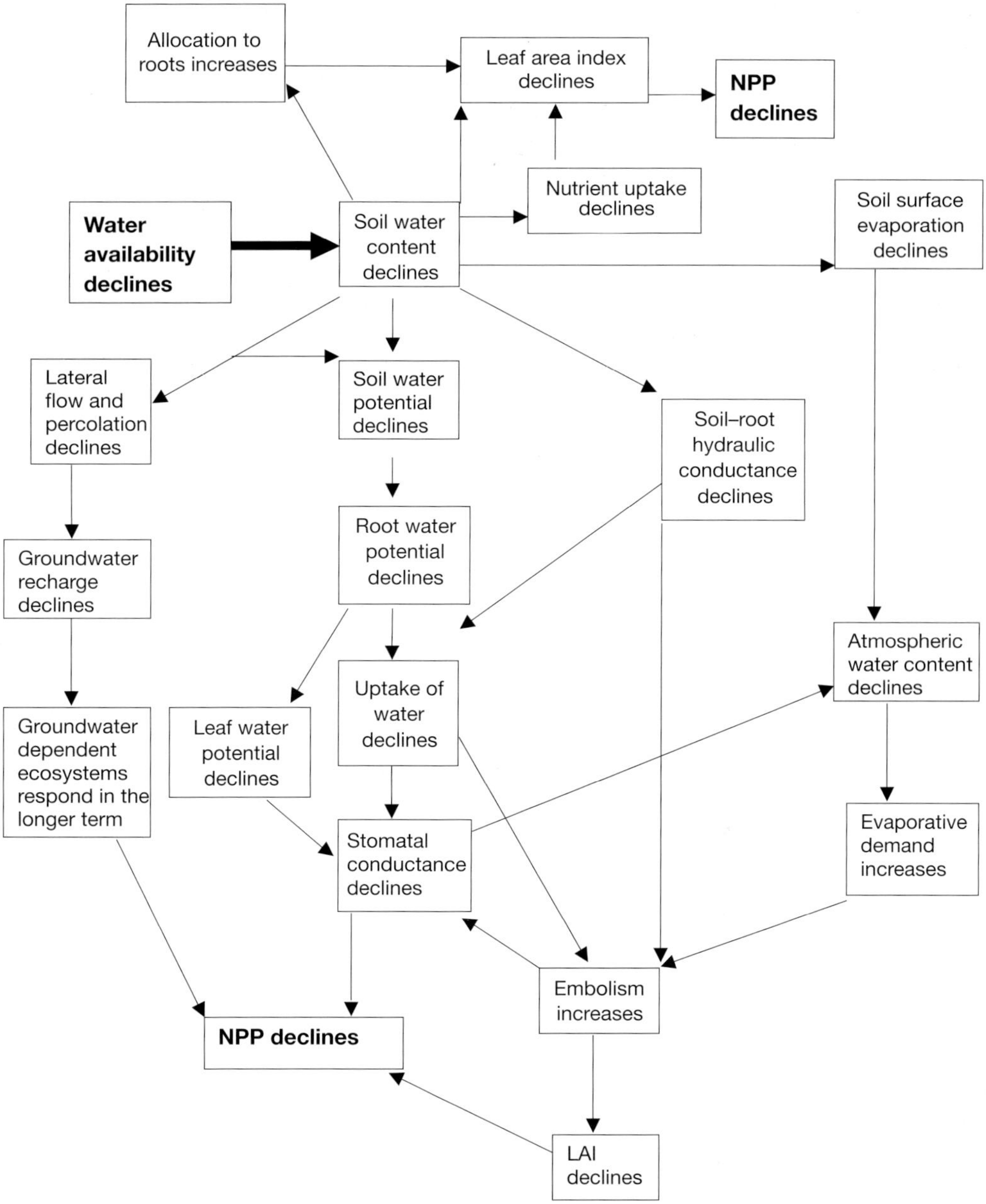

Figure 6.18 Schematic of some of the many inter-relationships among processes that link declines in ecosystem water status to plant behaviour and attributes. The same schema can be viewed as representing a comparison of two ecosystems differing in ecosystem water status by replacing 'Water availability declines' with 'Low ecosystem water status'. NPP = not primary productivity

Species of the Myrtaceae are a significant component of the vegetation of the Gnangara Mound. Myrtaceae shrubs including shallow-rooted *Astertea fascicularis* and *Pericalymma ellipticum* often dominate the lower-lying, winter-wet depressions of the Gnangara Mound. Populations of these species have declined significantly since 1966, attributed to declining groundwater and soil water availability (Davidson 1995, in Groom et al. 2000b). To determine whether this is indeed the case, recent studies of the water relations and stable isotope content of vegetation in and around these depressions has been undertaken (Groom et al. 2000b).

Vegetation on low-lying sites (groundwater depth <1 m) was compared to that on an embankment (groundwater depth 2–3 m) and a mid-slope site (groundwater depth 3–4 m).

Rainfall in the autumn of 2000–01 was 38% of the long-term average (98 mm compared to 258 mm). Rainfall is highly seasonal, with a winter dominant pattern (May–September). Groundwater depths were lowest at all sites between March and July 2001 and groundwater was lower (by approximately 1 m) in June 2001 than in October 2000 (end of wet season).

The water content of soils in all three sites (low-lying depressions, embankment, mid-slope) decreased between spring and summer (8–20% water content) and increased from summer to winter (to 20–80% water content), reflecting the seasonality of rainfall. Mid-slope sites maintained the lowest water content, reflecting lateral drainage to the depressions. The capillary fringe was evident for the embankment and mid-slope sites as an increase in soil moisture at depth. Differences in pre-dawn leaf water potentials (an integrated approximation of soil water availability) between summer and winter corresponded to seasonal differences in soil water content, with lower water potentials observed in summer than in winter.

Generally, pre-dawn leaf water potential declined (plants more stressed) as the depth to groundwater increased. In species growing on the embankment and mid-slope sites (*H. angustifolium* and *Eremaea pauciflora*), water potential and stomatal conductances declined during the summer as the upper soil profile dried. This was associated with declines in soil–leaf hydraulic conductivity, which was apparent in almost all species. The relatively low values of pre-dawn water potential (–3.3 MPa) for *Eremaea pauciflora* at the embankment and mid-slope sites indicates that access to groundwater was minimal (see below). However, this conclusion is in contradiction to comparisons of δ^2H values, which indicated that groundwater was being accessed. However, the lack of discrimination in δ^2H values with depth make the δ^2H values difficult to interpret (Groom et al. 2000b).

Stomatal conductance and rates of transpiration of representative plants did not differ between spring and winter. However, in the summer, species growing in the depressions had larger conductances and transpiration rates than species growing at other sites, presumably because of the larger availability of water in the depressions.

Groundwater δ^2H values were less negative than soil water and became less negative as the seasons progressed from spring to winter. Comparisons of branch xylem water δ^2H values showed that *P. ellipticum* (found in the low-lying depressions) were using surface soil water in spring but both *P. ellipticum* and *A. fascicularis* used deeper soil water (0.8–1.0 m) in winter. In the summer both species in the depression (*P. ellipticum* and *A. fascicularis*) used soil water throughout the soil profile. At the embankment site, *H. angustifolium* was using water in the top 1 m of soil in spring and summer and probably winter. This species is shallow-rooted and probably did not access water from deeper than 1 m.

Eremaea pauciflora had **multiple sources of water**. At the embankment site, this tree species used water from both shallow (<1.0 m) and deeper (>2.0 m) soil in spring and summer and may have used groundwater too, but lack of discrimination of isotope signal with depth makes this conclusion debatable (Groom et al. 2003). In winter, when soil water content was largest due to recent rainfall, *E. pauciflora* did not appear to use groundwater. In the mid-slope site, *E. pauciflora* may have used groundwater during spring and summer (when upper soil water was minimal) but used soil water in winter (when rainfall was largest).

Simple guidelines for managing *Banksia* woodlands as groundwater dependent ecosystems

Ecological water requirement (EWR) is the water required by an ecosystem to maintain ecosystem heath. The term 'ecological water requirement' is synonymous with the term 'environmental water requirement' but has the added benefit of highlighting the fact that it is principally the ecology (i.e. the living component) of the landscape that requires the allocation of water.

Different workers and pieces of legislation use different terms for the same thing. If possible, EWR guidelines should:

- be based on best available information;
- be simple to use;
- use an easily measured parameter;
- provide a measure of protection for the community as a whole rather than protection for a population or a limited number of species;
- be applicable to a wide range of communities.

To date, we are not in a position to provide EWRs that meet all these criteria (Froend 2002). This is principally because we have an insufficient number of detailed studies from which to infer guidelines that can be applied to a wide range of communities. Almost all studies have produced site-specific recommendations. The EWRs for the Gnangara Mound, like other such studies, are based on a series of ecophysiological measurements of one or two species of *Banksia* at a range of sites differing in groundwater availability. Furthermore, depth to groundwater is identified as the sole principal component of importance to groundwater dependent ecosystems, ignoring the timing, quality, quantity and location of groundwater availability.

Despite these limitations, EWR recommendations have been made by defining the acceptable level to which groundwater can be allowed to fall. However, problems inherent with this approach include the following:

- different life stages of a single species have different root depths and hence what is appropriate for a mature tree is unlikely to be appropriate for a sapling. It is preferable to protect the most vulnerable life stage;
- using current or recent hydrograph data to define acceptable limits for fluctuations in groundwater depth can be difficult without knowing whether these data represent average or above or below average years for recharge;
- the potential for new root growth probably declines with tree age and thus the rate of increase of groundwater depth and life stage probably determine the degree to which an adaptive response in root growth can be given;
- defining an acceptable level of groundwater extraction without defining an acceptable degree of change in the vegetation in response to that extraction;
- the relationship between groundwater extraction and soil texture (and hence soil water holding capacity) has not been defined (see the discussion of soil overburden in case study 1).

With these limitations in mind, the management guidelines defined by Froend and co-workers are thus:

- Three classes of depth of groundwater are identified. These are 0–3 m, 3–6 m and 6–10 m.
- It is assumed that a greater depth to groundwater in the reference period at a given site will result in a greater tolerance to groundwater decline of trees at that site because of the larger volume of soil (unsaturated) available for exploitation by roots.
- From ecophysiological data on growth rates and other plant attributes, as a function of groundwater decline, three response curves are suggested as describing plant responses (see Fig. 6.19).
- From the three response curves, **three categories of response** are identified:
  1) category 1 (0–3 m depth) – can tolerate up to a 0.75 m decline in groundwater depth;

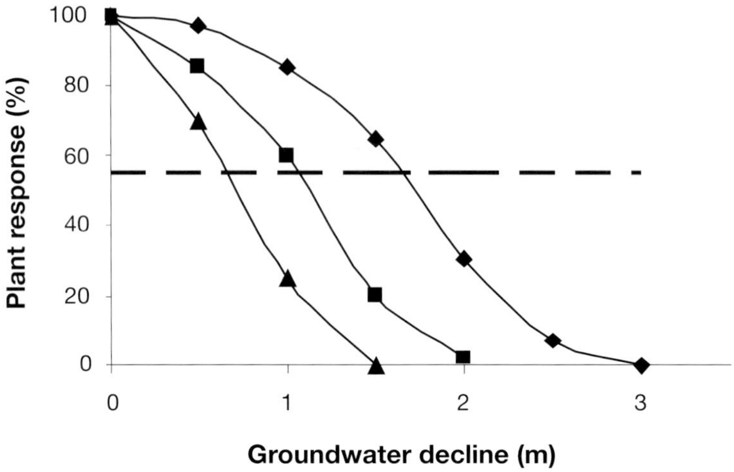

Figure 6.19 A threshold plant response (user defined; set by the dashed line) for the three response curves that characterise plant responses to changes in groundwater depth for the three depth categories

Source: R. Froend, pers. comm.

2) category 2 (3–6 m depth) – can tolerate up to a 1.25 m decline in groundwater depth;
3) category 3 (6–10 m depth) – can tolerate up to a 1.75 m decline in groundwater depth.

A **threshold** plant response (e.g. reduced growth) is defined by the manager and this value on the plant response axis is then used to define, for each of the three response curves, the upper bound (maximum allowable) for groundwater extraction. From the choice of acceptable plant response to increased depth to groundwater, three upper limits for the decline in water table are set and these are used to manage abstraction from the Gnangara Mound.

In response to the decline in groundwater level and changes in vegetation composition on the Gnangara Mound, several **management strategies** have been adopted, including:

- changing the location, rate and timing of pine plantation thinning to reduce plantation water use;
- reduced public consumption by applying water use restrictions;
- artificial supply of water to springs and wetlands to ensure continued health of those ecosystems;
- public education programs about the need to conserve water;
- increased scrutiny of water abstraction licence conditions and recovery of unused allocations.

Conclusion

At the start of the study of the Gnangara Mound vegetation, the following aims were stated:

1 to quantify the extent of vegetation changes on the Gnangara mound;
2 to identify the sources of water being used by vegetation;
3 to produce some management guidelines for protecting vegetation on the Gnangara Mound.

All these aims were met. In addition, three attributes were found to be important in determining the response of Gnangara Mound vegetation to changes in groundwater availability (depth). First, location (and hence depth to groundwater) in the landscape is important. Where groundwater depth was shallow, some species relied on that source for at least part of the year. Second, rooting depth (a species dependent attribute) determined the potential to access groundwater to different depths. Finally, time of year (and hence the water content of the upper soil profile and the depth to groundwater) influences the source of water accessed by different species. Because of the nature of the interactions among these three attributes, predicting which species are most likely to suffer first from a decline in groundwater availability is not straightforward.

Case study 4: Great Artesian Basin – a regional aquifer system supporting arid zone ecosystems

This case study is presented as a narrative of the issues and importance of a unique groundwater dependent ecosystem in Australia – mound springs of the Great Artesian Basin. Although they don't contain trees, they are iconic unique terrestrial ecosystems worthy of discussion. This section summarises work undertaken over many decades, rather than over 3–5 years in a single study. The aim is to show that recognition of a problem and the work required to understand and rectify it can take very many years. It also highlights the importance of integrating hydrological and ecological knowledge in order to understand ecosystem function. For a recent review of vegetation patterns in permanent springs wetlands of arid Australia, see Fensham et al. (2004).

Background and issues

Desert oases provide extreme and obvious examples of ecosystem dependence on water of remote origins. Where these oases arise with no connection to surface water systems (i.e. they are groundwater dependent), the biological isolation can lead to high local endemism. Perhaps the most notable examples of such ecosystems in Australia arise within the Great Artesian Basin, where remote desert springs discharge groundwater whose origins may be many hundreds of kilometres distant and from rainfall perhaps a million or more years ago. The remoteness and antiquity of these springs gives rise to local desert ecosystems of immensely high biological, cultural and economic significance. The post-European settlement management of this water resource has had major implications for regional and national development as well as for conservation. These desert springs are known as **mound springs**.

A groundwater system is said to be artesian if the local pressure level in the aquifer is higher than the level of the land surface. Such situations arise where the aquifer is **confined** over a distance such that the levels at the intake (recharge) end of the aquifer are higher than the ground level some distance away. This is illustrated in Figure 3.17 (Chapter 3).

The Great Artesian Basin (GAB) covers 1 711 000 square kilometres, about one-fifth of Australia. It is bounded on the east by the Great Dividing Range, on the west by the Central Australian Plateau, on the north by the Gulf of Carpentaria and on the south-west by the Lake Eyre Basin (Fig. 6.20).

Most of the GAB is **semi-arid to arid**, with mean annual rainfall increasing from less than 200 mm in the west to 600 mm in the eastern highlands, although rainfall in the far north-eastern portion may exceed 1200 mm. Nowhere does mean annual rainfall exceed potential evaporation, and groundwater is a more reliable source of water than surface resources. The surface water systems are highly variable in flow and their catchments are not contained by the

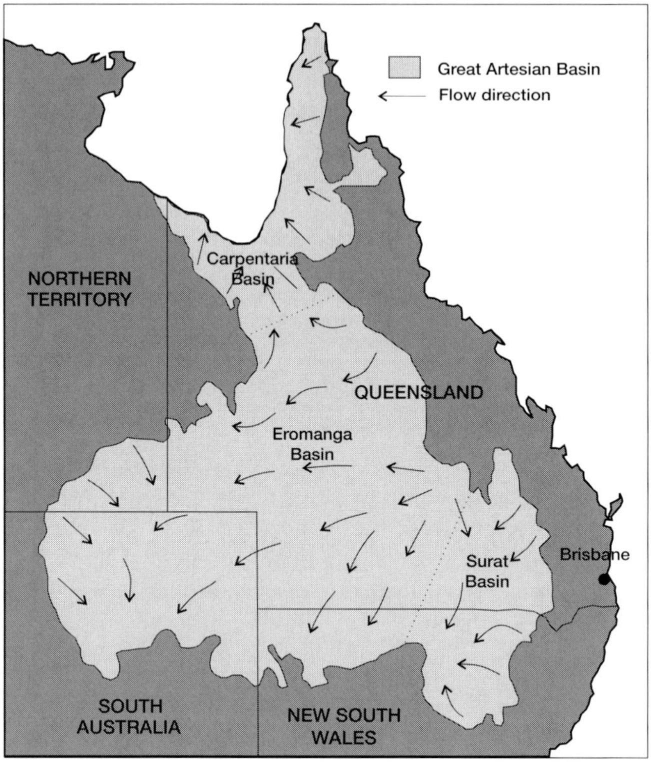

Figure 6.20 The Great Artesian Basin, Australia. Groundwater recharge occurs in the eastern highlands and along the edge of the Central Australian Plateau. There are three distinct zones of flow (sub-basins); by far the largest is the Eromanga Basin, whose discharge waters travel over 1000 km and give rise to the desert mound springs

boundaries of the GAB. Surface drainage in the region of desert springs discharge only rarely finds its way to its ultimate terminus at Lake Eyre in the south-west. Most importantly, the surface flows of water, even in relatively large and persistent rivers such as the Darling, do not interact substantially with underground waters once they flow out onto the plains.

The GAB is made up of a series of sedimentary **strata** (layers) deposited in depths of up to 3000 m. Some strata consist of sandstones that are apparently hydraulically continuous over the whole of the sub-basin (i.e. water under pressure can flow from one end to the other). These relatively transmissive sandstones are interbedded (sandwiched) with thicker layers of finer cemented materials with very low permeabilities. These serve as **aquacludes** to the sandstone **aquifers** that they hydraulically **confine**.

The confined aquifers flow as a result of a net fall of some 500–700 m over a distance exceeding 1000 km (Fig. 6.21). Flow rates are predictably low (1–5 m y^{-1}). Thus the water, upon discharge, is of great age (several thousand to 2 million years old; Mudd 2000). Discharge results from faults that provide a path for upward leakage through the confining layers, or by rapidly thinning aquifers along the line of flow.

The GAB comprises one of the largest underground stores of fresh water in the world, holding some 8700 million ML. Modern recharge volumes are estimated to be about 100 million ML per year, most of which occurs along the uplifted eastern margins but some also along western margins where aquifers are exposed to the surface or overlain by sandy sediments (Mudd 2000). More than half the water in the aquifers has a salinity of less than 1000 mg/L, making it a relatively fresh source upon discharge (or abstraction).

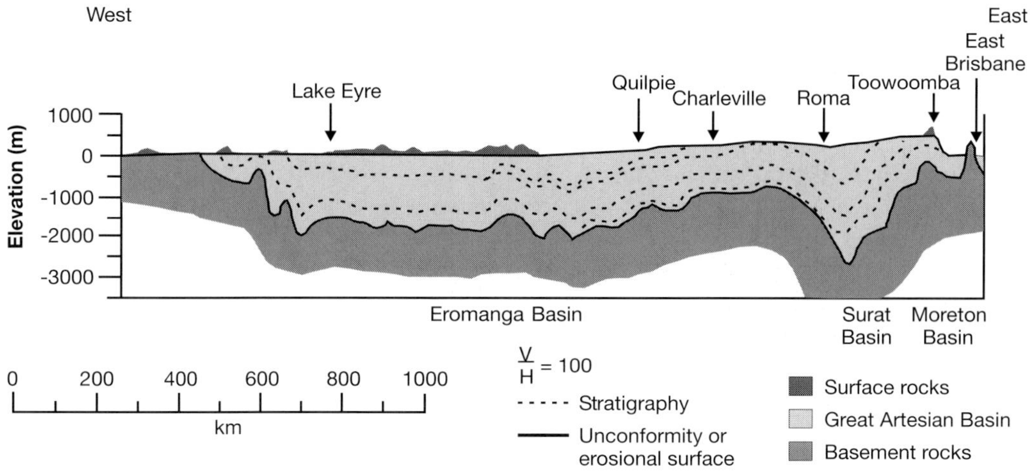

Figure 6.21 A roughly east–west cross-section showing the hydrogeology of the GAB. Note the extreme vertical exaggeration of the figure; an undistorted depiction of scales would represent the aquifer as a very thin continuous sheet

Source: Adapted from GAB Consultative Council (1998)

Mound springs

The aquifers of the Eromanga sub-basin partially discharge in localised features known as **mound springs**. Mound springs arise because of a connection between an artesian aquifer and the surface, through which water flows (see Fig. 3.17). There are over 600 groups of such springs in 11 or 12 complexes (area groupings). Hatton and Evans (1998) summarised the literature on these features. The most notable springs occur in a 1000 km arc from Maree to Oodnadatta, South Australia. This zone discharges about 40% of the current annual spring discharge of the GAB (50 000 ML; Ponder 1986) and includes the large Dalhousie Springs complex. Springs vary from 20 000 to 1 million years old. Mound springs can be up to 8 m tall and 30 m in diameter, although older extinct springs can be 40 m tall, suggesting higher artesian pressures than are currently observed (Mudd 2000). Springs often have wetlands surrounding them which are sustained by spring flows.

There are three major groups of springs in South Australia: the Dalhousie supergroup; the Lake Eyre supergroup including the Bubbler, Blanch Cup, Coward, Hamilton Hill Strangways and Elizabeth Springs; and the Lake Frome supergroup which includes springs east of Marree and those in Lakes Frome and Callabonna. Elsewhere in the region there are eight more aggregations: the Mulligan River, Springvale, Flinders, Barcaldine, Springsure and Eulo groups in Queensland and the Bogon River and Bourke supergroups in New South Wales. Many of the springs outside South Australia have declined or are extinct, but there are active sites in the Springvale, Barcaldine and Springsure supergroups.

The rate of discharge varies, with a maximum at Dalhousie of 160 L s^{-1}. Many discharges are sufficient to form outflow channels with associated wetlands. The springs are the only source of permanent fresh water within the desert environment, and the surrounding ecosystems are entirely reliant on the discharge of groundwater and have been since the late Pleistocene. This geographic isolation and dependence has given rise to unusual ecological and evolutionary phenomena.

Most of the springs support sedgelands and grasslands, but some of the larger pools at the Dalhousie complex support *Melaleuca glomerata* swamp woodlands or swamp shrublands. Vegetation at the mound springs tends to be of low species diversity, and high similarity with springs with other wetland systems. Any dissimilarity in composition may be related to variations in water chemistry among springs. There are few endemic or relict plant species.

Fish fauna at Dalhousie consists of six species in five genera from five families; four of these are endemic. The fish fauna of Dalhousie may be the most diverse assemblage of native fish species without exotics in desert mound springs worldwide (Kodric-Brown & Brown 1993). However, all these families and genera are represented elsewhere in the Lake Eyre Basin, although there is evidence to suggest that they have been isolated at Dalhousie for at least 10 000 years and it is unlikely that they could now disperse naturally in floodwaters.

It is among the invertebrates that the most unique features of mound spring fauna are found (Morton et al. 1996). The only known desert-dwelling isopod, the crustacean *Phreatomerus latipes*, is found only in the springs of the Lake Eyre Basin. The ostracod *Ngarawa dirga* is the only member of its subfamily and occurs only in the Lake Eyre supergroup of mound springs; other ostracod species have yet to be described taxonomically but are present. One copepod among the 12 species present at Dalhousie may be relictual (Morton et al. 1996). There is an endemic blind amphipod *Phreatochiltonia anophthalma* at Dalhousie. There is evidence of differentiation in the yabby *Cherax* spp. and the shrimp *Caridinia* spp., suggesting long isolation in the mound springs in which they occur. The mound springs are also rich in previously undescribed species of hydrobid snails, with two endemic genera *Fonscochlea* and *Trochidrobia*.

Lothian and Williams (1988) listed nine mound springs in South Australia as having special scientific significance. The mound springs were also listed as one of four special wetland areas in Australia with respect to biodiversity (Mummery & Hardy 1994). Harris (1992) provided an excellent review of mound spring values, their history of management and study, and their conservation, including a discussion on their immense importance to Aboriginal and European communities. Morton et al. (1996) concluded that two aspects of the aquatic systems of the Lake Eyre Basin are sufficiently unusual within Australia to warrant further assessment against World Heritage criteria: the nature and scale of the endorheic drainage systems reaching Lake Eyre from the east and the ecological responses to that system, and the most unusual nature of the evolutionary radiations among the scattered and isolated artesian springs of the Basin. They also concluded that while the Australian mound springs are not globally unique, the spatial scale over which they are distributed and their abundance is unusual at the global level.

Water resource development and the mound springs

Development of the GAB's groundwater resources in the 1880s opened the region to pastoralism, mining and settlement. Deep drilling yielded abundant artesian water, and in some cases local hydrothermal energy. The development of hundreds of bores in the ensuing decades led to peak abstraction of about 1 100 000 ML in 1918 (Mudd 2000) from about 1500 free flowing wells. Most of the water went to waste down natural watercourses or, later, constructed bore drains designed to reticulate water but losing perhaps 90% of the water in transmission. As the total number of flowing artesian bores drilled into the GAB has increased, the discharge rate has declined because of the drop in pressure within the aquifer (Fig. 6.22; Mudd 2000). The realisation that resources were finite prompted early legislative attempts to control abstraction in Queensland and New South Wales. Currently the rate of discharge from bores is about 1500 ML d^{-1} (Mudd 2000) but there are 34 000 km of bore drains. This flow has environmental consequences of its own, with contributions to wetland degradation, salinisation, spread of weeds, and overgrazing by livestock and wildlife concentrating around permanent water.

Many spring wetlands in New South Wales and Queensland have been destroyed by the lowering of groundwater pressure from extraction and development of unregulated bores. The single most important threat to mound spring ecosystems is this aquifer drawdown, which at its worst can eliminate the flora and fauna altogether. Habermehl (1988) believed that although a new steady state between recharge and discharge for the GAB was reached in the 1970s,

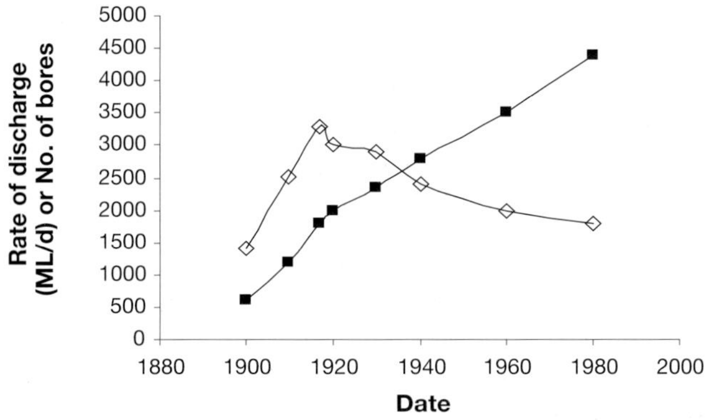

Figure 6.22 The number of bores drilled into the GAB (closed squares) has increased since the early 20th century but the rate of discharge (open diamonds) has declined because of the drop in artesian pressure

Source: Redrawn from Mudd (2000)

potential impacts of local groundwater extraction remain. This conclusion is underscored by concerns expressed in the environmental impact assessment associated with the Olympic Dam mining development and subsequent monitoring studies to ensure compliance with specified limits (Olympic Dam Operations 1990). Among the recommendations for mound spring conservation noted in Harris (1992), is a call for continuing research into the ecology and dynamics of plant and animal populations and the systematic monitoring of spring flow rates. Morton et al. (1996) also identified gaps in knowledge about mound springs, including the distribution of aquatic organisms among the springs, a detailed inventory of springs, and detailed studies of water budgets for outlets of the GAB (human-induced and natural). The third issue was considered urgent. Ponder (1986) suggested that the mound spring systems of Queensland required detailed scientific investigation.

Unregulated artesian flow from bores has led to the development of desert wetlands similar, if not equivalent to, mound spring ecosystems. Examples of such human-induced wetlands are associated with Nunn's, Angas' and McEwin's bores and Purni Springs in the Dalhousie supergroup. While the biological significance of these features is questionable given associated heavy grazing pressure and probable lack of endemics, they can have appreciable cultural significance. For example, Purni Springs is a popular outback destination for campers. Local resistance to regulating these bores and thus drying up these wetlands is one gauge of their cultural value.

Conclusion – a plan for recovery

Annual abstraction of groundwater from the GAB is 570 000 ML, unsustainable from an environmental point of view (GAB Consultative Council 1998). Legislation has progressively restricted access to groundwater, and Australia has undertaken an extensive program to cap free flowing bores in an attempt to reduce the rate of abstraction. The GAB Bore Rehabilitation Program, started in 1989, aims to place control valves on bores; the program is almost complete in South Australia. Associated with bore rehabilitation is the piping of water to usage points to eliminate transmission losses and increase efficiency. By 1998 the program had saved 79 000 ML of groundwater per year and eliminated 1000 km of bore drains. There is evidence that aquifer pressures are recovering in the vicinity of some rehabilitated bores.

The Great Artesian Basin of Australia is a very large sedimentary system with an immense volume of relatively fresh groundwater. This water flows slowly through confined aquifers under very low pressure gradients, and arises naturally as springs in the arid zone. These discharge features, or mound springs, have immense biological and cultural significance. The mound spring biota are entirely dependent on groundwater. They were put at risk by development of the region's groundwater supplies for industry; many have since disappeared. Regional drops in aquifer pressure are being redressed through a program of bore rehabilitation (capping) and the piping of water to reduce wastage. There is emerging evidence that this program has begun to reverse the drop in pressure, securing the future of the dependent ecosystems.

Conclusion

The existence of groundwater dependent ecosystems has only been identified by ecologists and hydrologists since the mid 1970s. The importance and value of such ecosystems has been recognised by science, the broader community and land and water managers since the mid 1990s. The legislative requirements for allocating water to maintain ecosystem health, including groundwater dependent ecosystems, is in many cases more advanced than scientific understanding of the actual water requirements of those ecosystems. This chapter has offered several case studies focusing on various issues pertaining to groundwater dependent ecosystems. It has shown that defining the degree of groundwater dependency is not easy, and quantifying the amount of water required for groundwater dependent ecosystems remains associated with large error terms. However, significant progress has been made, and continues to be made. The next chapter provides five additional examples of the application of ecohydrological studies to real world issues.

Further reading

Amlin NM & Rood SB (2002). Comparative tolerance of riparian willows and cottonwoods to water table decline. *Wetlands* **22**: 338–346.

Bell RW, Schofield NJ, Loh IC & Bari MA (1990). Groundwater response to reforestation in the Darling Range of Western Australia. *Journal of Hydrology* **119**: 179–200.

Clifton CER (2001). *A framework to assess the environmental water requirements of groundwater dependent ecosystems.* 3rd Australian Stream Management Conference, Brisbane.

Cook PG, Hatton TJ, Pidsley D, Herczeg AL, Held A, O'Grady A & Eamus D (1998). Water balance of a tropical woodland ecosystem, Northern Australia: a combination of micrometeorological, soil physical and groundwater chemical approaches. *Journal of Hydrology* **210**: 161–177.

Cooper DJ, D'Amico DR & Scott ML (2003). Physiological and morphological response patterns of *Populus deltoides* to alluvial groundwater pumping. *Environmental Management* **31**: 215–226.

Davis J, Froend R, Hamilton D, Horwitz P, McComb A, Oldham C & Thomas D (2001). *Environmental water requirements to maintain wetlands of national and international importance.* Canberra.

George RJ, MacFarlane DJ & Speed RJ (1995). The consequences of a changing hydrologic environment for native vegetation in south-western Australia. *Nature Conservation* **4**: 9–22.

Hayashi M & Rosenberry DO (2002). Effects of groundwater exchange on the hydrology and ecology of surface water. *Ground Water* **40**: 309–316.

Healy RW & Cook PG (2002). Using groundwater levels to estimate recharge. *Hydrogeology Journal* **10**: 91–109.

Chapter 7

Ecohydrology in action: case studies

This chapter focuses on six 'thematic studies' (case studies) to illustrate how knowledge of the behaviour of vegetation at plant, canopy and stand scales, in conjunction with knowledge of hydrology and micro-meteorology, can aid in understanding the behaviour and function of landscapes. Foresters and hydrologists interpreted and interrogated landscapes using their own language, paradigms and methodologies; this chapter will examine and interpret the behaviour and functioning of landscapes using both hydrological and plant functional approaches.

Throughout these studies information relating to plant function and hydrology, and the links between them, are presented. The first case study shows that the change in vegetation structure and function resulting from fire affects the hydrology of streams, **water yield** (the amount of water leaving the catchment as stream flow) of catchments and water quality. In the second study (Mountain Ash), a mechanistic understanding of how forests function and grow is combined with a hydrological understanding of a catchment to develop insight into how catchment water yield changes with forest age and disturbance (fire or logging). The third study examines the impacts of water extraction for mining in the arid zone on local hydrology and how these impacts are being managed. This contrasts to the fourth and fifth thematic studies (wetlands and floodplains), where it is the change in hydrologic regime caused by human activities that has caused a change in function. Three different wetland/floodplain sites are examined, each with contrasting issues (especially the degree of linkage between groundwater and wetland). The sixth case study (Lake Toolibin) shows how deforestation of a once forested landscape can threaten a freshwater lake and associated wetlands.

At the end of reading this chapter readers should be familiar with the following issues:

- the adaptations Australian vegetation exhibit in response to the frequent fire regime prevalent in Australia;
- how fire influences the ecohydrology of Australian landscapes;
- how catchment water yield varies with forest age, and what explains this pattern of change;
- the impact of clear-felling on catchment yield;
- the relationships among rainfall, tree size, micro-climate, tree water use and catchment water yield;
- how long it takes a burnt mature forest to recover following fire, in terms of catchment water yield;
- the links between mining and mound springs;

- how wetland function responds to changes in the hydrological balance of a site;
- what needs to be done to protect the future of Lake Toolibin in Western Australia.

Introduction

Vegetation and water resources face many threats. Some are obviously and immediately anthropogenic in origin, such as mining. When mines are created there are many potential ecological and hydrological impacts, including large tailings dams containing highly polluted water and the collection and pumping out of groundwater in mine shafts and open-cast mines. Others might be less obviously and immediately anthropogenic in origin, but upon inspection arise from human activities upstream of a site. Wetlands and floodplains can be impacted by changes in river flow, river trajectory and river quality that occur a large distance away. Other impacts are natural and arise from natural cycles of drought, fire and old age. Managing forests to maintain a city's supply of drinking water must be able to account for these stochastic (random) and developmental natural events.

Ecohydrological studies of the impact of anthropogenic and natural factors on vegetation structure and function and catchment hydrology can provide significant insight to the successful management of water and vegetation resources. This chapter gives broad-scale (fire and management of forests for water yield) and site specific (wetlands, mining and Lake Toolibin) examples. This chapter will discuss the following themed studies:

1. the impact of fire on Australian landscapes;
2. Mountain Ash forests – a city's water supply understood from an ecological and hydrological perspective;
3. the influence of mining on local hydrology;
4. links between hydrology and ecology of wetlands;
5. floodplains;
6. Lake Toolibin of Western Australia – an exercise in restoration ecohydrology.

Fire is a natural feature of Australia and has been for many many thousands of years. Large and intense fires impact both on vegetation and hydrology of catchments. Understanding these impacts is central to management of many Australian catchments. Mountain Ash forests cover many catchments around Melbourne and are subject to cycles of recruitment, middle age and death. They are also subject to logging and fire. Understanding how these influence catchment hydrology is important for the management of water yield. Mining, as has been stated, can influence both local ecology and regional hydrology, while diversions of water flows through rivers impact on water and nutrient flows into and through wetlands and floodplains. Finally, a narrative of Lake Toolibin in WA is presented to show how long-term studies (similar to the mound springs case study in Chapter 6) are often needed to gain sufficient information for understanding the causes and solution to the problems of managing inland lakes.

A primer is provided to introduce new terms and concepts required to understand the issues involved.

Case study 1: fire in Australian landscapes

Background and issues

Fire is a recurrent and large-scale factor that has influenced the ecology and hydrology of Australia (and New Zealand and subtropical Africa) for millions of years. Even before the arrival of humankind in any country, lightning strikes caused fires sporadically, stochastically and patchily. The arrival of Aborigines 40 000–60 000 years ago on Australian shores changed the fire regime substantially. In pre-European times, fires were used to manage the landscape for specific purposes, including making passage across the landscape easier, producing new flushes of growth and flushing animals out for hunting. The arrival of Europeans led to a second, perhaps more extensive and intensive change of the fire regime of Australia. Generally, Europeans have attempted to reduce fire frequency, especially throughout southern Australia. However, fire is increasingly being used as a management tool to reduce fuel loads and weed invasions and to enhance resprouting of native species (Thomson & Leishman 2005). Despite this, uncontrolled bushfires continue to have devastating effects. The fires of 2002–03 in Victoria, New South Wales and the Australian Capital Territory were widespread and associated with the largest and most severe drought in 100 years. These fires were probably more severe than the Ash Wednesday fires in 1983 and equalled the Black Friday fires of 1939. This section reviews the impact of fire on the ecology and hydrology of the Australian landscape, with particular reference to northern Australia where fire frequency is the highest of all Australia. Much of the savanna of northern Australia burns every 1–2 years.

Aims of the study

The aims of this themed case study are to answer the following questions.

1 What is the impact of fire on vegetation and how has Australian vegetation adapted to recurrent fire?
2 How does fire influence the hydrology of a site?

Fire: a primer

Obviously, bushfires are characterised by flames but it is not only the **heat** of the fire that has ecological impacts. The **smoke** (particulate matter), the **ash** deposited and **gases** released also have ecological and hence hydrological impacts.

The amount of heat released by a fire is determined by the **heat of combustion** of what is being burnt, and the **fuel load** (how much dry biomass is present at the site being burnt). Fuel for a bushfire mostly consists of grass, tree leaves, fallen branches and trees, and the stringy or papery/flaky barks of many Australian tree species. Because of variation in the composition of the fuel, fuel load at different sites can vary enormously. A typical range of values is 10 000–30 000 kJ kg^{-1}, with a value of 20 000 kJ kg^{-1} often assumed when site specific data are unavailable. In addition to variation in the heat of combustion (energy released by burning per kg of material burnt), there are large differences in the amount of fuel available for burning (fuel load) (Atwell et al. 1999).

Cattle grazing of grasses, land clearing of trees, a high frequency of fires and a large abundance of termites all act, individually or in combination, to reduce the accumulated fuel load. The accumulation of fuel load is highly variable in space (over distances of tens of metres) and time (inter-annually). A fuel load of 0.5 t ha^{-1} is deemed low, a large fuel load is 20 t ha^{-1}.

Fuel has to be suitably dried (**cured**) before a fire can properly take hold; consequently fires are more prevalent at the end of the dry season in monsoonal Australia or after a drought in southern and eastern Australia. When fires are absent from savannas of Australia for three to five years, fuel loads increase by a factor of 2 or more. Introduced grasses such as gamba grass can cause large increases in the amount of fuel accumulated at a site, compared to sites with native grasses.

The amount of heat released by a fire depends not only on the fuel load but also on wind speed, the degree of curing of the fuel and humidity of the atmosphere. A 'cool' fire, occurring

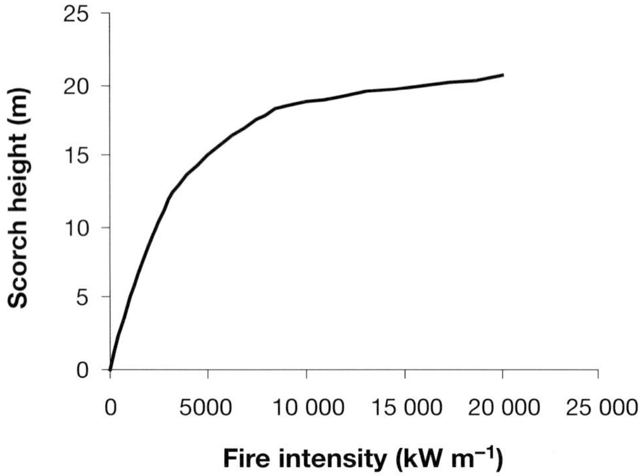

Figure 7.1 Char height and scorch height increase asymptotically with increasing fire intensity

in the early dry season of monsoonal Australia or in misty, low wind conditions in eastern and southern Australia, might release 1000 kW m^{-1} of energy. A 'hot' fire occurring late in the dry season or after a drought and after a fire-free period of a year or two, might release 10 000 kW m^{-1} and a 'very hot fire' can release up to 20 000 kW m^{-1}.

The amount of heat released is an important factor in determining the impact of the fire on the plant function and hydrology of a site. Cool fires are low-lying and do not exert a long-lasting effect on the canopy of vegetation taller than about 5 m. Consequently, 30 m tall tree canopies are relatively unharmed by frequent cool fires because the **scorch height** (the average height to which the fire's heat has an impact on leaves (Fig. 7.1) is much lower than the canopy). Similarly, the **char height** (the height to which tree stem blackening occurs) is low in low intensity (cool) fires. A fire intensity of 1000 kW m^{-1} will char to about 0.5–1 m. Hot fires can scorch a canopy to about 20 m and directly char a stem to about 3–4 m (Fig. 7.2, Colour Plate 6). However, even in cool fires the stringy and flaky barks of eucalypts and paperbarks can carry a fire rapidly up to the canopy, where it will spread fast. The high concentration of oils in the **phyllodes** (leaf-like structures) of many tree species facilitates this.

As the intensity of a fire increases, the proportion of the fuel that is combusted tends to increase. Low intensity (cool) fires tend to burn smaller debris, with less impact on large trees than hot fires which can burn almost all the fuel load present. This determines the amount of ash deposited after the fire, which has an important ecological effect. Thus the impact of a fire is very much a function of its intensity, which in turn is very much influenced by the vegetation type and the frequency of fire. Infrequent fire allows fuel loads to accumulate over time, making the eventual fire hotter and more severe.

Coping with fire: Australian plant responses to recurrent fire

Australian vegetation has evolved several adaptive features in response to the fire regime experienced over the past several million years (Atwell et al. 1999).

- Seeds are released from **woody capsules** when exposed to heat and smoke of a fire. Seeds of many Australian genera, including *Casuarina*, *Hakea*, *Eucalyptus* and *Banksia*, are enclosed in woody capsules which can remain on the tree for many years and release their seed after a fire has passed through the forest. This response is beneficial for two reasons. First, the ash on the forest floor is rich in nutrients from which newly germinated seedlings can benefit. Second, there is likely to be reduced competition from saplings as most will have been burnt by the fire.

- Seeds in the soil are stimulated to germinate after fire. Both heat and smoke can act together or independently to stimulate germination. Species that use this tactic (e.g. *Acacia* spp.) are called **fire ephemerals** (Atwell et al. 1999). Fire ephemerals are characterised by producing large numbers of long-lived seeds and very high rates of seedling growth.
- **Epicormic buds** are stimulated after fire. These buds are present in **lignotubers** (see below) and under the bark of stems, protected from all but the most intense fire. Eucalypts are good examples. These plants are known as **resprouters** because they resprout from buds after fire (or after being chainsawed). Resprouters are most dominant in locations that are frequently burnt.
- Flowering can be stimulated after fire so that new seed is released when the next rainy season occurs. Grass trees (*Xanthorrhoea* spp.) and some orchids show this behaviour.

A lignotuber is an underground storage organ that stores water, inorganic nutrients and carbohydrates. It is a swelling of the root immediately beneath the stem of a tree and can be more than 1 m in diameter. Lignotubers contain large numbers of latent buds which sprout if the main stem is killed (by fire or chainsaw). It is possible that the low above ground growth rate of seedlings and saplings of many eucalypts arises because of the large allocation of resources to the growth of lignotubers. Once the lignotuber has attained sufficient resources, the sapling often undergoes a sudden large growth spurt, taking it rapidly above the scorch height of the average fire.

Vegetation responses to fire in northern Australia

Savannas of northern Australia consist of a mosaic or patchwork of different vegetation communities. Within the matrix of eucalypt dominated open woodland or forest lies a large number of distinct vegetation types differing in species composition and structure (height, leaf area index, mix of grasses, shrubs and trees) (Williams & Cook 2001). Open savanna woodlands and forests are highly resistant to fire, but most of the other communities are fire sensitive and even infrequent fires (one every 5–10 years) over a period of 30 or more years can eliminate those communities from the landscape. Fire regime (a combination of frequency and intensity) influences the spatial distribution, composition and tree size distribution of savannas (Williams et al. 2003)

There are thousands of isolated rainforest patches within the savanna matrix of northern Australia and there are many forms of rainforest in monsoonal Australia, including dry sclerophyll forest, monsoon vine forest and *Allosyncarpia*-dominated riparian forest. In addition, paperbark forests (*Melaleuca* spp.), lancewood forests (*Acacia shirleyi*), Cypress Pine (*Callitris intratropica*) forests and **riparian** forests (forests growing along rivers and streams) occur within savannas. There are other community types, such as those found on sandstone escarpments, including the highly adaptable yet fire sensitive *Allosyncarpia ternata* and low-lying shrub vegetation. These communities are very sensitive to fire. Even a low frequency of fire (once every five years) will destroy them. We now discuss how these forests are maintained in the landscape.

Fire sensitive forests are maintained in the landscape through one of two mechanisms. First, by growing in perennially wet locations such as along rivers, or in low-lying positions in the landscape where soil moisture is maintained at high levels for most or all the year through surface flow and lateral subsurface flow. Because of this year-long availability of soil water, there is a dense and closed tree canopy so the understorey is very sparse and grasses absent.

Consequently the fuel load on the ground is low and the litter remains moist all year. Humidity at ground level is high and little solar radiation reaches the ground, so any fuel rarely cures fully. Fire is effectively excluded from these sites and therefore the forest, although fire sensitive and growing in a fire prone region, survives. Occasionally fire does penetrate these forests and the effect is severe. Tree death is pronounced and a long fire free period (5–10 years) is required for regeneration from the seed bank to occur and for the seed bank to be replenished from newly mature trees.

Some fire sensitive species survive by growing on rocky outcrops and sandstone ridges (Russell-Smith et al. 2001). In these places, fire is excluded because of the absence of a continuous layer of fuel. Large patches of bare rock and tall rocky outcrops provide fire shelters for very sensitive species. In these locations surface run-off is a major component of the hydrological balance, canopy interception of water is minimal and water is stored deep in rock fissures. The rare, patchy and cool fires that do occur have a minimal impact on the hydrological balance of these catchments.

When fire sensitive rainforests in low-lying regions are burnt too frequently, regeneration of rainforest species is inhibited, the canopy remains open for a long time and gradually grasses and eucalypt woodland species become established. Consequently fuel loads increase, more of this fuel is cured better and the probability of fire increases. As fire frequency increases, more of the rainforest tree species are killed, causing a feed forward response. Thus, a downward spiral of loss of rainforest species occurs and the site becomes dominated by eucalypt dominated, fire resistant savanna woodland and forest. Conversely, on eucalypt dominated savanna sites that are protected from fires for 20 or more years, the canopy closes, tree density increases, grasses become excluded from the understorey and fuel load is reduced. Consequently fire frequency declines, rainforest species become established (seed can be blown in or deposited by birds and bats) and savanna is replaced by a closed forest when water availability is not limiting. Water use by the rainforest is larger than that of the eucalypt forest it replaced and consequently interception losses increase, surface flow is reduced and **catchment water yield** will decline.

Hydrological impacts of fire

Determining the effect of fire on catchment hydrology is difficult for many reasons. It is generally difficult to obtain a data series that is of sufficient duration (decades) to provide statistically robust analyses. Studies of paired catchments, although providing a powerful approach, cannot overcome the temporal and spatial variability of rainfall, vegetation cover and slope, which are major sources of variation in stream flow and surface run-off (Liu et al. 2004). Analyses of stream flow duration curves need to take into account rainfall duration data since this, along with fire, can cause changes in flow duration curves. This is often difficult to achieve. Because of these problems, Liu et al. (2004) were unable to observe any effects of fire on stream flow in the Goulburn River in the Hunter Valley of Australia. Townsend and Douglas (2004) found few significant changes in catchment hydrology following fire in tropical Australia, a result they ascribe to the low intensity of the fire and the gently undulating, low slope terrain.

Despite the preceding discussion of the problems inherent in determining the effect of fire on catchment hydrology, some generally applicable results can be discerned.

For a low to moderate fire intensity, where the death of mature trees is usually minimal, the major impact on vegetation is removal of the understorey and some leaf scorching of the mid-storey, with little impact on the upper-storey canopy. This has several effects on catchment water balance. **Interception losses** are reduced when the leaf area index of a site declines, as occurs after leaf scorch and subsequent leaf drop. Consequently the amount of water lost from the site as surface flow often increases, and recharge to soil water stores can increase. This

Table 7.1 Effects of a bushfire on a small catchment in South Africa – significant increases in stream flow, storm flow, peak discharge and annual stream flow

	Pre-fire	Post-fire	% change
Weekly stream flow (mm)	9.9	14.1	+12
Storm flow (mm)	3.7	6.4	+62
Peak discharge (mm d^{-1})	6.6	22.5	+290
% change in total annual stream flow			+70

Source: Scott (1993)

water may be available to the remaining tree canopy or may increase the amount of groundwater recharge. Increased surface flows are associated with increased loss of soil and reduced stream quality as sediment and nutrient loads to the stream increase (discussed below). Increased evaporation from the soil surface also occurs, partly because of the increase in soil moisture resulting from increased rainfall reaching the ground and partly because of the increase in solar radiation and wind speed which occur because of the loss of the understorey and mid-storey leaf layers.

A **short-term** (months to a few years) increase in stream flow and storm flow following cool fires is often observed (Table 7.1), the result of four factors. First, the reduction in total leaf area index of the site causes reduced transpiration from the canopy; second, increased surface flow occurs because of reduced interception losses; third, increased subsurface flow may occur because of increased infiltration of water to the soil, but such increased infiltration is not universally seen because of the fourth factor – **soil hydrophobicity.** Fires can cause some soils to become **water repellent** (soils are hydrophobic) and therefore infiltration is reduced and run-off is increased (Scott 2004).

A thick layer of mulch tends to assist in maintaining good soil structure and high rates of infiltration. Fires remove this mulch and tend to increase surface flows of water. In addition, soils with a high clay or organic content can develop a hydrophobic characteristic, possibly through the activity of fungi in the soil. Fires remove the mulch layer, can kill surface fungi and can increase the occurrence of hydrophobic patches. Such patches may persist after a fire for many days or weeks so surface run-off of rainwater, especially during and after low intensity rain storms, can increase, causing short-term increases in stream flow and soil loss. The time required for the leaf area index of the site to return to pre-fire levels, plus the time for a layer of mulch to develop, determine the duration of the short-term response of stream flow to fire.

Longer-term changes in catchment hydrology occur following hot or very hot fires because of the larger impact on upper-storey **canopy cover.** If the fire is widespread, hot and in a mature old-growth forest, such as the Ash Wednesday fire in Victoria, natural regeneration occurs from the seed bank, forest regeneration can take 5–30 years and maturity is not reached for 100 or more years. Because of the ash bed effect (see above) and because of the very high initial density of saplings in the early stages of regrowth (discussed in detail below), the leaf area index of the site can increase to values much larger than those observed in the mature old-growth forest. Consequently interception losses and transpiration by vegetation are substantially increased. Surface and subsurface flows are reduced and water yield of the catchment declines for many years. This is similar to what happens when much of a catchment is extensively logged and regrowth occurs. Following the Black Friday fire of 1939 around Melbourne, water yields of Mountain Ash declined by up to 600 mm y^{-1} (6 ML ha^{-1}). An annual reduction of stream flow of about 25% occurred for many decades following the fire, and may take up to 120 years to recover.

The impact of fire on catchment hydrology is partially determined by the type of forest burnt. Fire sensitive species, such as Mountain Ash (*E. regnans*), Alpine Ash (*E. delagatensis*) and Shining Gum (*E. nitens*) are more likely to be killed by moderate and hot fires. Therefore the impact on catchment hydrology is more pronounced and longer lasting than in ecosystems dominated by fire resistant species (savanna species in general; dry sclerophyll forests), which are better able to survive such fires through a range of adaptations.

Impact of fire on stream water quality

Management of **water quality** is as important as water yield because quality determines the amount of treatment required to make the water acceptable for use.

The quality of water flowing from a catchment is influenced by several features, some of which change following fire (Townsend & Douglas 2000). Parent rock and hence soil type influence water quality but do not change following fire. However, fire does cause the loss of forest litter and replaces it with ash and bare soil which, along with increased run-off, cause decreased water quality. Run-off following fire contains increased amounts of sediment (soil), ash, nutrients and organic material (including charcoal) which appear in streams and rivers (Table 7.2). Large slopes, high rainfall intensity and severity and rains that quickly follow a fire increase the amount of low quality water entering streams and rivers. Early dry season fires in northern Australia have less impact on soil erosion and stream sediment load than late dry season fires because late season fires are hotter, cause more vegetation and litter loss and are soon followed by intense rains which wash soil and ash into streams before vegetation has had a chance to recover from the fire.

Low intensity fires throughout Australia can cause **leaf fall** immediately after the fire. The first rains after such a fire leach large quantities of organic material from these leaves, as well as moving ash, soil and leaf material with the surface flows into streams and rivers. This material is rapidly broken down by bacterial activity in the stream, thereby consuming large amounts of dissolved oxygen, causing **anoxia** in the water and associated **fish kills**. Anoxia in streams increases the amount of dissolved manganese and enhances the generation of reduced sulphur compounds. These factors reduce the potability of the water, which therefore requires more treatment. High concentrations of dissolved organic compounds in water require increased treatment with chlorine, which increases the potential for increased formation of **trihalomethanes**, which are toxic to humans. High nitrate concentrations are also toxic and may occur after fire.

High intensity fires throughout Australia burn most of the fuel load, including the upper canopy. Consequently much of the organic material in the fuel is burnt off (**volatilized**). Ash is the principal component left after the fire. The first rains after a high intensity fire carry soil and ash into the streams but less organic material than after a low intensity fire. **Eutrophication** (characterised by blue-green algal blooms) is a likely result because of the increased nutrient input to the receiving water (streams, rivers and lakes). Eutrophication is a major problem for water managers because of the toxicity of many blue-green algae and the enhanced **biological oxygen demand** (BOD) that can kill fish.

The hydrological impacts of fire differ between flowing streams/rivers and static lakes and reservoirs. Free flowing rivers and streams tend to be affected for a shorter period of time than reservoirs and lakes because of the **flow through effect**. In contrast, deep reservoirs (deeper than 10–20 m) are temperature and oxygen **stratified** (separated into layers). If the water entering the reservoir is cold and has a high dissolved organic carbon content, it will settle to the deeper parts of the reservoir and remain isolated from the oxygenated surface waters, posing a longer-term problem to water managers. If the water entering the reservoir is warm and nutrient rich, eutrophication of the upper layers is likely and an algal bloom may develop,

Table 7.2 A burnt catchment in Kakadu National Park (NT, Australia) with significantly reduced canopy cover, reduced riparian tree density, increased stream sediment loads and increased manganese content in the stream

	Unburnt catchment	Late burnt catchment
Canopy cover (%)	75	40
Riparian tree density (tree ha^{-1})	3000	600
Suspended sediment (kg ha^{-1})	40	80
Stream manganese content (g ha^{-1})	30	58

Source: Townsend & Douglas (2000)

with major impacts for water managers. The depth at which water is removed from deep reservoirs is adjusted to maximise the quality of water extracted at any one time, but the degree to which this can occur decreases as the depth of water in the reservoir declines.

Large increases in peak flows following fires (because of reduced canopy cover, increased surface flows and reduced vegetation water use) have significant in-stream effects apart from changes in water quality. Increased peak flows may occur at times normally associated with low rates of flow (usually driven by low **base flows;** flows arising from groundwater input), such as in the dry season in northern Australia. Large increases in peak flows can also cause increased **siltation** at some points and increased erosion of the stream bank and stream bed at other locations. Flooding of surrounding vegetation, including riparian and wetlands, is often observed. This can be beneficial: River Red Gums of southern Australia require occasional flooding for regeneration and wetlands require freshwater inflows to sustain them; see Chapter 6).

Conclusion

Fire has an impact on the structure and function of vegetation and hence on the hydrology of Australian landscapes. Australian plants show a range of adaptations to cope with recurrent fire. Decreased canopy cover, increased soil and ash movement into rivers, increased run-off and reduced vegetation water use all contribute to the changes in catchment hydrology occurring after fire.

The consideration of fire and its impacts on ecohydrology serves as a useful conceptual primer for the next section, which discusses the management of forests of Mountain Ash around Melbourne in relation to maintaining the city's water supply. These forests were extensively and severely burnt in 1939, which spurred much of the research that tried to link forest age, forest structure and catchment water yield.

Case study 2: Mountain Ash forests control a city's water supply[1]

Background and issues

Many cities of the world, including Melbourne, receive their water from surrounding **forested catchments**. The catchments in the Central Highlands north and east of Melbourne cover approximately 155 000 ha, much of the area dominated by **Mountain Ash** (*Eucalyptus regnans*). Mountain Ash grows at altitudes of 200–1000 m above sea level. Annual rainfall for much of the forested region is 1200–1800 mm and because the forests do not use all this rainfall the yield of water from the catchments into stream flow is high (Vertessy et al. 1994). Up to 80% of

1 The Mountain Ash case study derives much of its structure and content from Vertessy et al. (1993, 1995, 1998). For ease of reading *Ecohydrology* is not heavily referenced but the major contribution of these reports is acknowledged.

total stream flow through these catchments is derived from the Mountain Ash forests and the water is of high quality with little treatment required to make it suitable for human consumption. Therefore the provision of potable water to Melbourne is relatively cheap.

The relationships between **catchment management** and **water yield** was poorly known until the 1990s, although it became apparent that forest age and water yield were linked after a massive fire killed almost 80% of the old-growth Ash forest (Black Friday).

Seedlings of Mountain Ash are shade intolerant and therefore do not establish well beneath a closed canopy mature forest. When the canopy is killed by fire, seed is released from woody capsules and a dense regrowth of hundreds of seedlings per square metre compete for light and water. Density dependent competition occurs fiercely for the first 10–50 years with the number of seedlings declining from hundreds of thousands per hectare at germination to 10 000–20 000 per hectare at 5 years, to 200–300 per hectare after 50 years. In 2004, the 1939 regrowth cohort still dominates the forest and is 65 years old. Approximately 140 years after the fire (in 2079) tree density will be almost at equilibrium, with about 100 trees per hectare, although further slight reductions will occur over the subsequent 100–200 years. As the forest ages, from about year 15 to 20, an understorey of shrubs, ferns and medium sized trees (e.g. *Acacia*) will develop.

The principal aim of management of the catchments around Melbourne is to ensure that a sufficient quantity of water, of sufficient quality, is available all year to meet all needs. However, there is considerable pressure to allow tourism and logging interests to influence management practices. In order to manage sustainably and effectively, it is important to have a **mechanistic understanding of the ecohydrology** of these catchments and to understand how different management options will influence catchment water yield.

The following principles have been established in forest hydrology since the mid 20th century:

- an increase in the amount of forest in a catchment generally decreases the water yield of the catchment (yield defined as the amount of water available for human consumptive use, assuming that the forest first takes all that is required to maintain its own health);
- there is a proportional relationship between the amount of forest added or removed from a catchment and the change in water yield of a catchment;
- as mean annual rainfall increases, the net increase in catchment yield resulting from a decline in forested area within catchment increases (i.e. there is little point in changing the amount of forest in very low rainfall areas because the absolute benefit in yield will be very small). Thus as rainfall increases, run-off increases (Fig. 7.3a);
- forests use more water (per hectare of land) than grassland;
- as annual rainfall increases, annual tree water use increases (Fig.7.3b).

Such generalisations, although useful, have limited value when managing specific catchments at specific locations in specific years. The following describes the results of specific studies undertaken in the Mountain Ash forests around Melbourne to allow a scientifically sound basis for forest management.

Aims of the study

The case study aims to address the following questions:

1. how catchment water yield varies with forest age;
2. how forest water use varies with climate;
3. the impact of clear-felling and selective logging on catchment yield;
4. the relationships among rainfall, tree size, micro-climate, tree water use and catchment water yield;

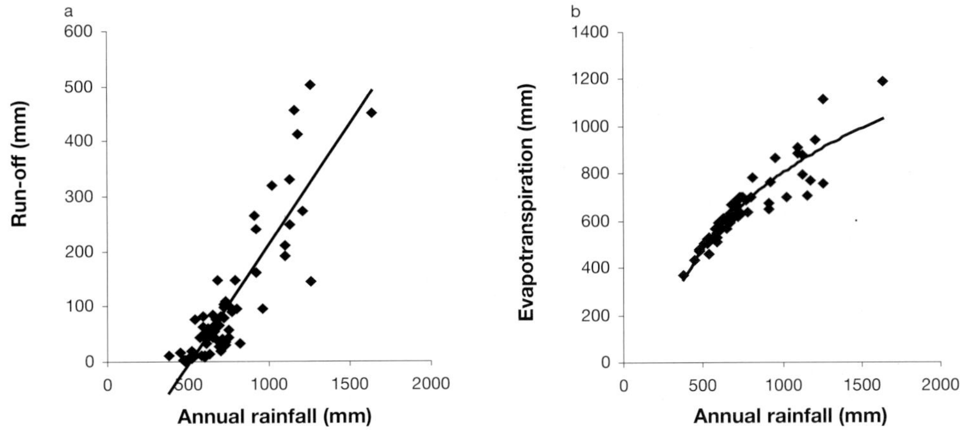

Figure 7.3 As annual rainfall increases in a range of Australian catchments, the amount of (a) run-off and (b) evapotranspiration from the catchment increases

Source: Zhang et al. (1999)

5 how long it takes for a burnt mature forest to recover following fire, in terms of catchment water yield;
6 whether it is possible to generate an empirical model linking water yield and forest ecology, at both a small (1 km^2) and large (100–1000 km^2) scale.

To address these six questions, field measurements, in conjunction with modelling, were used to develop small and large scale empirical models of relationships between water yield and catchment disturbance (Cornish & Vertessy 2001; see Watson et al. 2001 for a review). In particular, an annual **catchment water balance** for forested catchments around Melbourne was developed, and the causal links among tree density, canopy cover and principal components of the water budget (run-off, transpiration, infiltration, interception losses) were investigated.

A catchment water budget can be most simply stated as:

Annual rainfall = loss from the catchment + change in catchment storage of water

More specifically:

Annual rainfall = (transpiration from vegetation + interception losses + surface run-off + lateral flow to streams) + (change in soil and vegetation water content + deep percolation to groundwater reserves)

Changes in soil and vegetation water storage were assumed to be zero (as is usually the case over an annual time frame) and ignored. Therefore the equation simplifies to:

Annual rainfall = transpiration from vegetation + interception losses + surface run-off + lateral flow to streams + deep percolation to groundwater

To establish the **catchment water budget**, each element in this equation had to be measured, estimated or modelled.

Correlation of forest age and catchment water yield

Forest age and catchment water yield show a significant non-linear relationship (Haydon et al. 1996). This curve, idealised as the **Kuczera curve** (Fig. 7.4) shows that mean annual stream

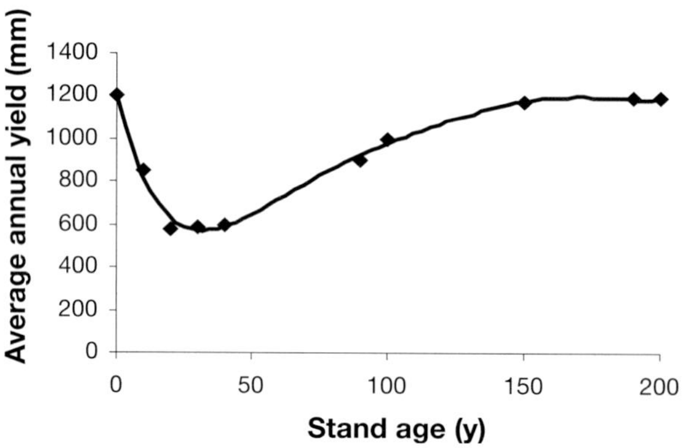

Figure 7.4 An idealised relationship between stand age and mean annual catchment run-off from Mountain Ash forest north of Melbourne

Source: Redrawn from Kuczera (1985), cited in Vertessy et al. (1998)

flow through a catchment declines substantially in the first 20–30 years following a major fire or other disturbance (such as clear-felling). Water yields increase for the next 100–150 years and reach a more or less stable value. This curve was synthesised using data from several large forested catchments in Victoria and is a hypothetical response. Two principal problems are associated with the curve. First, the error associated with the curve is substantial (up to $\pm$30%) and largest for the period 30–80 years of age. Consequently, accurate predictions of actual water yield are not possible for a large fraction of the forest life time. Second, the hypothetical curve is a composite, derived from patchy data from multiple and differing catchments and therefore any understanding about variation between catchments has been lost. Because of these limits, new experiments were undertaken in the 1990s, and are discussed below.

Developing a Mountain Ash catchment water budget

Methods

Sixty kilometres north of Melbourne lie the Maroondah water supply catchments. Seventeen experimental catchments were identified to address the six questions listed above, using a combination of the following techniques:

- sapflow measurements of tree water use;
- lysimeter estimates of soil and litter evaporation;
- rainfall interception measured with rainfall gauges;
- saturated soil hydraulic conductivity measured with a constant head borehole permeameter;
- testing and parameterising a small-scale catchment model (**TOPOG**) which simulates interactions among soil water, groundwater, canopy micro-climate plant growth and catchment topography;
- developing a large-scale model (**Macaque**) to simulate large-scale water balances;
- applying **GIS** to map large catchments.

Self thinning

Mountain Ash stands **self thin** very rapidly in the first decade, but the entire thinning process can take almost 100 years. Figure 7.5 shows a typical pattern of **tree density** as a function of **stand age** in Mountain Ash forests of Victoria (Vertessy et al. 1998). It is not clear from this relationship exactly how tree and stand water use vary with stand age (or density) since it is

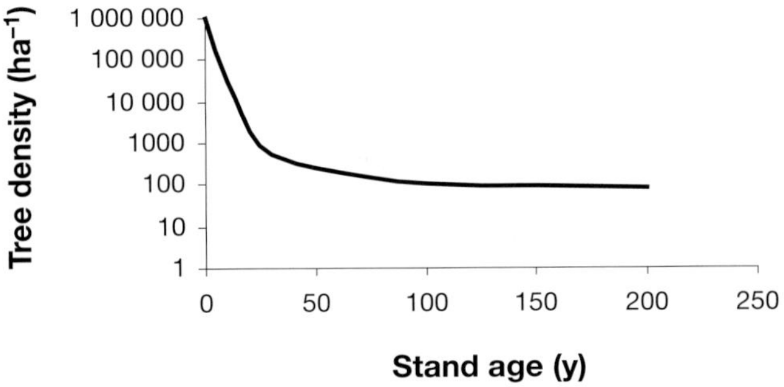

Figure 7.5 Relationship between stand age and tree density for Mountain Ash forest north of Melbourne

Source: Redrawn from Ashton (1976), cited in Vertessy et al. (1998)

leaf area and sapwood area, in conjunction with climate, that determine tree and stand water use. Therefore we need to know how leaf area and sapwood area vary with stand age and tree density if we want to predict how water yield will vary over time.

Determination of leaf area index and sapwood area

Leaf area index (LAI; the ratio of leaf area to ground area) is difficult to measure (because of spatial and temporal variability and the height of the canopy) but knowledge of LAI is vital to any mechanistic understanding of forest water use and hydrology. A low LAI (0.05 to 0.2) is found in arid zones and means that the canopy is very open and usually low-lying. Very little leaf area is present to intercept radiation and transpire. A high LAI (5–8) is found in humid tropical zones with high rainfall and tall canopies (rainforests). Eucalypt woodlands and forests generally have an LAI of 1–4. The larger the LAI, the more interception of rainfall occurs and the more water transpired in total by the vegetation. In addition, a high LAI causes shading of the ground and understorey and thus reduced evapotranspiration from soil, litter and understorey. An understanding of how LAI varies with stand age contributes significantly to the development of a mechanistic model of forest water use (Vertessy et al. 2001).

From knowledge of the diameter of all trees in a subplot of a forest and a regression of leaf area against tree diameter, it is possible to calculate the LAI of a forest. The LAI of Mountain Ash forests peaks at around 15 years of age (Fig. 7.6) and declines gradually for the next 200+ years. This trend is similar to an inversion of the Kuczera curve and clearly helps explain at

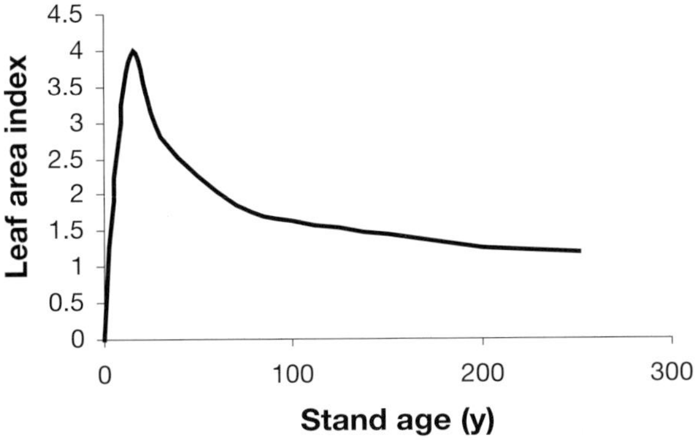

Figure 7.6 Relationship between stand age and LAI of Mountain Ash forest north of Melbourne

Source: Redrawn from Watson & Vertessy (1996), cited in Vertessy et al. (1998)

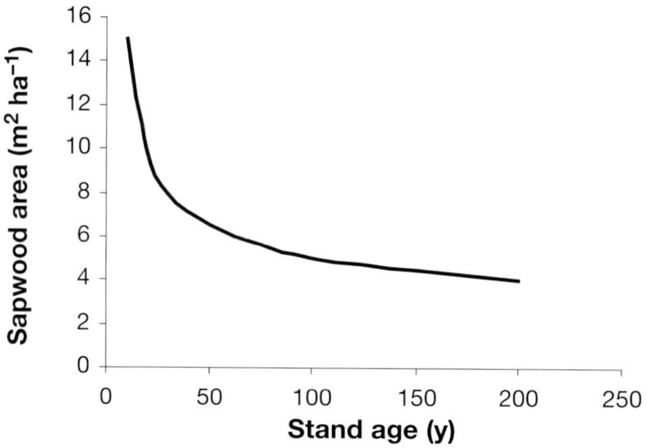

Figure 7.8 As a Mountain Ash forest ages, the sapwood area per hectare of forest declines

Source: Vertessy et al. (1998)

least some of the changes in catchment water yield with age. As the LAI increases to a peak, water use by the vegetation increases and therefore water yield declines. As LAI declines thereafter, water use declines and water yield increases (Vertessy et al. 2001). However, this is not the full story.

Sap velocity, sapwood area and tree water use

The **velocity of sapflow** up Mountain Ash (and other species, see Roberts et al. 2001) stems does not vary with tree age. It ranges from about 12 cm h^{-1} in the summer to about 6 cm h^{-1} in winter. Because these velocities do not change with stand age but sapwood area per hectare of forest does vary with age, it is possible to predict the rate of forest water use as a function of stand age. Clearly we need to know the rate of water use by a forest as a function of forest age to generate a catchment water budget for different age forests.

Sapwood area of trees is determined by **coring** a large number of trees. The inner and outer boundary of sapwood is generally easily seen by a distinct colour change at the cambium–sapwood boundary and the sapwood–heartwood boundary (Fig. 7.7, Colour Plate 7). Knowing the inner and outer diameter of the sapwood allows calculation of the area of sapwood for each individual tree. From dozens of cores in as many trees, a relationship between sapwood area and DBH and stand age can be established. Sapwood area declines with stand age, from about 15 m^2 ha^{-1} for a very young forest to about 4 m^2 ha^{-1} for a 200 year old forest (Haydon et al. 1996; Fig. 7.8). Since sap velocity is constant with age, the almost 4 fold decline in sapwood area is associated with an almost 4 fold decline in the rate of water use by the forest (Vertessy et al. 1995).

It is clear that both LAI and sapwood area decline as the age of the stand increases past about 15–25 years of age. Both contribute to the gradual increase in water yield observed in the

Table 7.3 Relationship between increasing stand age, declining sapwood area per hectare and declining mean annual stand water use

Stand age (years):	**14**	**45**	**160**
Sapwood area (m^2 ha^{-1})	11	6.5	3.1
LAI	3.6	4.0	3.4
Stand water use (mm d^{-1})	2.2	1.4	0.88

Source: Data from Roberts et al. (2001)

Kuczera curve from about age 25 years. Similar results have been observed elsewhere (Roberts et al. 2001), as shown in Table 7.3.

Understorey transpiration, litter and soil evaporation

Water is transpired by trees and the understorey in a forest. In addition, water is evaporated from wet soil and litter following rain. Estimations of understorey water use can be made using **lysimeters, open top chambers** and **sapflow sensors** (Chapter 4). The LAI of the understorey in a Mountain Ash forest was found to increase from about 0.1 in a 5 year old forest to about 3 in a 300 year old forest. As the forest ages, the understorey represents an increasingly significant pathway for the discharge of water from the catchment.

Water use by understorey plants tends to be less than that of overstorey tree canopy. This is because of the higher humidity (lower vapour pressure deficit), the lower temperatures and lower levels of solar radiation and wind speeds in the understorey. In Mountain Ash forests, understorey water use is approximately 50–80% of overstorey water use, but on cloudy or cool still days understorey water use was as low as 10% of overstorey water use.

Soil and litter evaporation can be measured using micro-lysimeters and open top or closed top chambers (Chapter 4). Although soil and litter evaporation tend to be larger in young stands, soil and litter evaporation account for about 10% of total annual evapotranspiration of the Mountain Ash forest (Vertessy et al. 1998).

Interception losses

Interception of rainfall by vegetation and litter and subsequent evaporation to the atmosphere represents an unproductive loss of water from the catchment. Interception losses vary between ecosystems (Fig. 7.9) and age (Haydon et al. 1996) because of differences in structure of the ecosystem (LAI and branch angle) and differences in rainfall characteristics (e.g. lots of short intensity light rainfall, compared to long duration heavy rainfall). It is possible to estimate canopy interception losses by measuring the amount of rainfall above a canopy and subtracting the amount of rainfall that reaches the ground beneath the canopy. It is also important to measure the amount of water flowing down the trunks of trees (**stem flow**) and the amount dripping from the canopy after the rain has stopped (**throughfall**) to accurately estimate

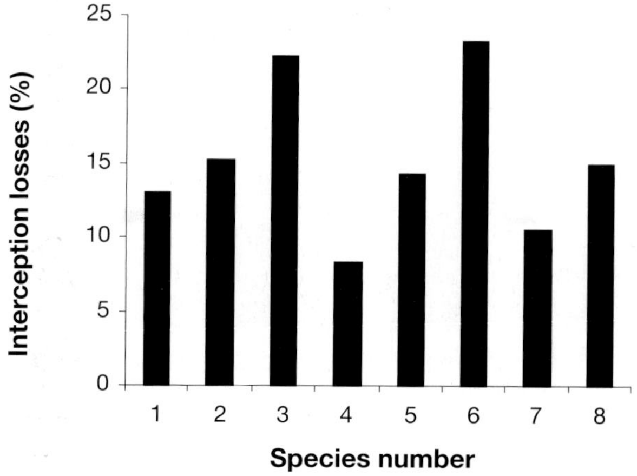

Figure 7.9 Species and ecosystems differ in the amount of rainfall lost as interception. Species 1 = *Acacia aneura*, 2 = *A. harpophyla*, 3 = *Agropyron koeleria*, 4 = *Eucalyptus* spp., 5 = *E. camaldulensis*, 6 = *E. regnans*, 7 = *E. rossii*, 8 = *E. belangeri*

Source: Data from Zhang et al. (1999)

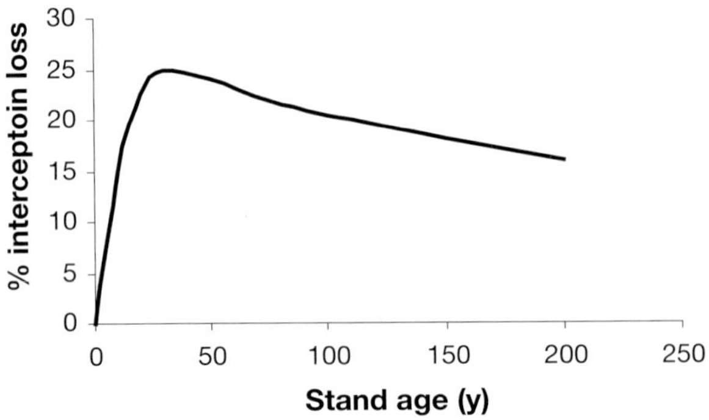

Figure 7.10 Canopy interception losses increase rapidly in the early years of forest establishment but decline once the forest reaches about 30 years of age

Source: Redrawn from Vertessy et al. (1998)

canopy interception. Interception losses can vary from 0% (no canopy present) to about 25%, and vary with stand age (Fig. 7.10). Presumably the decline in interception loss with age contributes to the increase in catchment water yield with age.

Preliminary water balance for a Mountain Ash forest

Annual overstorey water use was highest in young stands and declined with stand age, from about 750 mm y^{-1} at 15 years to about 250 mm y^{-1} at 240 years (Vertessy et al. 1998) As overstorey water use declined understorey water use increased, from about 50 mm y^{-1} to about 300 mm y^{-1}. Because of this imbalance, and changes in interception (which peaked at about 30 years of age) run-off increased from about 500 mm y^{-1} to about 900 mm y^{-1} over the same age span (Fig. 7.11).

Small-scale catchment modelling

A small-scale **catchment model** called **TOPOG** (Vertessy et al. 1993; Chapter 5) was used to simulate interactions among rainfall, soil water, groundwater, micro-climate, vegetation

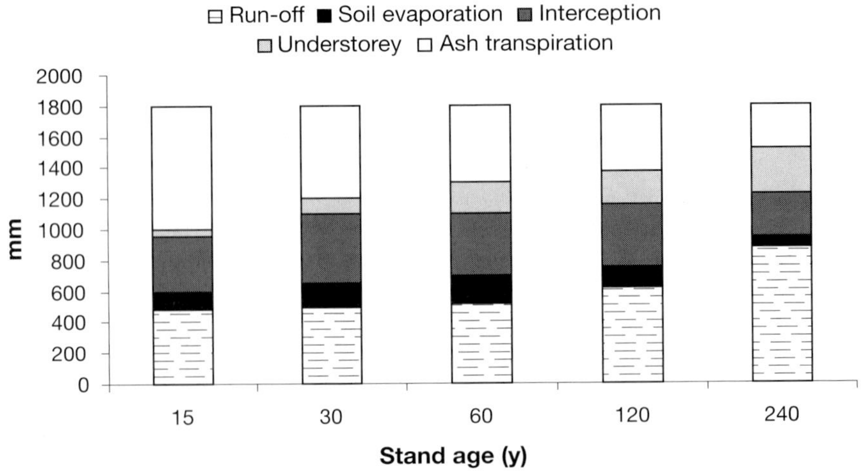

Figure 7.11 The proportion of run-off, tree transpiration, understorey transpiration and soil evaporation changes as a stand of Mountain Ash ages

Source: Data from Vertessy et al. (1998)

growth and catchment topography (e.g. slope). Inputs to the model include daily radiation, rainfall, temperature and atmospheric humidity and the model runs on a daily time step. From these inputs, plus information about vegetation structure, the model calculates transpiration for both overstorey and understorey vegetation, plus surface evaporation, soil moisture content and stream flow. Vegetation growth (above and below ground biomass) is also calculated.

A small catchment (the Myrtle 2 catchment) of 0.32 km^2 was chosen to test TOPOG. A 12 year period for the catchment was modelled, during which time the catchment was covered by old growth forest. A regression of predicted and observed monthly run-off yielded a slope of 1.06 with an intercept of 0.1; that is, the model and observations showed very close agreement and the model explained 87% of variation in monthly stream flow throughout the 12 year period. By the end of the 12 year period, modelled run-off was within 5% of actual run-off.

The ultimate aim of the TOPOG model was to be able to predict what happened to stream flow after logging or fire caused massive loss of canopy. Consequently, to test the model's ability to cope with disturbance to a Mountain Ash forest, the Picaninny catchment was chosen for further study and modelling. This catchment covers an area of 0.52 km^2 and was chosen because data were available for a pre-logging and post-logging period. A 3 year pre-logging period was modelled, followed by a 20 year post-logging period when vegetation dynamics (germination, regrowth, competition and thinning) were substantial.

The model was able to predict throughfall to within 1% of observed values for an 18 year period. Soil moisture fluctuations over a 4 year period were very close to observed values and modelled leaf and stem biomass accumulation were close to observed values, for both overstorey and understorey. Finally, the model accounted for 76% of variation in monthly stream flows.

Large-scale modelling

TOPOG is not able to deal with large catchments (100–1000 km^2) and a new model, **Macaque**, was developed (Vertessy et al. 1998). Although many similarities with TOPOG exist, a major difference is the simplification of the soil moisture subroutine. Daily inputs of rainfall, solar radiation, temperature and humidity are required and the large catchment is divided into a number of hillslopes, each of which is treated individually. Because of the larger scale of application of Macaque, with concomitant increased variation in soil properties, slope and elevation (which influences surface run-off characteristics, temperature, humidity and rainfall), and variation in forest structure and age, it was necessary to use a GIS (geographic information system) which allows representation of such variation across large spatial scales.

The Macaque model was applied to the Maroondah catchments to estimate daily run-off over an 82 year period (1911–92). This period included several years before the massive 1939 fire which removed most of the old growth forest on the catchment and many years after the fire when regeneration was occurring.

The model is able to predict run-off at daily, weekly and yearly time frames and accounted for 66% of the variance in observed yearly run-off. Poor resolution of daily rainfall input data limited the accuracy of the model outputs.

Predicted annual run-off for three hillslopes within the Maroondah catchments are shown in Figure 7.12. It is clear that the response curve closely resembles the Kuczera curve.

Conclusions

The Mountain Ash case study employed a wide range of ecohydrological techniques to determine the behaviour of small and large forested catchments. By applying knowledge and techniques from several disciplines, including plant ecophysiology, hydrology, forestry and micro-meteorology, it was possible to accurately simulate the changes in forest growth, water use and catchment water yield (stream flow) on short (weekly) and long (yearly) time frames.

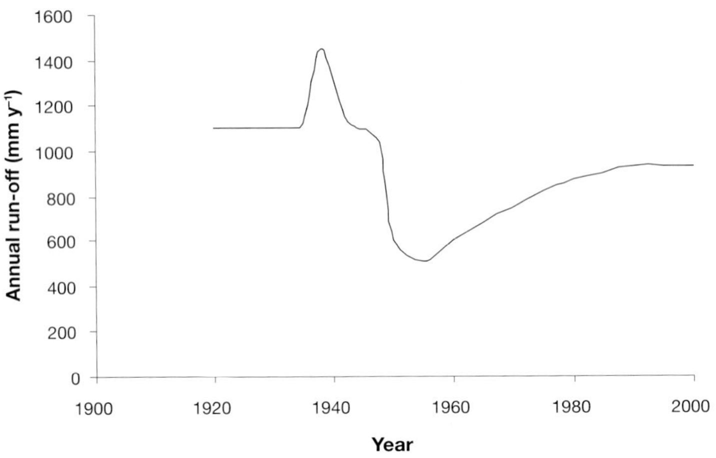

Figure 7.12 Predicted annual run-off before and after a wildfire in a Mountain Ash forest

Source: Redrawn from Vertessy et al. (1998).

Such information is being used to aid sustainable management of those forested catchments on a sound scientific basis.

Case study 3: mining and ecohydrology

Background and issues

Mining occupies a pre-eminent place in the Australian economy, returning large economic dividends to the nation from relatively small (at a regional scale, although large at a local scale, in some cases) demands on water resources, particularly in comparison to the agricultural sector. However, impacts of mining on local hydrology can be high, and often involve changes in water quantity and quality.

Mining activities can result in changed ecohydrological conditions because of the need to develop a large and reliable water supply for the processing of ore and because mines often sink lower than the upper level of the water table and therefore have a tendency to flood. Activities associated with mining that impact on local ecohydrology include:

- development of mine water supply;
- clearing of land;
- mine dewatering;
- storage of mine tailings;
- dealing with mining voids when mining ceases.

Much of Australia's known mineral resource is in the arid or semi-arid zone where water supplies can be limited and from sources with only limited recharge and high ecosystem dependence. Many mining operations make use of 'fit-for-purpose' water resources, recognising that some parts of their operations can utilise relatively low quality water (e.g. saline groundwater). Abstraction of fresh water from local or remote sources is generally accompanied by some level of trade-off with the environment.

The initial development of a mine site can involve the removal of native vegetation, which in turn can change the local water and salt balance. This is usually on a highly localised scale and thus of little impact to regional ecohydrology. However, some mineral resources are sufficiently extensive in nature (coal, bauxite) and their mining on a large enough spatial scale to significantly change land cover, and thus evapotranspiration, over large areas. This can translate into increased stream flow and increased salt loads.

Ore is often extracted from below a regional water table, and local **dewatering** is necessary to maintain operations through to mine closure. Western Australia alone has about 150 mines operating below the water table. Dewatering an aquifer in the vicinity of a mine can translate to reduced or redirected flows of groundwater with respect to local groundwater dependent ecosystems. Disposal of the water into the environment over the period of dewatering can also alter the **hydroperiod** of downstream ecosystems. In some cases these changes will adversely affect ecosystems through oversupply of water, or cause the development of dependency that will cause vegetation to suffer following mine closure.

Ore processing usually involves either caustic or acidic agents and the liquid waste that is generated is normally sequestered in **tailings dams** or ponds. There is a fairly extensive history of leakage or overflowing associated with such structures, with resulting downstream contamination with toxic metals, organics, acid and salinity. Solid waste material is stored in large piles of tailings from which **acid mine drainage** is a recurrent problem. Acid mine drainage (AMD) is a worldwide problem and can occur at coal and metalliferous mines. It has detrimental impacts on groundwater, surface water and soil quality, instream flora and fauna and terrestrial flora and fauna (especially microbial populations). The biological and chemical oxidation of pyrite (FeS_2) or other sulphide minerals, in the presence of water, results in sulphuric acid formation. This is leached from tailings piles over a period of years to decades. Streams that receive significant quantities of AMD become more acidic, often contain elevated concentrations of iron and other metals and have reduced biodiversity and abundance. Increased precipitates of iron reduce the visual depth of the water and can block gills and feeding of stream fauna. One tool for minimising acid mine drainage is to put store-release covers on the storage pits. Such covers consist of soil, clay and rubble with a vegetation cover that, it is hoped, will discharge water through transpiration, thereby preventing water reaching the solid waste and preventing acid solutions leaching out.

Disposal of tailings water, which is usually contaminated with moderate to high levels of heavy metals and is generally acidic, is similarly problematic. Disposal into rivers and streams during periods of high flow (to maximise the dilution effect) and spraying onto large areas of native vegetation has been used in the past, although evaporation is the most frequently used method of disposal.

At the end of mining operations mine pits often fill with surface or groundwater, which in turn acidifies, becomes saline or contaminated with heavy metals. The subsequent flow of contaminated groundwater into downstream environments poses an ecological hazard similar to that of AMD. Western Australia has about 1000 mine voids with the potential to become pit lakes. These may have ecological impacts if they intersect aquifers that can transmit contaminated or salinised water through the landscape.

Olympic Dam Mine

Mining operations at the Olympic Dam site have impacted on the mound springs of the Great Artesian Basin close to the mine. This case study is presented as a short narrative that summarises work undertaken over several decades. Although not dominated by woody vegetation (the principal focus of this book), mound springs have significant cultural, biological and conservation value. The case study exemplifies how human activities can have unforeseen consequences.

The mound springs of the Great Artesian Basin were described earlier in this book. Of the many pressures on the groundwater resources of the GAB, few provide so large a financial return per unit water as Western Mining Corporation's Olympic Dam copper-uranium mine in South Australia. The plant began commercial production in 1988, developing the world's largest known uranium ore body. In 1998, the mine produced 85 000 tonnes of copper, 1600 tonnes of uranium oxide, 13 tonnes of silver and 850 kg of gold. Expanded annual production rates were approved to levels of 200 000 tonnes of copper (eventually increasing to 350 000 tonnes), and the mine is expected to operate for at least another 50–100 years.

A great challenge to developing this arid land resource is ensuring a consistent and adequate water supply. The water supply for the mine and associated township relies on two remote well fields (A and B) on the southern edge of the GAB near Lake Eyre, 80–200 km distant. These bore fields are in the proximity of the Lake Eyre supergroup of mound springs (Fig. 7.13), and from the beginning some impact from mining was expected.

Water demand associated with the initial level of production averaged 15 ML d^{-1}, which increased to 34 ML d^{-1} with increased production levels. The anticipated production levels may require 42 ML d^{-1}, with 36 ML d^{-1} of this coming from well field B. Table 7.4 summarises the regional water balance implications of this development at a time when Olympic Dam Mine was producing 85 000 tonnes of copper per annum (Power 2000). Natural discharge,

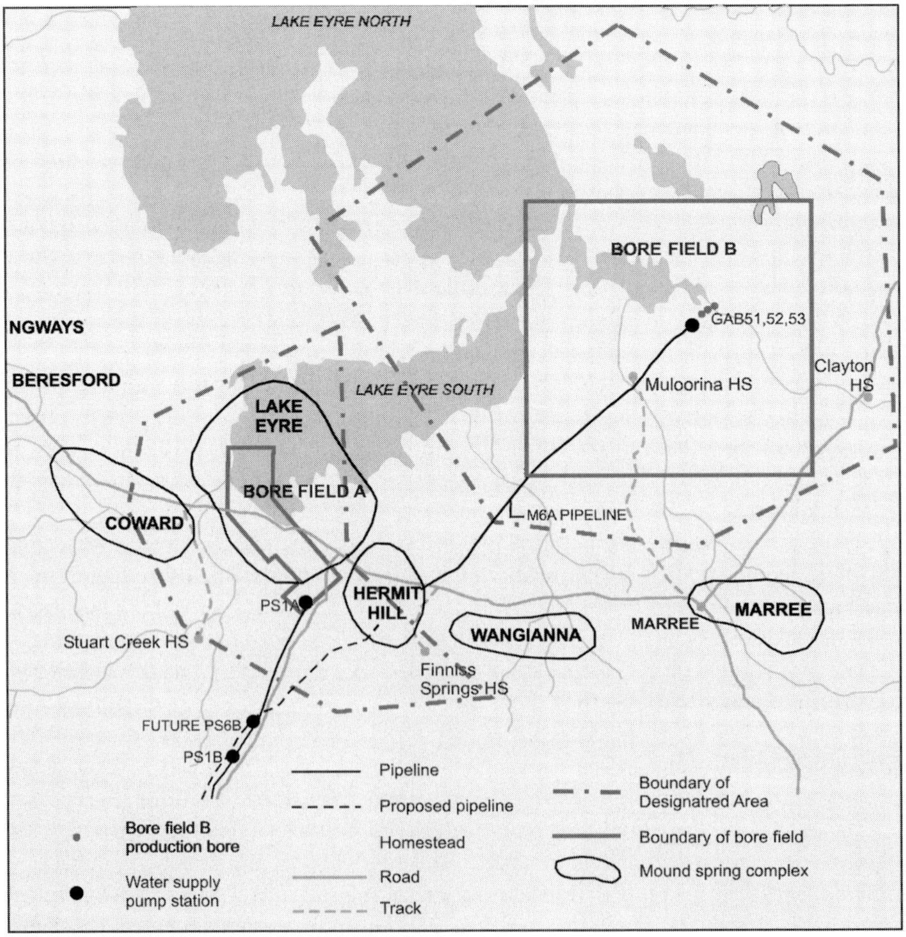

Figure 7.13 Location of well fields A and B and nearby mound springs of the Lake Eyre supergroup, supplying the Olympic Dam Mine and Roxby Downs townsite to the south (off map)

Source: Kinhill (1997)

Table 7.4 Water balance for the South Australian sector of the Great Artesian Basin

Water balance components	ML d^{-1}
Recharge	425
Inflow into South Australia	
Discharge	
Pastoral use	132
Flow associated with gas and oil production	22
Olympic Dam Mine (1996)	15
Mound springs	66
Vertical leakage	190

either through springs or as diffuse leakage, still dominates the discharge. Water abstraction for the mine and associated town is about 10% of that used by pastoralists, but is highly localised. The water contains approximately 2000 mg L^{-1} of total dissolved solids (mildly saline). Part of this supply is desalinated for potable water. While the total inflow to South Australia is 425 ML d^{-1}, the inflow into the Lake Eyre region is only 76 ML d^{-1} (Berry & Armstrong 1995), and therefore the 15 ML d^{-1} abstraction in 1996, and the projected abstraction of 42 ML d^{-1}, becomes a large fraction of the total flow.

Average extraction from well field A reached 15 ML d^{-1} before well field B was commissioned. Within a few years of pumping it was clear that the impact on the mound springs had been underestimated. By 1990, the flows at the Venebles and Priscilla springs had ceased, and there were reductions in flow at Hermit Hill, Beatrice and Bopeechee springs. These impacts generally exceeded those predicted when water allocations were initially made (Table 7.5), and it is widely conceded that the estimates were based on very limited data. With the availability of water from well field B, extraction from well field A was reduced to 6 ML d^{-1}. Since that time, pressures in the latter region have recovered but there has been a slower recovery of mound spring flows (Power 2000).

The groundwater resource is managed by a criterion of a general default pressure fall of not more than 5 m at an agreed boundary around well fields A and B; for a particularly sensitive portion of the boundary of well field A near one of the impacted mound springs (Hermit Hill complex) the maximum permissible pressure drop is 2.2 m. Other existing users of groundwater in these designated areas (e.g. pastoralists) maintain their right to use the same resource for proper management of existing developments; where the drawdown of groundwater due to mining development resulted in reduced flows for these other industries, the mine operator has provided piped distribution systems to maintain stock water supply. There is some evidence that continuing abstraction at present levels in well field B is inhibiting recovery of pressures in well field A (Mudd 1998).

Clear ecological impacts on mound spring ecosystems have resulted from abstraction of water for mining, and there is concern in many quarters regarding future impacts (summarised in Keane 1997). In such isolated ecosystems with complete dependency on groundwater flow, cessation of flow leads to local removal of endemics (Table 7.5). Restoration of flow cannot necessarily restore this fauna.

Conclusions

Mechanisms to prevent further collapse of mound springs in the region have been identified. The first is to increase water efficiency at the mine (water used per unit tonne of ore processed). In recent years this has been increased by 25%. The second is the completion of the bore capping program for any remaining free flowing pastoral bores and the piping and control of stock water (this conserves the resource for both environment and industry). The third

Table 7.5 Mound spring flow reductions resulting from groundwater abstraction in well field A (increasing from 1 ML d^{-1} in 1983 to 13 ML d^{-1} in 1995) for mining as predicted in 1984 and as recorded in 1995, with observations on faunal impacts at mound springs.

Spring complex	Spring group	Flow reduction as predicted in 1984 (%)	Actual flow reduction in 1995 (%)	Observed change in endemic fauna
Hermit Hill				
	Venable (bore)	100	100	Total loss of endemic fauna
	Bopeechee	20–30	43	Decline in all endemic fauna
	Bopeechee (bore)	80	80	
	Dead Boy	10–17	no record	
	Sulphuric	8–15	insufficient data	Decline in all endemic fauna
	West Finniss	10–13	insufficient data	Decline in all endemic fauna
	West Finniss (bore)	no data	20	Decline in all endemic fauna
	Hermit Springs	5–33	insufficient data	
	Hermit (bore)	no data	36	
	Old Woman	<3	insufficient data	Decline in all endemic fauna
	Old Finniss	<2	insufficient data	Decline in all endemic fauna
Lake Eyre				
	Goose	5–25	insufficient data	
	Priscilla	60–75	100	
Wangiana				
	Davenport	1–10	~0, one record	Decline in all endemic fauna
Coward				
	Blanch Cup	no data	no data	Decline in all endemic fauna
				Ostracod present; no trend data
				Total loss of endemic fauna
				Decline in all endemic fauna

Source: Olympic Dam Operations Expansion Project EIS (1997) as summarised by Keane (1997) option involves redesign and relocation of bores or well fields to locations more removed from mound springs. Finally, local groundwater pressures or flow regimes in the vicinity of mound springs at risk can be artificially maintained until longer term solutions arise. This strategy has already been attempted at springs in the Hermit Springs complex (Venables, Beatrice) although success is uncertain.

The key issues arising from this case study are the nature of trade-offs between groundwater dependent ecosystems in the arid zone and water-dependent industry, and the need for monitoring and adaptive management in the face of poor and limited data for initial planning.

Case study 4: three wetlands

Background and issues

The development of water resources (rivers, lakes, groundwater) is a major cause of loss of wetland habitat. Diversion of flow by dams, extraction of river water for consumptive use, and increased drainage of low lying areas all have the potential to alter local hydrologic regimes and hence the health of wetlands. The most famous international case is the diversion of water for irrigation from the Amu-Darya and Syr-Darya Rivers in Uzebekistan and Kazakhstan. These rivers feed the Aral Sea, which used to cover an area of the size of Ireland but now is only 25% of that area. Extraction of water has been so severe that the level of this inland sea has dropped by 27 m. The water is also 2.4 times more salty than seawater. Wetlands in the region have declined by 40% in area, with concomitant loss of habitat and biodiversity.

Aims

The aims of these wetland case studies are to provide a brief description of wetlands in Australia, to describe the functional impacts on altered hydrological regimes on wetlands, and to provide contrasting examples of wetland sites that have been influenced by altered hydrological impacts.

The Environmental Water Requirements of wetlands in Australia were recently reviewed by Davis et al. (2002) as part of the National River Health Program. We do not deal extensively with this aspect of wetlands; rather we provide a summary of the interaction between wetland function and hydrology of three wetlands.

A wetlands primer

What are wetlands?

The definition of 'wetland' is problematic. The **Ramsar Convention on Wetlands of International Importance** (1971) uses a broad definition which includes mangroves, peat bogs, paperbark swamps and marshland. Under the Convention on Wetlands wetlands are defined as:

> **Article 1.1**
>
> For the purpose of this Convention wetlands are areas of marsh, fen, peatland or water, whether natural or artificial, permanent or temporary, with water that is static or flowing, fresh, brackish or salt, including areas of marine water the depth of which at low tide does not exceed six metres.

Article 2.1 provides that wetlands:

> may incorporate riparian and coastal zones adjacent to the wetlands, and islands or bodies of marine water deeper than six metres at low tide lying within the wetlands.

These definitions include habitats that are periodically wet with free standing water for some of the year and habitats that are perennially wet with up to 6 m of free standing water. We accept this definition but wish to exclude marine, coastal or estuarine systems. For the purpose of this book, therefore, we define wetlands as a terrestrial ecosystem that periodically or permanently contains free standing fresh water. Thus we follow Wheeler (1999) in discriminating between **aquatic wetlands** which contain shallow or deep fresh water (lakes, streams) for all or most of their existence, and **telmatic** or **paludic wetlands** which are wet terrestrial systems. We acknowledge that this is not a wholly satisfactory distinction – for example, what is an ephemeral stream? However, it does allow us to ignore mangrove and instream components of river systems and to include marshes, swamps, bogs and fens.

Sources of water for wetlands

Wetlands are deemed to be wetlands when the substrate is saturated for some or all of the year. Keeping the substrate saturated requires an input of water at a rate equal to the rate of discharge of water through evapotranspiration, drainage and surface run-off. The sources of water include rain, groundwater, surface run-on and subsurface lateral flows. Groundwater fed wetlands (spring fed wetlands) generally remain wet all year because of the constant input from the spring. However, in many wetlands the height of the water table fluctuates markedly because of seasonal changes in rainfall, run-on and evapotranspiration. These fluctuations in water table can be sufficient to cause wetlands to periodically flood and dry out. Consequently three classes of wetland can be identified.

1 Permanent wetlands have vegetation that is relatively stable and perennial species dominate. Fluctuations in water table depth are relatively small.
2 Seasonal wetlands come and go according to season. Species are often ephemeral, although in *Melaleuca* swamps, trees (perennial) dominate.
3 Fluctuating wetlands are characterised by long term fluctuations in the water table which cause cyclic changes in vegetation, according to whether the water table is rising or falling.

Plant survival in wetlands

Plants living in wetlands must survive periods of time when their roots, at least, are inundated with water. Soil saturated with water soon becomes **anoxic** (**anaerobic**) and therefore roots, which require oxygen for respiration, must have some mechanism to cope. In addition, the redox potential of soils declines under anoxic conditions and this can increase the availability of Fe^{2+}, Mn^{2+} and S^{-}, all of which can become toxic to roots.

Some species avoid the negative impacts of periodic flooding by surviving as seeds or tubers. Others have a dense mat of shallow roots close to the surface, where oxygen availability is higher. Finally, some species develop **aerenchyma** – tissue in roots and shoots that has a large internal volume that is not taken up with cells, but is full of airspaces which facilitate the diffusion of oxygen from the atmosphere into the roots. Some species counter the reducing conditions of the saturated soil by producing oxidizing conditions at the root surface. This is achieved by allowing rapid diffusion of oxygen laterally through the roots to the root surface (via aerenchyma) or by enzymatic oxidation at the root surface. Such a tactic reduces the uptake of phytotoxic compounds by the root and may increase the availability of nitrate, both of which aid in plant survival in saturated soil.

Distribution and persistence of temporary wetlands in arid Australia

Many wetlands in Australia are temporary (**ephemeral**), partly because rainfall in Australia is highly variable. Consequently those wetlands that are dependent on rainfall are likely to show significant variation in depth and areal extent over time. This, coupled to the increasing extraction of water from rivers, lakes and groundwater stores, means that wetlands in Australia are under increasing threat. Many species of endemic waterbirds utilise wetlands in arid zones as breeding and migratory habitats and therefore the loss of wetland habitat has implications for many such birds.

The many thousands of temporary wetlands in arid Australia range in size from less than 1 hectare to thousands of square kilometres. These wetlands result from one of three processes.

1 They occur as terminal water bodies filled from major drainage systems. Lake Eyre is one of these.
2 They occur as terminal water bodies filled by very local drainage systems (e.g. Lake Gairdner and Lake Torrens).
3 They occur as a result of stochastic flood events from major drainage systems, such as the flooding that occurs along the floodplains of the Darling and Warrego Rivers.

Surprisingly, large areas of the arid zone are flooded at frequent but highly irregular intervals (Roshier et al. 2001). The strongest determinant of whether wetlands fill in any given year in the arid zone is the pattern of tropical weather systems, which extend their influence into southern Australia. A role for the El Niño/La Niña oscillation (see Chapter 1) is also likely. Using models to backcast, Roshier et al. (2001) showed that significant increases in the frequency and extent of wetland filling were likely to have occurred in the 1910s, 1950s and 1970s, with less frequent filling in the 1920s, 1930s and 1960s. Importantly, there was no period longer than 12 months when wetland filling was predicted to have not occurred. However, increased groundwater abstraction, regulation of river flow and drainage of wetlands represent significant threats to both the flora and fauna of these fragile habitats. The maintenance of the mosaic of large and small, temporary and permanent wetlands in arid Australia is central to the maintenance of viable populations of waterbirds.

Naturally flowing floodplain rivers are extremely dynamic and complex, with high spatial and temporal variability. The volume, timing and frequency of flooding is highly variable and determines the biotic responses of the river, the floodplain, wetlands, fresh and saltwater lakes, swamps and billabongs located on the plain.

Flooding of floodplains and associated ecosystems results in a suite of responses. Some of these are schematically outlined in Figure 7.14, which illustrates how flooding events represent important drivers for the ecology, structure and function of floodplains and associated ecosystems. Unfortunately, surface water resource management in the past has not understood the value of these flood events. Indeed, flooding is viewed by many as wasted water and many surface water managers are have been mandated to minimise 'waste'.

Ecological impacts of altered hydrological regimes on floodplain wetlands

Wetlands have important conservation, ecological, aesthetic, biodiversity and economic value. They are increasingly being viewed as having significant environmental value because of their ability to act as filters of water moving through the landscape. The construction of **artificial wetlands** to assist in water treatment, for example around mining activities and as pre-treatment for **urban stormwater** run-off, has intensified the attention given to the ecohydrology of wetlands. It is therefore unfortunate that globally and in Australia wetlands and floodplains are being drained or starved of water through poor water management activities around them (Kingsford 2000). Wetlands have a high biodiversity of waterbirds, native fish, invertebrates, aquatic plants and microbes which is being threatened by poor water management.

Most river flows are highly **regulated** in Australia, which precludes the stochastic flood events of the past thousands of years and to which floodplain ecology has evolved. The hydrology of rivers and floodplains has been managed through a range of interventions, including (Kingsford 2000):

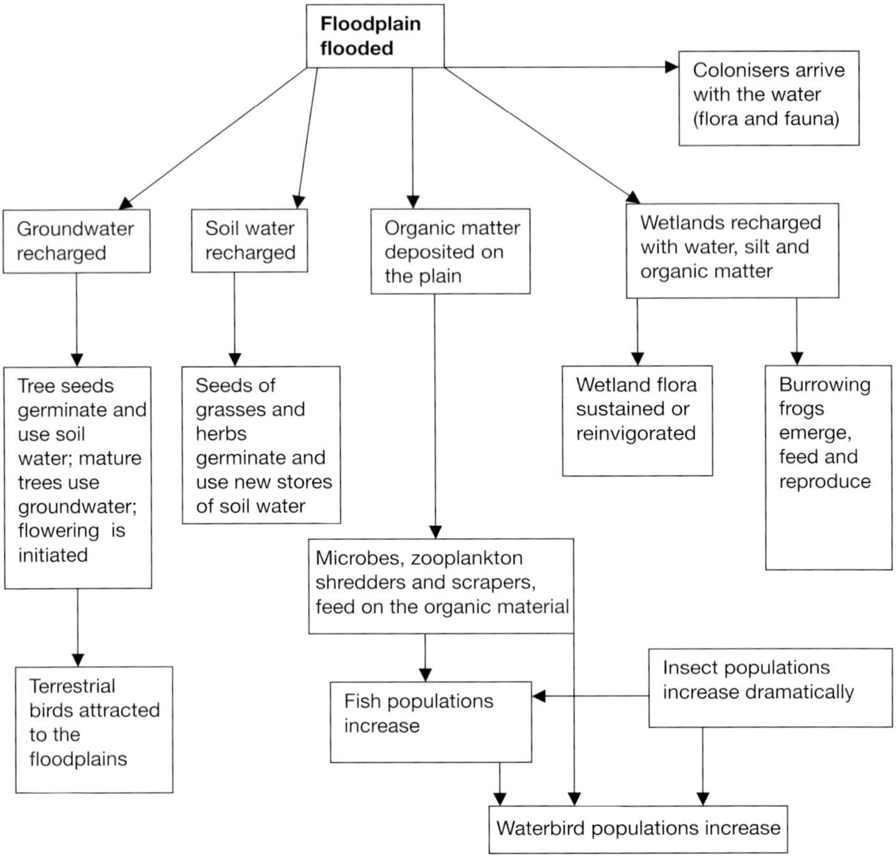

Figure 7.14 Schematic diagram outlining some of the interactions and effects of a flood event across a floodplain

Source: Derived from Kingsford (2000)

- levee banks to prevent flooding;
- channels to divert water to irrigation users;
- cutting – changing the course of a river by excavating soil to provide new pathways;
- weirs to measure stream flow and generate ponds;
- farm dams to provide water for livestock and small-scale irrigation;
- locks to improve navigation and access by boats;
- pumps to move water out of the river into irrigation channels;
- reservoirs and dams, to provide large-scale storage for irrigation and domestic supply.

These structures change the **flow regime** (volume, frequency, timing and duration) of stream and river flow and hence flooding of floodplains. The floodplains become **disconnected** or divorced from the river and eventually are lost as the wetland dries out and terrestrial species begin to colonise what was a wetland.

Case study 5: responses of three floodplains to an altered hydrological regime

In the first example within this case study, the construction of two dams and water extraction has caused a reduction in the frequency and extent of flooding of a floodplain and its attendant wetlands. Similarly, in the second example, reduced flooding with fresh water because of extraction of river water for irrigation reduced the frequency, intensity and extent of flooding. The principal result was the accumulation of salt in the root zone because of saline groundwater interception. The final example differs from the first two in having a significant groundwater dependency.

Example 1: Barmah-Millewa forest and Moira marshes

The Barmah-Millewa forest covers almost 70 000 ha on the upper Murray River, in Victoria and New South Wales. A small part is listed as a wetland of international importance under the Ramsar Convention. It also contains the largest River Red Gum *(E. camaldulensis)* forest in Australia (approximately 60 000 hectares). The Moira marshes lie to the south of the forest. A clear pattern of **vegetation zonation** in response to the hydrologic regime is apparent. Frequently flooded areas are dominated by rushes and sedges, red gums dominate in the less frequently flooded zones and Black Box (*E. largiflorens*) occurs in the least flooded areas.

The floodplain receives low rainfall (400 mm per year) and experiences high evaporation rates (>1000 mm y^{-1}). Consequently the wetland is highly dependent on river flows to replenish water lost through evapotranspiration. The natural pattern of river flows was high flows in winter and spring but low flows in summer and autumn. However, because of the Hume and Dartmouth dams and concomitant withdrawal of water for irrigation, the rate of river flow is now less than half of what it was. Furthermore, water flows are now summer dominated rather than spring dominated. Because river flows are reduced and highly regulated, the percentage of years that the Barmah-Millewa forest has flooded has declined from 80% to 35% and the period of flooding has declined from almost 3 months in 78% of years to 1.3 months in 37% of years (Davis et al. 2002). In addition, the number of no-flood years has increased significantly. These hydrologic changes have had significant impacts on vegetation.

Areas previously flooded most often responded first. The species composition, growth and regeneration of the sedge and rush communities have changed, with the common reed and moira grasses declining. Red gums are beginning to move from the previously moderately flooded regions into the previously frequently flooded regions and black box is moving into areas formerly dominated by red gum. Many of the mature red gum trees are showing crown death, reduced regeneration and increased susceptibility to insect and fungal attack. These are classic symptoms of trees under stress. The already threatened parrot *Polytelis swainsonnii* is further threatened because it requires mature red gum to breed.

Faunal populations are also showing a response to the altered hydrologic regime. Populations of many fish, waterbird and snake species are declining substantially. Brolgas, glossy ibis and little egrets have disappeared from the forest.

The Murray-Darling Basin Commission now manages water releases from the dams in order to improve local ecology. The forest has been divided into management units which are managed independently. Pre-regulation flow patterns (timing, duration, volume) are being replicated wherever possible to enhance ecological health. An environmental water allocation of 100 GL per year has been made from the Hume Dam and Yarrawonga Weir.

Example 2: Chowilla floodplain

The Chowilla floodplain covers almost 18 000 hectares and includes lakes, billabongs, islands and over 100 km of anabranch creeks. It is located on both sides of the Murray River, west (downstream) of the junction of the Murray and Darling Rivers. Chowilla is listed in the Ramsar Convention and contains extensive floodplain forest. The lower Murray is regulated through the presence of weirs and barrages and through managed releases of water from Lake Victoria and the Menindee Lakes. Of the approximately 13 500 000 ML of flow per year that naturally occurred (with much inter-annual variation), nearly 10 000 000 ML is now removed, mostly for irrigation use. Median natural flows to the floodplain are now about half of what occurred previously. The floodplain previously flooded every 1.2 years; now it receives water about every 2.5 years and the area of land that was formerly flooded every 10 years has declined from 77% to 54%. Floodplain health has declined because of reduced flooding frequency and the rise of a saline water table, which is discharging into once ephemeral streams (Akeroyd et al. 1998).

The **Chowilla anabranch** system is made up of a network of streams that flows from the River Murray above Lock 6 of the river, across a c.10 km wide floodplain. These streams come together to form the Chowilla Creek which flows back into the Murray River downstream of Lock 6. The total length of the streams is about 100 km, and billabongs and lakes are present. Before Lock 6 was built, flow in these streams was ephemeral – it only flowed during floods. However, when Lock 6 was constructed a large weir pool (70 km long) was generated and flow from this pool through the streams occurs all year. About 50–75% of Murray River flow now occurs through the creeks.

The floodplain lies above a **saline aquifer.** Prior to construction of the weirs, groundwater moved at depth through a sandy aquifer into the Murray River, taking water and salt to the river. The floodplain, creeks and lakes were hardly affected by salinity because groundwater levels were sufficiently deep. However, construction of weirs has made groundwater levels rise by 2–3 m (through increased local recharge) and water resides in the clay layer above the sand (Jolly & Walker 1995). Surface evaporation has increased (potential evaporation is about 2000 mm per year) and the salt concentration of surface soils has increased significantly. In addition saline (20 000–60 000 mg L^{-1}) groundwater that previously flowed beneath the floodplain into the Murray River is now intercepted by creeks (because groundwater depth has decreased). Finally, floods that previously leached salt from the soil and supplied fresh water for vegetation no longer occur as frequently and saline groundwater is increasingly entering the floodplain.

Vegetation on the floodplain shows distinct zonation, determined principally by the frequency of flooding. Black Box (*E. largiflorens*) occupy the zone highest on the floodplain and receive very infrequent and short-lived floods. These communities occupy about 40% of the floodplain. River Red Gums (*E. camaldulensis*) occupy a region lower in the landscape and closer to the stream, receiving more frequent and longer-lasting floods. Cane grass and lignum are found between the River Red Gum and Black Box. River Red Gums can occur very close to the river on localised high points in the landscape where they can use river water and local bank recharge. Closer to the stream is a littoral and rush zone which requires frequent and long-lasting floods.

Black box trees are dying as a result of salt accumulation in the root zone and the too infrequent flooding, but not only vegetation is showing declining health. Populations of 18 species of snails have declined in the past 50 years; this is likely to be influencing fish populations and hence the size of waterbird populations. Native fish populations have also declined substantially in the Murray River. Red gum dieback in the summers of 2003 and 2004 affected over half the red gum trees, which are not dead or in severe decline. A combination of salt accumulation and drought is the cause.

Plant function and hydrology are coupled to floods

Rainfall is lower for Chowilla than for the Barmah-Millewa forest, averaging 250 mm per year and therefore the **ecology of the floodplains showed a heavy reliance on river flows and floods** before the river was regulated. Not only does the ecology rely on flooding but fresh groundwater depends on flood events. Groundwater recharge by floods occurs through three mechanisms:

- bank recharge, where water in the flooded stream recharges the aquifer;
- diffuse recharge, where floodwaters on the floodplain percolate down into the aquifer;
- localised recharge, where floodwaters accumulate in low points in the landscape and recharge quickly into the aquifer.

When floods recharge groundwater, they leach salt out of the soil profile and the fresh floodwater sits on top of the aquifer as freshwater lenses. Furthermore, the direction of the flow of water is reversed by floods. Prior to a flood, groundwater flows into the river. During floods, water flows from the river (and floodplain) into the aquifer. Consequently salt concentrations in the river are reduced for a short period as the initial impacts of the flood (increased bank recharge with freshwater and dilution effects due to more water flowing in the river) are experienced. However, for many months (12–24 depending on the size of the flood) after the flood, **salt loads** (flux of salt per day per km of river) into the river rise because increased groundwater levels result in an increased head difference with the river, which drives saline groundwater into the creeks. After small floods, the increase in salt load lasts only a few months as the river banks discharge into the river. After large floods where all or most of the floodplain is flooded, diffuse recharge and localised recharge occur at larger distances from the river and so the pulse of salt carried into the river takes longer to dissipate. For the Chowilla anabranch system, diffuse recharge appears to be relatively unimportant – bank and localised recharge are the most important processes (Jolly & Walker 1995).

Water use by river red gums and black box

Table 7.6 shows stand characteristics, rates of water use for black box and red gum and the sources of water used by these species growing on the Chowilla floodplain (Thorburn et al. 1993). Rates of water use by black box are lower than those of red gum, principally because of

Table 7.6 Stand characteristics, rates of water use for black box and red gum trees and their sources of water, Chowilla flooplain

	Black box site 1	**Black box site 2**	**Red gum**
Tree density (ha^{-1})	75	350	50
Leaf area index	0.6	1	1.5
Groundwater electrical conductivity ($dS\ m^{-1}$)	25	33	11
Soil chloride content ($mg\ L^{-1}$)	4000–12 000 in the 0.5–4 m depth range	2000–12 000 in the 0.5–4 m depth range	1000–7000 in the 0.5–4 m depth range
Transpiration rate ($mm\ d^{-1}$):			
Summer	0.3	–	1.3
Winter	0.3	0.2	1.8
Depth (m) of source of water transpired	1.7–3.3	Mixed: 0.2 and 3	Mixed: 0.1 and 2.8
% of water sourced from groundwater:			
Summer	100	65	79
Winter	100; 44	51	58
Groundwater discharge ($mm\ d^{-1}$):			
Summer	0.3	-	1.0
Winter	0.3; 0.13	0.1	1.0

Source: Thorburn et al. (1993)

their lower leaf area but also because of the higher soil and groundwater salinity of two black box sites (Table 7.6) compared to the red gum site. Despite its salinity, groundwater can be used extensively by these species. Not all trees at all sites switched their water use from saline groundwater to relatively fresh water in the upper soil profile immediately after rainfall (Thorburn et al. 1993) and patterns of water uptake as a function of depth were variable and complex (Table 7.6). It is clear that despite low (in absolute terms) use of groundwater, the low frequency of flooding of the black box sites (a decade or longer) results in the slow but significant accumulation of salt in the root zone. The higher frequency of flooding in the red gum site (4–7 y) results in a more frequent of leaching of salt from the root profile.

Red gum trees located closest to the streams rely on surface water for their transpiration needs, and more than 50% of their water is derived from the stream. As the distance from the stream increases, the reliance on surface water declines so that at 15 m distance no surface water (stream water) is used. For trees very close to an ephemeral stream, surface water use is less than 30% and transient (Jolly & Walker 1995). Surprisingly, where saline groundwater is available to the roots and saline soil water is available, both are used in preference to the less saline surface water in the stream.

The decline in health of black box on the Chowilla floodplain has been noted for the past 20 years. Today it is estimated that 55% of the black box trees are dead or in severe decline (Overton & Jolly 2004). Generally, black box health declines as the depth to the groundwater decreases and the salinity of the groundwater increases (shallow groundwater causes increased soil salinisation). A **critical depth** of 4 m was determined for black box. When groundwater depth is more than 4 m, the trees appear healthy (Jolly & Walker 1995). As flood frequency increases this critical depth is likely to decrease because frequent flooding will leach salt from the upper soil profile more frequently and therefore salt loads for the trees are reduced.

Water use by black box during drought is influenced by the depth to groundwater and groundwater salinity. During drought, the upper soil profile is soon depleted of plant available soil moisture. For trees growing above shallow (i.e. available to tree roots) groundwater with low levels of salt, water use continued at relatively high levels and trees maintained their vigour throughout the drought. However, for trees growing above saline shallow groundwaters, the groundwater could not be used and transpiration was much reduced. Tree vigour declined as the drought progressed and tree death occurs when salinity in the root zone becomes too high for the roots to survive.

Flooding is usually associated with anoxia of soil and as such would be expected to reduce root function. However, Akeroyd et al. (1998) showed that flooding did not reduce tree water use of Black Box on the Chowilla floodplain. During the flood, tree water use closely followed potential evaporation, indicating minimal control of water use through stomatal regulation. Following the recession of the floods tree water use increased significantly (Fig. 7.15a) at some sites but not others because of enhanced leaching of salt from the root zone (Fig. 7.15b) at some sites and not others. Leaching of salts was most pronounced on sites with a high sand content of the soil.

Management options for Chowilla

From this and related studies, four management options for salinity management of the Chowilla floodplain are apparent (Jolly & Walker 1995). These are:

- do nothing – not a sustainable strategy;
- intercept groundwater by pumping or drainage and subsequent evaporation of the water to extract the salt;

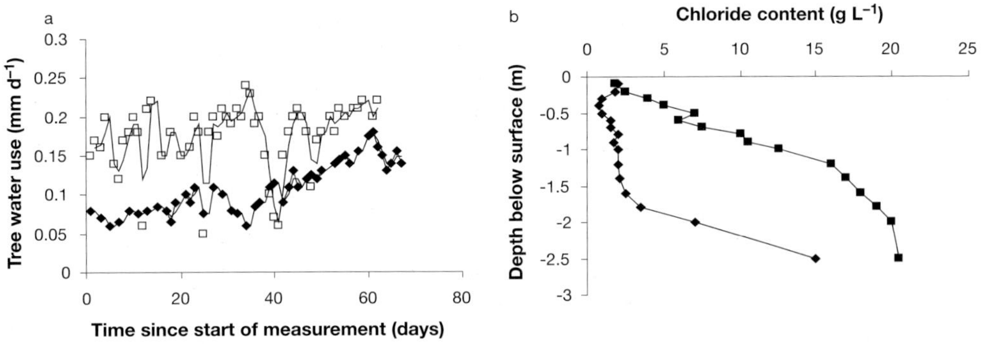

Figure 7.15 (a) Tree water use by Black Box growing on the Chowilla floodplain is low (<0.25 mm d^{-1}) partly because of the low LAI and partly because of the saline groundwater and soil water that is available. At some sites flooding does not result in increased tree water use (open squares) but at others it does (closed diamonds). The horizontal arrow indicates the period of flooding. (b). Closed squares are before flooding, closed diamonds are after flooding

Source: Akeroyd et al. (1998)

- increase the frequency and extent of flooding by releasing more water from storage or manipulating surface flows using control structures;
- artificial irrigation.

The first option is not discussed further! Doing nothing simply allows further environmental degradation and loss of ecosystem health.

Groundwater pumping can reduce local groundwater depths and allow increased recharge (with fresh water) during floods. The aim is not to freshen the groundwater, but to freshen the soil profile by lowering the groundwater, leaching the salt and reducing the rate of salt accumulation. A salt interception scheme has been proposed for the Chowilla floodplain. The system design has been based on field pumping tests and a model (see Chapter 5) of groundwater movement.

Artificial watering trials have been undertaken on Chowilla as a means to combat the decline in tree health (especially red gums). These trials have used pumps to move water from permanent creeks into ephemeral creeks and wetlands. A good response in the fringing tree health has been seen, and these artifical watering events may become permanent management options in drought periods.

In conclusion, as observed in the first case study, changes in hydrologic regime resulting from the construction of weirs on the river have influenced the movement of water and salt through the landscape. These in turn have had detrimental impacts on the ecology of the floodplain.

Example 3: water balance of a groundwater fed wetland in north-eastern Victoria

In the previous two examples groundwater input to the wetlands was unimportant (until the hydrologic balance was changed by human activity in the Chowilla example). Those wetlands are isolated from groundwater and are sustained by rainfall, surface flow and intermittent flooding from rivers. In contrast, the wetland of this third study is closely coupled to groundwater availability and changes in availability have a rapid and pronounced effect on the hydrology and ecology of the wetland.

The Reids wetland, located in the Kiewa Valley in north-eastern Victoria, consists of a reed swamp of 2 ha with a catchment of 50 ha. It is located at a break of slope on the edge of the floodplain of the Kiewa River. A small ephemeral stream supplies surface water to the wetland from the east and water exits the wetland on the west. *Phragmites* spp., *Juncus* spp. and *Paspalum* spp. dominate the vegetation. Rainfall is approximately 800 mm annually but in dry years can be as low as 500 mm.

Raisin et al. (1999) investigated the contribution of groundwater to water and nutrient flows from the Reids wetland. Surface water influx and surface water efflux were measured using a standard V-notch weir (inlet) or concrete weir (outlet), along with rainfall. Base flow was measured following a fire that burnt most of the wetland; aquifer depths were measured using bores. A water balance for the wetland was derived from:

$$P + SWI + GWI = ET + SWO + GWO + \Delta S$$

where P = rainfall, SWI = surface water inflow, GWI = groundwater inflow, ET = evapotranspiration, SWO = surface water outflow, GWO = groundwater outflow and ΔS is change in soil storage (Raisin et al. 1999).

During the 30 month study period 26 inflows of greater than 0.5 ML d^{-1} surface water occurred, caused by rainfall within the catchment. These inflows ranged from approximately 1–6 ML d^{-1}. During the period February–May 1994, surface and rainfall inputs were zero yet a daily discharge from the wetland of approximately 0.65 ML d^{-1} (ranging from zero to 2.9 ML d^{-1} but generally 0.5–1 ML d^{-1}) occurred. This flow represents **base (groundwater) inflow** and accounted for 97% of the water leaving the wetland. Even during wetter periods, groundwater represented a significant fraction of the water leaving the wetland.

Two aquifers were found below the wetland – a shallow (2–6 m) aquifer overlying a deeper aquifer, with an overall upward hydraulic gradient moving water from the deeper to the more shallow aquifer. However, the two systems appeared to operate as a single hydrological unit. Both the shallow and deeper aquifers contributed water to the wetland. Hydrographic data showed groundwater discharge to be approximately 0.65 ML d^{-1}, in complete agreement with the base flow values derived from the discharge weir. A water balance for the wetlands is shown in Table 7.7.

Evapotranspiration was estimated from pan evaporation (measured as 1373 mm for the study period). Groundwater influx represented 97% of total flow, with surface flow and rainfall each contributing about 1.5%. As groundwater levels declined because of a prolonged drought, the discharge of groundwater to the wetland declined. This study highlights the importance of maintaining groundwater levels for the maintenance of ecosystem (wetland) health and shows the complexity and variability of groundwater and surface water fluxes. The construction of artificial wetlands to intercept point and diffuse sources of pollution (e.g. to intercept water being discharged from a sewage treatment plant or mine tailings water; or water being discharged from a rural catchment containing pesticides and

Table 7.7 Water balance for the Reids wetland in the Kiewa Valley, north-east Victoria

	Entering the wetland	Departing the wetland	Residual (groundwater)
Surface flow (ML)	13.5	223.4	
Rainfall (ML)	14.4		
Evapotranspiration (ML)		22.0	
Groundwater input (ML)			217.5

Source: Raisin et al. (1999)

herbicides) requires a deep understanding of the processes and interactions within the wetland and surface and groundwaters.

Case study 6: Lake Toolibin – saving an inland lake

Background and issues

The inland area of the south-western corner of Australia, known locally as the Western Australian wheatbelt, is a hydrologically and biologically distinctive region of some 20 Mha. Remarkably high biodiversity arises from rich assemblages of terrestrial and wetland plant communities, including valley floor complexes comprised of chains of ephemeral lakes (playas) with characteristic and diverse salinities and **hydroperiods** (typical duration of inundation or waterlogging). Clearing native vegetation from 85–90% of whole catchments for agriculture over the last 150 years has changed both water and salt balances to greatly extend the periods of waterlogging and increase the salinity of previously fresh systems.

The extent of this phenomenon is so great that only one large example of a reasonably intact fresh lake ecosystem remains: **Lake Toolibin**. The natural heritage value of this Ramsar-listed wetland is so high that since the mid 1970s effort has been focused on understanding the lake's hydrogeology, hydrology and ecohydrology to identify the most effective protection and recovery strategy. The work done to save Toolibin highlights the complexity of natural ecosystems in this region as well as the size of the challenge involved in protecting or restoring natural assets subject to landscape-scale hydrologic changes.

Regional setting

Lake Toolibin (300 ha) and its catchment (48 000 ha) are situated near the headwaters of the Northern Arthur River, which drains into the Blackwood River (Fig. 7.16). Like other rivers of the Western Australian wheatbelt (see Hatton et al. 2004 for review of regional hydrology and hydrogeology), the first distinctive feature of the river system is that mean annual rainfall is lowest (less than 350 mm) at its headwaters. The second distinction is that the landscape is older in the headwaters than at the sea, and the headwaters flow with remarkably low gradients. Gradients in the upper reaches fall only fractions of a metre per kilometre. A key feature of this gradient is that it is interrupted by large, essentially flat playas that drop water from one to another when they (occasionally) overflow. It is particularly important to appreciate that these drainages do not all flow as one linked system except in the most extreme rainfall events. Parts of the upper Blackwood drain into Lake Dumbleyung, which is a large permanent salt lake that is said to have been dry more often than wet before land clearing occurred. Lake Dumbleyung is thought to have overflowed into the Lower Blackwood only three times since the 1870s. The chains of (usually dry) wheatbelt lakes form a series of local storages that in most years are not overtopped by surface flows from upstream. Toolibin is part of a small complex of playas, the largest of which is Lake Tarblin, a much larger playa a few kilometres downstream whose biology and hydrology has already been radically affected by the changed hydrology.

The failure of these extensive, relatively dry and flat headwaters to frequently purge, combined with the great age of the landscape, means that immense amounts (up to 1 tonne of salt per square metre of land surface) of cyclic salt have naturally accumulated over millennia. Prior to clearing, this salt was generally concentrated below a depth of 10–15 m in unsaturated regolith. Increased groundwater recharge (by factors of 10 or even 100) resulting from tree clearing has brought a sea of salty water to the land surface and to the surface waters (see Chapter 8).

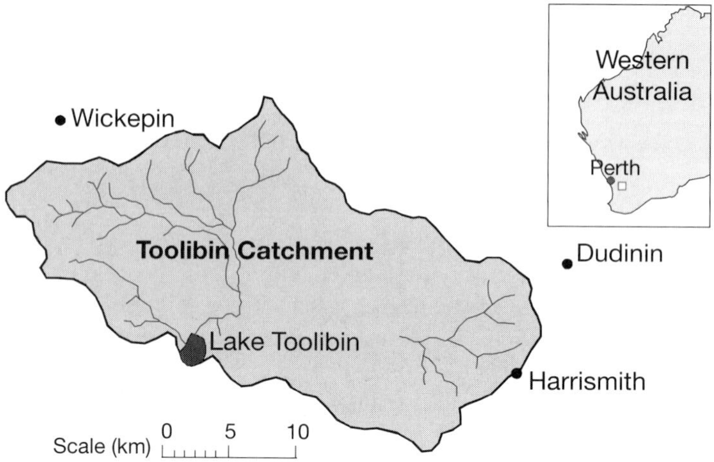

Figure 7.16 Location of Lake Toolibin, Western Australia

Climate

Lake Toolibin receives an average 400 mm of annual rainfall, predominantly in winter. Actual annual rainfall can vary by a factor of two or three. Potential annual evaporation is 1900 mm, and mean monthly values exceed mean monthly rainfall. Summer rainfall does occur from north-west cloud bands and (less frequently) from tropical depressions; summer events are typically the most extreme rainfall events. The aridity and unpredictability of the timing and amount of rainfall has had a major impact on evolutionary adaptations of Australian biota generally and the productivity of Lake Toolibin's ecosystems in particular.

Vegetation, hydrology and hydrogeology

The vegetation of Lake Toolibin is dominated by stands of *Casuarina obesa* and *Melaleuca strobophylla*. By virtue of its freshness, the lake is an important breeding habitat for native fauna and is listed under the Ramsar Convention as a Wetland of International Importance, and the Toolibin Nature Reserve is considered one of the most important single sites in the conservation estate in Western Australia. Recently, however, the combination of dry years, salt inflows and rising saline groundwater tables has led to declining cover of mature vegetation, greatly reduced recruitment of seedlings and a decline in system health. Because most similar systems in Western Australia have already been highly impacted or lost, the fate of the Lake Toolibin Nature Reserve is of great relevance.

Our hydrogeological understanding of Lake Toolibin is based on a long series of investigations, summarised by George and Dogramaci (2000) and Dogramaci et al. (2003). Toolibin has the common wheatbelt palaeochannel stratigraphy of alluvial sands and gravels at its base, trending upwards into lacustrine clays and finally into recent alluvial and aeolian sands and clays (Fig. 7.17). A deep and relatively transmissive palaeochannel exists in the lower catchment and extends at least 5 km upstream of the lake. The periphery of the lake is comprised of deeply weathered granite regolith and very recent sediments.

The groundwater under Lake Toolibin is within 1–1.5 m of the lakebed, and was almost certainly much deeper prior to clearing of the catchment. Groundwater salinity exceeds 30 000 mg L^{-1}. The groundwater system is to varying degrees confined by layers of low permeability clays, but these do not extend uniformly across the lakebed. A portion of the recharge (and the hydraulic head) to the system arises from within the surrounding granite regolith. The lakebed provides additional recharge to the system when inundated, then becomes the discharge point

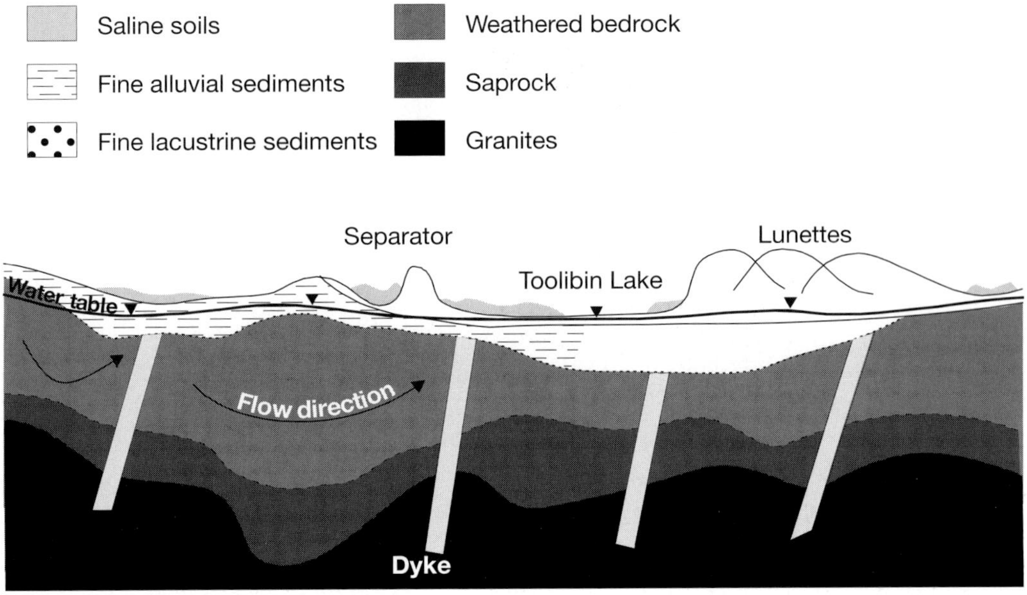

Figure 7.17 Hydrogeological cross-section of Lake Toolibin

Source: George & Dogramaci (2000)

when it dries out. It is believed that the groundwater system has not yet reached a new equilibrium with post-clearing recharge and that, without intervention, hydraulic pressures and levels will continue to rise for some time. It is important to note that the majority of groundwater movement in this system is thus vertical (on average, groundwater levels are rising five times faster than they are flowing laterally).

Determining the annual water balance of the Toolibin catchment is straightforward due to the low gradients, low rainfall and high potential evaporation. Lateral groundwater losses are negligible, and surface discharges out of the lake do not occur in most years. Thus evaporation accounts for the majority of rainfall in the catchment. A secondary feature is the net accumulation of groundwater, which on present trends may result in about 25% of the catchment eventually having a shallow water table. The implications are that salinity and waterlogging are liable to have strong negative impacts on catchment vegetation. Chapter 8 discusses a range of options for ameliorating the development of saline catchments through extensive planting of trees.

Toolibin's salt

The dominance of vertical subsurface flow and shallow water tables indicate a high potential for salinisation from groundwater processes. However, the major source of modern salt inputs to Toolibin arise from **saline groundwater discharges** upstream that are washed down by surface flows then trapped in the lake. The salinity of the lake water ranges from 1800 mg L^{-1} at a volume of 2 million cubic metres (this would taste salty to drink and exceeds drinking limits for humans) to more than 10 000 mg L^{-1} (one-third seawater) at a volume of 0.5 million cubic metres (although the lake has been dry since about 1993). Inflow salinities range from 240 mg L^{-1} to 26 000 mg L^{-1}, depending on the volume of stream flow (Table 7.8). Clearly, the freshwater lake that Lake Toolibin once was is now very saline because of the changed ecology of its catchment.

The large differences between the mean and median values shown in Table 7.8 reflect the greater influence of smaller events per unit salt load, arising from more constant annual salt

Table 7.8 Stream flow and salinity inflows to Lake Toolibin

	Stream flow (ML)	Salt load (t)	Salinity (mg L^{-1})
10 percentile	0.03	0.05	400
Median	272	489	2110
Mean	1405	1290	3500
90 percentile	2490	2980	4200

Source: Dogramaci et al. (2003)

discharge upstream relative to the variation in annual rainfall and run-off. Inflow salinities have been trending upward since at least 1984, a rise of about 1000 mg L^{-1} over 20 years. Modelling suggests that the lake and (lakebed) has accumulated 35 000 tonnes of salt over this period. This observation is the reason for the major intervention now on trial at Toolibin, described below.

A plan for recovery

The two **key physical parameters** to restore to Lake Toolibin are a minimised input of surface salts and a lower groundwater table. Three interventions were introduced in the 1990s in an integrated management strategy. First, **trees** have been planted around the lake. Hydrological modelling has demonstrated that this strategy alone cannot be effective. However, it will reduce the need for or extent of pumping, as well as delay the onset of the most severe impacts.

The second intervention has been extensive **groundwater pumping** to increase the depth to the water table. In the longer term, when groundwater levels are sufficiently deep and with more commercial revegetation options for landholders, revegetation that is now confined to around the lake may extend across most of the catchment and groundwater pumping can be abandoned.

The third intervention has been to install a **surface diversion** which intercepts and redirects downstream all run-off from the beginning of the wet season until the stream salinity is reduced below a threshold of 1000 mg L^{-1}. In the relatively dry years since the diversion was built, this threshold has rarely been reached. Operating since 1994, the installation of this surface input control has resulted in the diversion of 3500–4000 tonnes of salt away from the lake.

The largest **costs** associated with the recovery of Lake Toolibin are the installation and running of the pumps. Groundwater modelling was used to assess the design of the well field. Starting in 1996, pumps were installed in the western half of the lake and yielded approximately 200 kL d^{-1}. Location of a higher yielding palaeochannel on the eastern side provided an opportunity for more efficient water extraction, and since 2002 some 500 kL d^{-1} have been pumped from three bores on that side of the lake. Thus, 700 kL d^{-1} is withdrawn and discharged for evaporation downstream at Lake Tarblin. This translates to the export of some 8000–10 000 tonnes of salt per year from under the lake, thereby reducing the threat of this salt moving up into the lake.

Conclusions

The cost of the attempt to recover Lake Toolibin has exceeded $1 million and success is not certain. Groundwater pumping is not only an uncertain activity but expensive and fraught with maintenance problems, such as iron fouling the well screens. The hydrogeological modelling upon which the strategy is based is also uncertain, even in a locale such as Lake Toolibin where intensive field investigations have been undertaken. The response time to such processes is slow, and the success of the outcome remains to be seen. Assessment of the effects of the surface diversion, recharge control measures and groundwater pumping is being confounded by a run of years with low rainfall. There is some indication that the surface soil is freshening over the first few centimetres, and some recruitment of juvenile tree species is evident in places. However, overstorey trees are still dying off. It will be most interesting to see the physical and biological responses to wet years, when they come.

Lake Toolibin must be seen as an important experiment in ecohydrological management. Key lessons for the management of up to 6 Mha of land and associated built and natural assets at risk of salinisation in Western Australia are being learned.

Conclusion

Vegetation structure and function of all landscapes is influenced by many factors, including climate, soils and disturbance history. Disturbance history includes the impact of natural events such as storms and fire, and anthropogenic effects such as mining and diverting river flows across the landscape. Almost all large disturbance events can alter vegetation cover and hence hydrology. This chapter has shown some of those impacts.

There is one final and major anthropogenic event that has affected large areas of temperate Australia since European settlement. Removing trees from the landscape is a major change in land use and a major change in vegetation cover. It is therefore not surprising that large-scale changes in the hydrologic balance have occurred. Because the spatial and economic magnitude of the impacts of this change in land use and land cover – and its amelioration – are so large, it is discussed separately in the following chapter, which considers salinity in Australia.

Further reading

Anon. *Olympic Dam operations.* Roxby Downs, Environmental Management Program.

Bosch JM & Hewlett JD (1982). A review of catchment experiments to determine the effect of vegetation changes on water yield and evapotranspiration. *Journal of Hydrology* **55**: 3–23.

Brock MA, Smith RGB & Jarman PJ (1999). Drain it, dam it: alteration of water regime in shallow wetlands on the New England Tableland of New South Wales, Australia. *Wetlands Ecology and Management* **7**: 37–46.

Burch GJ, Bath RK, Moore ID & O'Loughlin EM (1987). Comparative hydrological behaviour of forested and cleared catchments in south-eastern Australia. *Journal of Hydrology* **90**: 19–42.

Cramer VA & Hobbs RJ (2002). Ecological consequences of altered hydrological regimes in fragmented ecosystems in southern Australia: impacts and possible management responses. *Austral Ecology* **27**: 546–564.

Dogramaci S (2000). *Toolibin: a life and death struggle for the last freshwater wheatbelt lake.* Hydro 2000: 3rd International Hydrology and Water Resources Symposium, Canberra, Institution of Engineers.

Dogramaci S, George R, Mauger G & Ruprecht J (2003). *Water balance and salinity trend: Toolibin Catchment, Western Australia.* Conservation and Land Management, Perth.

Eamus D, Hutley LB & O'Grady AP (2001). Daily and seasonal patterns of carbon and water fluxes above a north Australian savanna. *Tree Physiology* **21**: 977–988.

Hughes FMR & Rood S (2003). Allocation of river flows for restoration of floodplain forest ecosystems: a review of approaches and their applicability in Europe. *Environmental Management* **32**: 12–33.

Kingsford RT (2000). Ecological impacts of dams, water diversions and river management on floodplain wetlands in Australia. *Austral Ecology* **25**: 109–127.

Large ARG & Prach K (1999). Eco-hydrology: plants and water in terrestrial and aquatic environments. In *Ecohydrology* (eds AJ Baird & RL Wilby). Routledge, London, pp. 237–268.

Pettit NE & Froend RH (2001). Variability in flood disturbance and the impact on riparian tree recruitment in two contrasting river systems. *Wetlands Ecology and Management* **9**: 13–25.

Pettit NE, Froend RH & Davies PM (2001). Identifying the natural flow regime and the relationship with riparian vegetation for two contrasting Western Australian rivers. *Regulated Rivers: Research and Management* **17**: 201–215.

Reid MA & Brooks JJ (2000). Detecting effects of environmental water allocations in wetlands of the Murray-Darling Basin, Australia. *Regulated Rivers: Research and Management* **16**: 479–496.

Sundaravadivel M & Vigneswaran S (2001). Constructed wetlands for wastewater treatment. *Critical Reviews in Environmental Science and Technology* **31**: 351–409.

Williams R, Duff G, Bowman D & Cook A (1996). Variation in the composition and structure of tropical savannas as a function of rainfall and soil texture along a large scale climatic gradient in the Northern Territory. *Journal of Biogeography* **23**: 747–756.

Wullschleger SD, Hanson PJ & Todd DE (2001). Transpiration from a multi-species deciduous forest as estimated by xylem sap flow techniques. *Journal of Hydrology* **143:** 205–213.

Chapter 8

Salinisation: an ecohydrological perspective

This chapter discusses saline landscapes and focuses on the relationships among tree function, groundwater and salt. It does not intend to discuss the ecological impacts of salinity in Australian landscapes. The chapter is divided into three sections.

The first section describes the **causes, extent** and **impacts** of salinity in Australia and differentiates between **primary** (natural) salinity and **secondary** (anthropogenic; human induced) salinity. It further differentiates between **irrigation** salinity and **dryland** salinity – the latter is the sole focus of this chapter. Dryland salinity arises because the hydrological balance of landscapes has been shifted through removal of trees. A note on units for measuring salinity and a brief discussion of the physiological impacts of salinity and waterlogging are also presented.

The second section discusses the principles underlying the use of trees in the landscape to ameliorate (offset, repair) salinisation of landscapes in Australia. It starts by comparing the ecohydrology of Australia before and after European settlement, then presents a hypothetical water balance for pre-and post-clearing to show how groundwater recharge has increased post-clearing. It also shows how replacing trees in the landscape *may* be a way to tackle Australia's salinity problem. Guiding principles are provided for small, intermediate and large-scale aquifers and there is a discussion of the information needed to design agroforestry (mixtures of trees and agriculture in the landscape) systems to ameliorate salinity.

The third section discusses the **ecohydrological implications** and **requirements** of using trees to ameliorate salinity in the landscape. It discusses the possibility of planting trees on shallow saline aquifers to increase groundwater depths and thereby reclaim degraded land. Three scenarios for planting trees above saline aquifers are included. Trees may be planted on a regional scale above a regional scale aquifer of high transmissivity; trees may be planted on a regional scale above a regional scale aquifer of low transmissivity; or trees may be planted at a local scale on a local scale perched aquifer. This section also considers how to decide where (**break-of-slope, points of discharge/recharge**) to plant trees and includes a discussion of the optimal formation (**belts, blocks** or **alley farming**) for trees in the landscape.

After reading this chapter readers should have developed an understanding of the following:

- the cause of dryland salinity;
- the two principal sources of salt in landscapes affected by dryland salinity;
- the two units commonly used to express salinity of groundwater;
- the concept of 'leakage' of water past the rooting zone in landscape ecohydrology;

- the reasons why regions with high rainfall (>900 mm y^{-1}) tend not to develop salinity problems;
- the relationships among annual rainfall, tree leaf area index and total ecosystem water use;
- the differences between local (small) aquifers and intermediate size aquifers in their response to revegetation;
- the factors that influence the establishment of tree belts in the landscape and the interception of groundwater by those belts;
- the differences between highly transmissive aquifers and aquifers with low transmissivities in their response to revegetation;
- the comparative effects of planting trees in alleys or rows or on shallow water tables as a means of addressing dryland salinity.

Causes, extent and impacts of dryland salinity

The development of dryland salinity in Australia is so large in its spatial, economic and ecological impact, and the amelioration of the impacts so complex, that an entire chapter of this text is devoted to this one ecohydrological phenomenon.

The development of dryland salinity in Australia is, perhaps, the largest and most important ecohydrological problem facing the continent (Australian Dryland Salinity Assessment 2000; White et al. 2002). Dryland salinity is best interpreted as a change in the **hydrological balance** of the landscape arising from changes in the ecology of the landscape. Briefly stated, replacing **deep rooted native trees and shrubs** that use water throughout the year with **shallow rooted crops and pasture** that use much less water over a limited growing season each year, has changed the hydrological balance of the landscape (Dunin et al. 1999). In particular, more water is draining through the soil and past the root zone (**leakage**) so that groundwater levels have risen. This rise in the water table brings dissolved salts within the rooting depth of the vegetation and even up to the surface of the soil. Salts are also moving laterally in subsurface flow into streams and rivers, as evidenced in the Blackwood River in Western Australia (MacFarlane & Williamson 2002; Fig. 8.1). This chapter explores this simple synopsis and outlines possible responses to the challenge of dryland salinity.

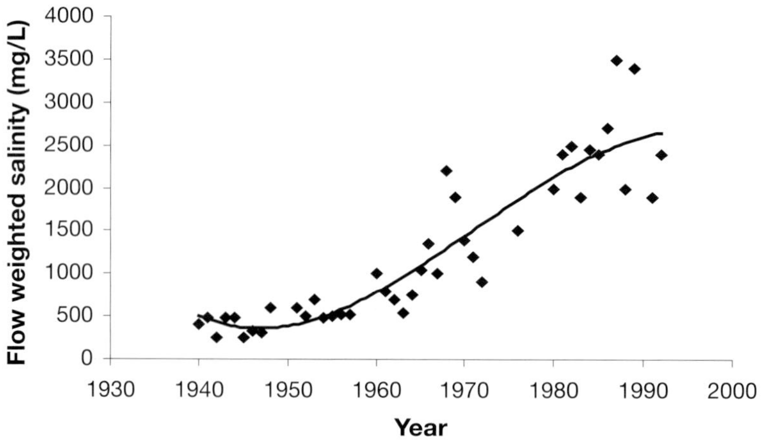

Figure 8.1 Changes in stream salinity for the Blackwood River in Western Australia

Source: Adapted from MacFarlane & Williamson (2002)

Figure 8.2 A saline seep from rising groundwater in New South Wales. Picture by M. Zeppel, UTS

Primary and secondary salinity

There are two kinds of salinity confronting Australia. The first, called **primary salinity** or **natural salinity**, arises from the natural accumulation of salt in the landscape through non-anthropogenic inputs. There are three sources of this salt. First, deposition of sea salt occurs through the action of salt spray and wind (Hingston & Gailitis 1976). Salt from the oceans surrounding Australia can travel significant distances inland. Second, salts arising from parent rock also accumulate in soils. In most places the input of salt from these two sources over millions of years has not caused problems because rainfall was sufficient to leach the salt from the surface of the soil down into layers of soil below the root zone. A third natural source of salt in the Australian landscape is the salt that was deposited when ancient inland seas evaporated and formed large salt pans in arid and semi-arid Australia.

In some locations, lateral flow of salty water to low points in the landscape, followed by evaporation, has led to the natural formation of salt scalds (Fig. 8.2) and salt lakes, especially in arid and semi-arid zones. These may be shallow (common), deep (rare), permanent (less common) or temporary (**ephemeral**, most common) features in the landscape. The depth, areal extent and salinity of these saline lakes are governed by the balance between inputs and outputs of saline water. In all cases, the major discharge pathway for water is evaporation from the surface into the hot dry atmosphere of arid and semi-arid Australia. For temporary salt lakes the most important water inputs are rain and ephemeral stream and surface flows following localised rainfall. For permanent salt lakes, groundwater flows play a more significant role, in addition to rain, subsurface lateral flow and overland flow.

Three patterns to the characteristic hydrology of saline lakes can be distinguished:

- permanent salt lakes that are large enough to contain water all year – volume and areal extent increase after seasonal rains and decline during the dry season, but lakes do not have a history of drying out;
- ephemeral salt lakes that increase in volume as a result of seasonal rains, but dry out in the dry season because they are too small to remain wet. Salinity increases during the dry season and there is a clear pattern in the timing of increasing and decreasing volume;
- ephemeral lakes that exist only sporadically after unpredictable and rare large rainfall events. These lakes can be dry for many years then wet-up for a period of a few weeks or months, before drying out again for many years. Lake Torrens in South Australia is usually a dry salt pan, but was a salt lake of 0.5 m depth within two weeks of rains in 1989.

The accumulation of salt in the upper soil profile is also associated with the development of **chenopod** shrubland. Chenopod (saltbush) shrublands are found extensively in the arid zones of Australia, covering about 7% of the mainland. *Atriplex* (saltbush) and *Maireana* (bluebush) are two important genera. Chenopod shrublands are short (<0.5 m) and found on alkaline and saline soils or in saline depressions around salt lakes (Morton & Landsberg 2003). The key point is that there are natural saline ecosystems across Australia of biological and cultural significance; not all salinity is a problem. Ironically, these saline systems may be vulnerable to the same kinds of hydrological changes associated with human-induced salinity.

The second type of salinised environment is labelled **secondary** salinity, or **anthropogenic** salinity (salinity arising from human activities). Two types are clearly identified. First, there is the salinity that arises from using too much irrigation water and/or using irrigation water of poor quality (i.e. containing salt) without sufficient leaching and drainage. Such **irrigation salinity** is the cause of much of the salinity problem in the Murray-Darling Basin. The second is **dryland salinity**, arising because of a change in the vegetation cover of the landscape. In both cases, land and water salinisation is caused by a change in the hydrological balance of the landscape. In the case of irrigation salinity, increased input of water (as irrigation) has raised the water table and brought salts, previously located at depth, to the surface. These salts become concentrated in the root zone. Using poor quality (salty) irrigation water has exacerbated the problem.

In the case of dryland salinity, it is the removal of trees and their replacement with crops and pasture that has resulted in groundwater rising (Hatton & Nulsen 1999). When trees are returned to the landscape in sufficient area and in appropriate locations, increased depth to groundwater can be achieved (George et al. 1999; Fig. 8.3). It is important to recognise that while this generality holds true, the local pathways by which this cause gives effect to soil and

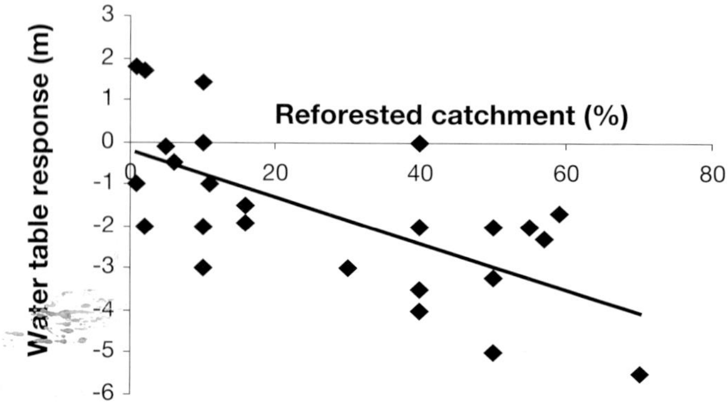

Figure 8.3 In a review of 80 sites in Western Australia, George et al. (1999) showed that reforestation could reduce groundwater depths substantially

Source: Redrawn from George et al. (1999)

water salinisation can vary greatly. In the simplest case, tree clearing increases local groundwater recharge, which causes a vertical rise in groundwater directly under the cleared land, bringing salts up to the surface. In a local hillslope system, increased vertical recharge as well as increased surface run-off may cause saline groundwater levels to develop vertically but also to flow and accumulate downslope (compounded by the accumulation of run-off), resulting in some displacement of cause (tree clearing and increased recharge) and effect (the surface expression of saline groundwater and waterlogging). In more regional systems, such as those associated with the bulk of salinised land in Western Australia, broad valley floors are salinising through complex combinations of all these phenomena at very large scales: enhanced local recharge due to clearing of valley woodlands, the vertical rise of saline water tables partly as a direct result of this increased recharge, large regional flood events arising from enhanced run-off from cleared hillslopes, and increased hydraulic gradients from evolving groundwater levels under cleared, adjacent hillslopes. The key message is that while the ultimate cause of secondary salinisation is clearing, the conceptual models for how this translates into salinised land and water vary greatly. In turn these difference have great implications for strategies to contain or remedy salinisation.

The rest of this chapter is concerned only with dryland salinity. However, before we can discuss this in detail we need to know how to express the degree (amount) of salinity in an aquifer.

Quantifying salinity

The salt content of water is usually expressed in units of µS cm^{-1} (microSiemens per centimetre). These are units of electrical conductivity, often abbreviated to **EC units**. For simplicity, we ignore the non-linearities that can arise in the relationship between electrical conductivity and salt content and can approximate the relationship as:

$$1000\ \mu S\ cm^{-1} = 1000\ \text{EC units} \approx 0.64\ g\ L^{-1} = 640\ \text{parts per million (ppm)} = 1\ dS\ m^{-1}$$

Seawater has a typical value of electrical conductivity of 54 000 µS cm^{-1} while rainwater is of the order of 100 µS cm^{-1} (typically higher near the coast and declining inland). Saline groundwater has salinity in the range 2000–50 000+ EC units (although the lower limit for 'saline' groundwater is arbitrary).

An alternative measure of salinity in water is the amount (mass) of salt present in a known volume (1 L) of water. Seawater contains, on average, approximately 34.6 g L^{-1} (or 600 millimolar). Clearly this is too salty to drink or to use as irrigation water.

A low salinity groundwater might contain less than 2 g L^{-1}, while moderately saline groundwater might contain 2–10 g L^{-1}. Highly saline groundwater contains more than 10 g L^{-1}, approximately equivalent to one-third the salinity of seawater.

The World Health Organisation recommends that water used for drinking by humans should not contain more than 0.5 g L^{-1} of salt (corresponding to an EC of 800 µS cm^{-1}). Water begins to start to taste salty to humans at about double this value.

Having established the causes and quantification of dryland salinity, we now assess the areal extent and the impacts of dryland salinity in Australia.

Extent of dryland salinity in Australia

The Australian Dryland Salinity Assessment 2000 report (National Land and Water Resources Audit 2001) compiled state by state assessments of the presence and potential for development of dryland salinity. The potential for the development of dryland salinity is based upon information about shallow groundwater tables and land use practices. From this assessment, the

Table 8.1 The area of land which is highly likely to become saline affected in Australia by 2050 (in hectares)

State/Territory	1998–2000	2050
NSW	181 000	1 300 000
Vic	670 000	3 110 000
Qld	Not assessed	3 100 000
SA	390 000	600 000
WA	4 363 000	8 800 000
Tas	54 000	90 000
Total	**5 658 000**	**17 000 000**

Note: The Northern Territory is considered to have extremely low probability of dryland salinity developing and is not included in this table

Source: Data taken from the Australian Dryland Salinity Assessment 2000

area of land with a high potential to develop dryland salinity totals more than 5.5 million ha. By 2050 this area could increase to 17 million ha (Table 8.1). Currently there are approximately 2.5 million ha of land affected adversely by salinity. About 70% of this is located in Western Australia, but this dominance in Western Australia is likely to change in the next 25 years.

The major regions currently showing and likely to show significant dryland salinity in the next 50 years occur in the highly productive agricultural land of south-western Western Australia, much of Victoria, central and eastern New South Wales and eastern Queensland (Fig. 8.4). Not only does salinity degrade productive agricultural land and streams, it also corrodes railways, roads, pipes and building foundations.

It is clear that loss of productive land, loss of infrastructure (roads, rail), loss of surface water quality (streams and lakes) and loss of genetic biodiversity pose a serious challenge to land and water managers at local and regional scales. The economic, environmental and social impacts of salinity across the most heavily populated regions of Australia cannot be ignored.

The financial costs of dryland salinity to Australia are vast. Within the Murray-Darling Basin alone, lost agricultural production, damage to infrastructure and other costs are estimated at $250 million per year.

Plant responses to salinity

Salt within the rooting zone of plants causes reduced seed germination, reduced growth, reduced yield and eventually plant death. Salinity at the landscape scale has impacts on ecosystem

Table 8.2 A summary of the assets located in areas of high risk from salinity development or shallow groundwaters

Asset	2000	2050
Agricultural land (ha)	4 650 000	13 660 000
Native perennial vegetation (ha)	631 000	2 020 000
Length of streams and lake perimeters (km)	11 800	41 300
Rail (km)	1600	5100
Roads (km)	19 900	67 400
Number of towns	68	219
Number of important wetlands	80	130

Source: Australian Dryland Salinity Assessment 2000

Figure 8.4 Areas at high hazard or risk of developing dryland salinity (indicated by dark shading) include the most agriculturally productive land in Australia

Source: National Land and Water Resources Audit, © Commonwealth of Australia 2001

species composition, ecosystem structure, net primary productivity and biodiversity (Murray et al. 2003).

Salinity exerts its negative impact on plant growth through two effects. The first is an **osmotic** effect. As the concentration of salt in soil water increases, the water potential of the water is reduced (see Chapter 2). As the water potential of soil water declines, the availability of water to a plant is greatly reduced and plant growth is reduced. A ready and freely available supply of water is a precondition for plant growth.

Figure 8.5 illustrates the relationship between the salt concentration of a solution, the water potential of the solution and the conductivity of the solution. An increased salt concentration causes solution conductivity to increase and water potential to become more negative, which means the water is less available to plants.

The second way in which salt reduces plant growth is through the toxic effect of sodium and chloride ions (especially sodium ions) on metabolism. The vast majority of plants do not need sodium for normal metabolic processes such as enzyme activity, respiration or photosynthesis. When sodium ions accumulate in cells, they disrupt normal metabolic processes by interfering with enzyme function. Growth is therefore impaired.

When plants absorb water that is saline, a large proportion of the salt accumulates around the roots. Consequently the actual concentration of salt around a root (and hence the osmotic effect and toxic effect) is much higher than the average concentration in the soil solution.

Crops and most trees show a poor tolerance to conductivities above about 1000 $\mu S\ cm^{-1}$ (or 1000 EC units or 1 $dS\ m^{-1}$). Some tree species, such as *Casuarina cunninghamiana* and

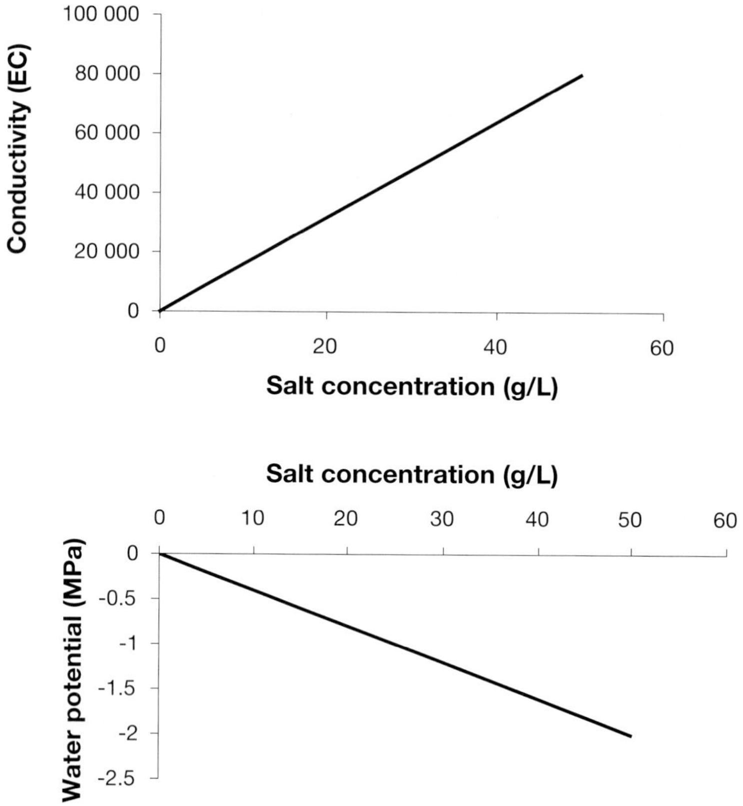

Figure 8.5 As the concentration of salt (NaCl in this example) in a solution increases, the electrical conductivity increases approximately linearly and the water potential of the solution decreases linearly. The exact relationship (slope) changes with temperature and type of salt. A decreased water potential means that water is less available to a plant. A soil water potential of less than –1.5 MPa is often called the wilting point of the soil as it can induce wilting in some (crop) plants. The wilting point of Australian trees is much lower than this value

Eucalyptus camaldulensis are frequently identified as being relatively salt tolerant (e.g. compared to annual crops like lettuce) and can grow well in soil salinities of almost 1500 $\mu S\ cm^{-1}$ (Sun & Dickinson 1995). Niknam and McComb (2000) and Bell (1999) reported on the comparative response of many Australian tree species to salinity. Marcar et al. (in Stirzaker et al. 2002b) presented a comprehensive evaluation of key tree species and their suitability across rainfall, soil and salinity parameters. It is important to note that relative tolerances in greenhouse trials may not be the same as those observed in field trials, although a salt concentration of 20 000 $\mu S\ cm^{-1}$ or more causes plant death in both field and glasshouse trials. In contrast, growth of saltbush, which can be used as a fodder crop, is stimulated by low levels of salt (it is a halophyte, a salt-loving species) and can tolerate very high levels of salinity (Fig. 8.6).

Waterlogging and saline landscapes

Dryland salinisation is associated with the development of high groundwater tables and this can cause significant **waterlogging** of soil and roots (MacFarlane & Williamson 2002). Saturation of soil with water (waterlogging) creates an **anaerobic** environment for roots and reduces root growth, ion and water uptake and eventually plant mortality if prolonged periods of waterlogging occur. Plants exposed to salty and waterlogged soils are more sensitive to the salt

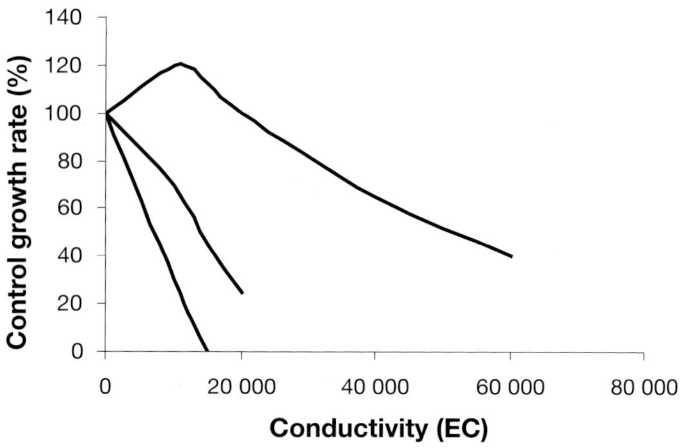

Figure 8.6 Rice and wheat (the lower two lines) are intolerant of salt in the environment. Such very sensitive plants are called glycophytes. Salt bush (upper curve) growth is stimulated by low levels of salt and it tolerates high levels of salt; such plants are called halophytes

than plants exposed only to the salt, that is, water logging reduces the plant's ability to tolerate salinity (MacFarlane & Williamson 2002).

Having established the causes, extent and impacts of salinity in this first part of the chapter, the second part explains the principles applied when considering the use of trees in the landscape to manage catchment hydrology and hence ameliorate dryland salinity.

Principles underlying the use of trees to ameliorate salinity[1]

Catchment water balance before and after clearing

The ecohydrology of Australian landscapes before and after European settlement has been discussed extensively (Dunin et al. 1999; Eberbach 2003; Hatton et al. 2004) and the idea of mimicking the pre-clearing pattern of water use through changes to modern land use management has gained significant momentum. The following summarises current thinking.

In most of temperate Australia, rainfall is highly variable within and between years and monthly potential evaporation exceeds monthly rainfall for most months (Chapter 1). Rainfall is highest close to the coast and lower inland; this gradient is opposite to that for potential evaporation, which is larger inland than at the coast. Topographic relief is minimal, especially in river headwaters, and so river flows tend to be slow. Soils have accumulated salt for millennia and stored salt ranges from 0.001 to 0.17 kg m^{-2} y^{-1}. Prior to European arrival, water tables in Western Australia, for example, were generally deep and aquifers were rudimentary (Hatton et al. 2004). Native vegetation had several adaptations to these conditions, including:

- sclerophylly, where leaves are tough and thick, with low water content and long-lived;
- perennial patterns of water use, with a dominance of evergreen woody species;
- stem flow of water was favoured so that deep infiltration into the ground immediately around tree stems was rapid;

1 The structure and content of this section of Chapter 8 draw extensively from the excellent work by Silberstein et al. (2002a, b), Hatton et al. (2002), Stirzaker et al. (1999) and Coram (1998). This book is not heavily referenced for ease of reading. However, the major contribution of these authors (and other works as cited) is acknowledged here.

- deep roots, with lignotubers often present;
- leaf area index (LAI) showed small seasonal change.

As a consequence of these characteristics of climate and vegetation, the following hydrological cycle was consistently observed in temperate Australia prior to the arrival of Europeans:

- almost all the rainfall that fell was either transpired by vegetation or evaporated;
- surface flows were minimal and local rainfall was captured and remained local;
- water use by vegetation occurred year round;
- groundwater recharge was minimal, typically less than 5 mm y^{-1}, and in regions where rainfall is <600 mm y^{-1} base flow in rivers was zero or very low;
- leaching of salt from the soil profile was extremely limited and salt accumulation in soil was large over time;
- trees of many ecosystems could use shallow soil water stores immediately after rain, switching to deeper soil stores as required and groundwater reserves when it was available, even using groundwater of low salinity (Chapter 6);
- very little liquid water left a region.

In contrast, much of the Australian landscape now has the following attributes:

- sclerophylly has been replaced by soft mesophytic leaves of crops with high water content;
- water use does not occur all year;
- the LAI of annual crops is larger than that of native vegetation but only for a short period of time;
- groundwater recharge has increased by up to 3 orders of magnitude;
- run-off and groundwater recharge have increased so base flow to rivers has increased, carrying salts to the streams and rivers, which flow more often;
- perched aquifers are developing, mobilising soil stores of salt.

We can formalise these generalised observations by considering a catchment water balance (Brooks et al. 1997):

$$P = I + RO + E + T + D + \Delta S$$

where P = rainfall, I = interception losses, RO = run-off, E = evaporation from soil and lakes, T = transpiration from vegetation, D = drainage below the root zone (which may become recharge of groundwater or stream flow) and ΔS = change in soil water store (and, technically, change in vegetation water store, which is usually ignored). Note that E is often taken to include evaporation from wet canopies and therefore can include interception losses, or interception losses can be measured separately.

For a typical eucalypt woodland in southern Australia, the following values can be used as an example: P = 900 mm y^{-1}, E + T = c. 800 mm y^{-1}, I = 70–80 mm y^{-1}, RO = 10–20 mm y^{-1} and therefore the amount of water draining and potentially becoming groundwater recharge is 10–20 mm y^{-1}. In the Murray-Darling Basin, rates of deep drainage have been estimated to be 5–30 mm y^{-1} (Allison et al. 1990) while Knight et al. (2002) cited estimates of recharge under remnant mallee in the Murray-Darling Basin as being <0.1 mm y^{-1}. Dunin (1992) estimated native vegetation in southern Australia to have a **leakage rate** (see below) of approximately 5 mm y^{-1}, although the true value is probably 0.1–10 mm y^{-1}. Most **recharge** beneath native vegetation reappears as base flow of streams and rivers, so groundwater volumes are static.

These rates of recharge are very low, especially since groundwater discharge through vegetation occurs and therefore base flow of streams is low when rainfall is <600 mm y^{-1}. Consequently groundwater tends to be deep, but more importantly, it is **in equilibrium and the catchment is in hydrologic equilibrium: the rate of input equals the rate of output**.

For a typical crop or pasture grassland growing at the previously wooded site, the numbers are quite different (Greenwood et al. 1985; Farrington et al. 1992). P does not change and remains at 900 mm y^{-1}. However, E + T = 600–700 mm y^{-1}, I = 60–100 mm y^{-1} and RO = 20–40 mm y^{-1}, therefore groundwater recharge is increased to about 60–220 mm y^{-1}. Some of this deep drainage is discharged as base flow of streams and rivers, carrying salt with the water flow, but there is also a **substantial rise in the water table** towards the surface, bringing salt to the surface where it is deposited when the water is transpired by vegetation or evaporated directly to the atmosphere. The catchment is no longer in hydrologic balance and dryland salinity results. In addition, the increased base flow of groundwater to streams carries with it some of the salt that has moved upwards. Consequently stream salinity increases (Fig. 8.1). This increased stream salinity (see below) has negative impacts on aquatic biota and the water's suitability for human consumption and irrigation use.

These principles are illustrated in the study by Bari and Schofied (1992) who applied a simple water balance approach to an experimental site south of Perth in Western Australia. They showed that following extensive clearing of native forest, land and stream salinity increased. Following reforestation of 70% of the cleared land at an average stocking density of about 350 stems per ha and a crown cover of about 40%, groundwater depths beneath the plantations at the site declined by about 7 m within 10 years of reforestation, while groundwater depth beneath pasture (control) sites increased by 2 m. Rainfall and pan evaporation rates showed no difference between control and reforested sites. The difference in groundwater depths are clearly the result of reduced recharge because of **reduced leakage past the root zone** (Bari & Schofield 1992). Eldridge and Freudenberger (2005) show how the sorptivity (the ability of soils to absorb water in the initial stages of infiltration) of fine textured soils and the rate of steady state infiltration was substantially lower in cropland and pasture than in woodlands (Table 8.3). Much of this effect was because of the large number of macro-pores present in the woodland site and the absence of such pores in the pasture and cropland. The ratio of sorptivity under ponded conditions to that under unsaturated conditions (sorptivity ratio) is a measure of the degree of macro-pores present in the soil. Large values of this ratio indicate that macro-pores contribute substantially to infiltration.

Drainage past the root zone and groundwater recharge

One key issue in managing landscapes to prevent or ameliorate dryland salinity is that of **drainage** past the root zone, also called **leakage**. Leakage is defined as the movement of water past the root zone of the vegetation and is the phrase used for the remainder of this chapter. Following clearing in salinisation prone landscapes, this enhanced volume of leakage (or a large fraction of it) reaches the water table (hence it is called groundwater recharge) and causes the water table to rise to the surface, bringing salts.

Table 8.3 Different rates of sorptivity and steady state infiltration for different land uses

	Woodland	Pasture	Cropland
Sorptivity (mm $h^{-0.5}$)	140	15	10
Steady state infiltration (mm h^{-1})	240	75	50
Sorptivity ratio	46	14	8

Source: Data from Eldridge & Freudenberger (2005)

Table 8.4 Leakage of water past the root zone occurs when the rate of input exceeds the rate of evapotranspiration of the ecosystem. Rates of leakage are lowest under woody vegetation

	Leakage rates (mm y^{-1})		
	Pasture	Crops	Trees
Summer rainfall	20	30	5
Winter rainfall	60	40	10

Source: Estimates from Silberstein et al. (2002a)

The rate of increased leakage following clearing is determined by several key factors. The total annual rainfall and its timing, soil type, rooting depth of the species planted, and duration of the fallow (uncropped) period are the main determinants of the rate of leakage. Measuring leakage rates is notoriously difficult and any values quoted are estimates only (Table 8.4).

Knight et al. (2002) estimated deep drainage below cropland in the Murray-Darling Basin to be 13–35 mm y^{-1} and Smettem (1998) estimated drainage to be 20–100 mm y^{-1} under agricultural land in southern Australia. Leakage from native vegetation is in the range 0.1–10 mm y^{-1} (Allison et al. 1990; Knight et al. 2002).

Asseng et al. (2001) recognised the complexity and variability inherent in measuring leakage in the field and used a **simulation model** (Agricultural Production Systems Simulator; APSIM) to analyse leakage under wheat crops in Western Australia. They also compared model outputs with field data wherever possible. They showed that deep drainage under wheat could vary between 0 and 386 mm per year in the high rainfall zone (460 mm y^{-1}) and between 0 and 234 mm per year in the low rainfall zone (310 mm rain y^{-1}). However, annual average drainage rates ranged from 36 mm (low rainfall site) to 134 mm (high rainfall site) on sandy sites and ranged from 4 mm (low rainfall site) to 57 mm (high rainfall) on clay soils, values that are about one order of magnitude larger than the rate observed under native vegetation (Asseng et al. 2001). Similarly, Hatton and George (2001) reviewed 60 case studies of recharge under native or agricultural land and showed that recharge is much larger under agricultural land, especially when annual rainfall is less than 1000 mm (Fig. 8.7). Petheram et al. (2002) reviewed 41 studies of recharge and concluded that rainfall, land use change (e.g. replacing wooded landscapes with annual crops) and soil type were the major determinants of recharge rates. They concluded that recharge under shallow rooted annual vegetation was much larger than under deep rooted perennial vegetation.

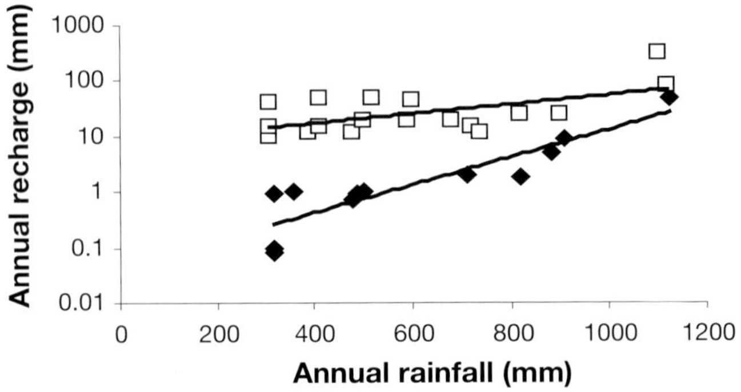

Figure 8.7 Annual rates of groundwater recharge increase with increasing rainfall and are generally larger under agricultural land (black diamonds) than land covered with native forest (open squares), especially where rainfall is less than 1000 mm y^{-1}

Source: Redrawn from Hatton & George (2001)

It is surprising how little water is required to cause groundwater levels to rise dramatically. This is because most of the space within the water table (within the aquifer) is actually taken up by soil and rock and the water already present (low specific yields). Thus, for an aquifer with a specific yield of 0.05, if 25 mm of water is added to the surface above the aquifer the water table will rise by 500 mm (25/0.05).

Clearly the removal of trees and their replacement with crops has significantly altered the hydrologic regime of Australian landscapes. The question that now occupies the rest of this chapter is: Can replanting trees return the landscape to a healthy, salt-free state?

Planting trees to regain hydrologic balance: first principles

There are two fundamental and distinct roles for trees in addressing dryland salinity. The first is to **reduce recharge** through interception of water prior to it reaching groundwater. This water is transpired and thus recharge is reduced. The second is to **enhance discharge**; this requires tree roots to access groundwater (or the capillary fringe above it) and discharge groundwater through transpiration. This distinction must be borne in mind during the rest of this chapter.

As the availability of water (crudely measured as rainfall but better assessed as the ratio of rainfall to annual pan evaporation (Eo)) increases, the LAI of native unmanaged woodland or forest will increase linearly before reaching a maximum value and thereafter showing no further increase. From this, we can deduce that if we wish to reduce leakage (i.e. the amount of water recharging groundwater) and we can't manipulate rainfall, we must increase the LAI of the vegetation **all year round**. Having a large crop LAI only in the summer will not suffice, especially where rainfall is winter dominant.

The rate of water use by native woodlands and forests in southern Australia tends to a maximum of approximately 1200 mm y^{-1}. This is because much of southern Australia receives less than 1600 mm y^{-1} and a fraction of rainfall is always unavailable to roots because of evaporation from soil and plant canopies. From this fact, plus the relationship between LAI and the ratio of rainfall to Eo (Fig. 8.8), we can generate some predictive rules about the probability of dryland salinity occurring (Stirzaker et al. 2002a).

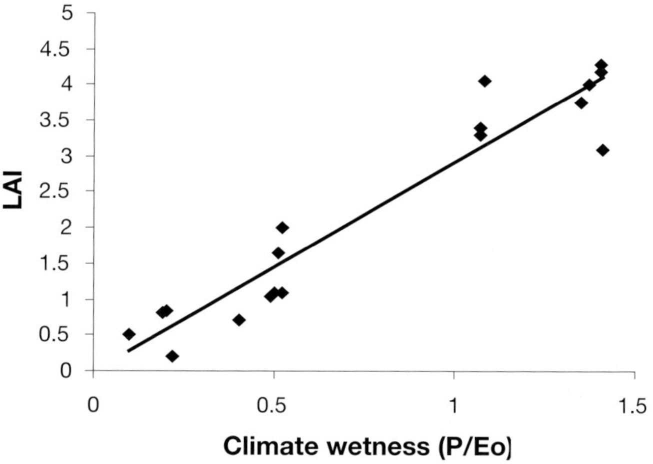

Figure 8.8 As the climate wetness index increases (site available water increases) the LAI of the site increases, for a range of Australian sclerophyllous woody vegetation types. P = precipitation, Eo = annual pan evaporation

Source: Redrawn from Ellis et al. (1999)

These rules can be formalised thus:

1 Where rainfall is less than about 1000–1200 mm y^{-1}, water use by vegetation will be similarly reduced from the maximum rate of approximately 1200 mm y^{-1}. However, if the LAI is at the equilibrium value predicted from Figure 8.8 we can predict that groundwater recharge is minimal and dryland salinity should be unusual, broadly speaking (specific local conditions can change this broad prediction). This is because the vegetation is in equilibrium with the water supply and leakage is very low (Dunin et al. 1999; Hatton & Nulsen 1999).
2 Where rainfall is much higher than 1200 mm y^{-1}, we can predict that some of this rainfall is recharging groundwater (which may be discharged as stream flow or cause the water table to rise). However, the probability of development of salinity in the landscape is lower than in low rainfall zones because the natural leakage rates would have prevented the accumulation of salts in the soil and aquifer. Changes in the amount of woodland and forest in the catchment have an impact on water yield of the catchment and less impact on salinity of the water coming from the catchment (Chapter 6).
3 If the average annual LAI is less than the equilibrium LAI predicted from Figure 8.8, groundwater recharge is likely to be occurring.
4 Where local topography produces surface or subsurface lateral flow, and this flow is reasonably fresh, the LAI of the vegetation downslope is likely to be larger than predicted from Figure 8.8 because it is the total water availability that is important and rainfall, in this particular case, is not the sole source of water for the vegetation. The same is true for groundwater dependent ecosystems (Chapter 6).
5 Low rainfall catchments (<900 mm y^{-1}) with an equilibrium LAI are likely to have more salt stored in the subsoil than high rainfall catchments (>900 mm y^{-1}) with a higher equilibrium LAI. This is because the high rainfall catchments have had water percolating down through the subsoil and leaching the salt to deeper layers. Consequently, if we wish to minimise the movement of salt into streams, planting trees is less effective in high rainfall zones than in low rainfall zones.

Not all aquifers are the same

For convenience, three types of aquifer associated with secondary salinisation may be recognised (Coram 1998). First, there are small (<3 km in one or more dimensions) **local flow** aquifers. These are found in locations with **high topographic relief** (hilly) and on edges of plateaus and ridges. Second, there are intermediate scale aquifers (3–50 km in one or more dimensions) which are often alluvial and occasionally in glacial valley fill in foothills and valleys (Coram 1998). Finally there are regional scale aquifers (such as the Great Artesian Basin) which are larger than 50 km in all dimensions. These tend to be broad riverine plains on depositional basins (Coram 1998).

In addition to the size of the aquifer, the average rainfall within the catchment must be considered when determining the value of trees in the landscape. In high rainfall zones (>900 mm y^{-1}), removal of trees tends to be less important in relation to causing salinity, although even high rainfall zones of northern Australia are not immune to dryland salinity. However, the enhanced water yield of these high rainfall zones after clearing can be important for diluting downstream river salinity arising in lower rainfall zones further upstream.

There are two basic approaches to revegetation in salinising catchments. The first is to attempt to enhance the discharge from the groundwater to keep the water table in low lying areas well below the soil surface. The second approach is to use revegetation to control leakage (and ultimately groundwater recharge) upslope of the salinising areas.

Discharge enhancement with trees is now widely recognised as a short-term strategy of only limited application. Once regional saline groundwater tables are at or near the surface,

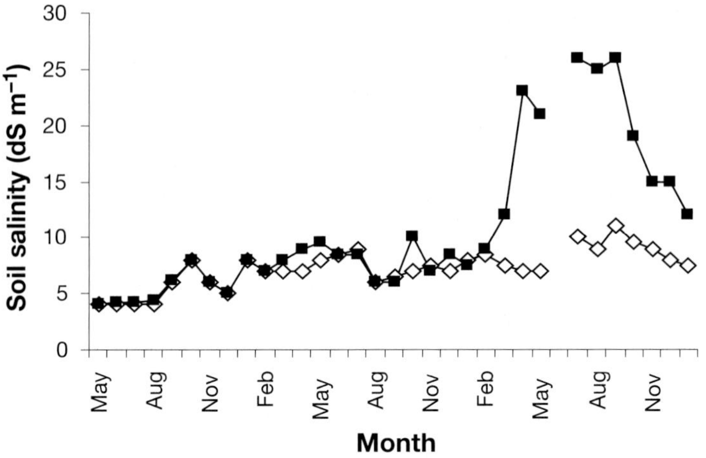

Figure 8.9 The uptake of saline water by roots causes a gradual accumulation of salt in the root zone because the roots exclude the salt. Salinity at 300 cm increased 5-fold (closed squares) for a 6-month period, compared to the salt content lower in the profile (open diamonds)

Source: Redrawn from Morris & Collopy (1999)

there is a real possibility that even if trees can be established their root zone can be expected to run a high risk of salinising as transpired groundwater leaves behind an accumulation of subsurface salts (Fig. 8.9). It should be noted, however, that under natural conditions the broad valley floor woodlands in what is now the wheatbelt of Western Australia were probably essential in keeping water tables and salt at depth. Prior to clearing, there was up to 70 m of unsaturated, relatively fresh sediment that would occasionally recharge as a result of regional flooding. Over subsequent years this water would be transpired, resulting in a dynamic equilibrium that accumulated brines at great depth below a deeply leached profile. Now that these saline water tables have risen to or near the land surface, trees cannot access this salty water. This hysteretic relationship between trees, salt and the environment precludes taking a 'natural' pathway back to the original condition via tree planting in these systems.

The modern emphasis on **recharge control strategies** with trees is informed by consideration of the following (Stirzaker et al. 2002a):

- The discharge capacity of the aquifer – can the trees be expected to transpire at sufficient rates to reduce catchment recharge to something like the discharge capacity? In other words, can leakage be reduced to something less than the rate at which the aquifer can discharge laterally? If not, then water tables will rise until new discharge mechanisms emerge to match recharge, typically discharge via evaporation at the soil surface with associated salinisation;
- Size of the aquifer (local, intermediate, regional). Small systems can respond quickly (within decades) and substantially to sufficient changes in vegetation type and land use practice. In contrast, regional aquifers may show slow (>100 years) and small responses to changes in vegetation and land use.

The following section outlines the ecohydrological considerations for small and intermediate aquifers.

Small aquifers with high discharge capacity

Small aquifers with high discharge capacity are often found on the edge of plateaus and ranges of hills. Where high discharge is occurring, it is usually as seeps on the side of hills arising from the presence of a layer of impermeable rock. Groundwater discharge is usually of relatively low salinity. Replanting of trees in **alleys or belts** or scattered throughout the recharge area can significantly reduce recharge (White et al. 2002), and plantings within the **discharge zone** can also be effective in reducing salt intrusion into streams. However, these are not always effective (Greenwood et al. 1995). In one extensive study near Katanning in Western Australia it was shown that, where discharge rates are high, groundwater depth is responsive to reductions in recharge arising from revegetation of the land. White et al. (2002) calculated that replanting 16% of the catchment with trees would reduce leakage to 5 mm y^{-1}, although Dunin (1992) concluded that 22% of the catchment would have to be planted with trees. Interestingly, if the annual crops were replaced with lucerne, which has deep roots, the area of land that had to be planted with trees was reduced to 8% (White et al. 2002) or 12% (Dunin 1992). Most importantly, a **tree/lucerne/crop rotation** system was found to be economically viable, despite the low rainfall (about 450 mm y^{-1}) of the site (Poole et al. 2002).

Small aquifers with low discharge capacity

When discharge is controlled by changes in the topographic gradient (break of slope catchments) discharge tends to be slow. Discharge can also be slow for other reasons, such as faults in the rock placing a layer with low transmissivity in the path of discharge. The salt content of the subsoil tends to be high and the groundwater being discharged tends to be very saline. These sites are only slowly responsive to revegetation and care must be taken when placing trees to avoid exposing the roots to excessively saline conditions. Extensive planting of trees in the recharge zone is typically required for there to be much effect on reducing saline discharge or salinised area. These aquifers are widespread in south-west Western Australia and usually exist where groundwater flows occur through deep regolith (Coram 1998). The Toolibin catchment (see Chapter 7) is one example in Western Australia. The tablelands of New South Wales and the Liverpool Plains exhibit this type of aquifer (Coram 1998).

McJannet et al. (2000) reported on the efficacy of **break-of-slope plantings** at a site in north-eastern Victoria. They showed that tree water use is substantially larger than pasture water use and that recharge can be reduced by tree plantings. However, they showed that the presumed access of the trees to relatively shallow groundwater was not occurring and that the plantation is susceptible to drought conditions, a result not expected from theory since it is always assumed that shallow aquifers will be accessed by tree roots.

Intermediate scale aquifers

Intermediate aquifers have groundwater flows occurring over distances of typically 5–20 km (but up to 50 km). These aquifers are found in fractured rock or shallow sediments. Discharge from these aquifers is controlled by reductions in hydraulic conductivity along the gradient in hydraulic head, or by a reduction in transmissivity or hydraulic gradient as the aquifer thins, or by the presence of impermeable layers. Discharge of groundwater is found in low points in the landscape.

In Western Australia there are many broad valley systems, associated with low transmissivity. Salinity is a pronounced problem in such systems and many thousands of hectares are affected by very saline groundwater discharge. Similarly, in South Australia, Victoria and New South Wales, salinity problems are often associated with such intermediate size aquifers and broad valleys (Hatton et al. 2002).

In intermediate aquifer systems, the equilibrium between recharge and discharge following vegetation clearing takes many decades and most of these systems in Australia are yet to reach equilibrium. Thus, the extent of salinity has not reached a plateau in Western Australia. Furthermore, discharge capacities are relatively low and consequently revegetation has only a small impact on regional water table rises. A large proportion of the entire landscape (rather than belts of trees) must be revegetated to have a significant impact on discharge. Successful planting on discharge zones is difficult because of the highly saline groundwater being discharged, which most trees are unable to tolerate. Further, the large influence of occasional regional flooding events can override any hillslope revegetation effects on valley floor water tables. Effective solutions for recovering these systems from salinisation usually involve an **engineering component** (groundwater pumping, groundwater drainage or surface water control), which may be enhanced through the inclusion of tree planting.

Groundwater response times to land use change

It is clear that land use change has had significant effects on the salt and water balance of Australian catchments. The large time lags between changes in land use and response in groundwater and salt movement has made it difficult to effectively manage landscapes and prevent the development of salinity problems. The long time lags also make it difficult to predict the outcomes of different management options (Jolly & Cook 2002). However, research conducted since the early 1990s shows time lags of 40–100 years in the Mallee region of south-east Australia. This is the time taken for the wetting front that arises when trees are replaced with pasture and crops to move through the soil profile to the water table. Once the wetting front reaches the water table, recharge occurs and there is increased flow of saline groundwater to the Murray River, for example.

Jolly and Cook (2002) and Dawes et al. (2004) used a relatively simple modelling approach to answer a question: If revegetation at a site can reduce drainage past the root zone (deep drainage) to zero, how long does it take for groundwater recharge to respond? Jolly and Cook (2002) modelled two different soil types for the Mallee region of south-east Australia and Dawes et al. (2004) applied their model to the mid-Macquarie catchment of New South Wales. Jolly and Cook (2002) showed that for a loamy sand soil profile, recharge begins to decline after 5–20 years, depending on the depth of the water table, and takes a further 25–130 years to reach an equilibrium of zero recharge. For a sandy loam profile, these times are extended by 5–50 years. The important conclusion is that the **cessation of deep drainage does not result in an immediate cessation of recharge** because of the volume of post-clearing water stored in the soil profile above the water table. This volume continues to move down and recharge the groundwater until water no longer moves downwards due to gravity. Dawes et al. (2004) estimated the equilibrium response time for groundwater to a change in land use and concluded that the impacts of land clearing in the second half of the 20th century will not reach equilibrium before the end of the 21st century and that, even with large scale planting on the mid-Macquarie catchments, it will take at least 50 years for salt input to streams to decline.

In conclusion, four key pieces of information are needed to design an agroforestry system to ameliorate salinity (Hatton et al. 2002). These are:

1. discharge capacity of the aquifer and the local constraints on that capacity;
2. groundwater salinity levels;
3. the size of the aquifer system;
4. the endpoint required (for stream or land salt contents).

The following generalisations (Hatton et al. 2002) can be made:

1. local (small) aquifers respond faster to revegetation (through reduced recharge) than intermediate or regional aquifers;
2. shallow water tables and relatively fresh groundwater will favour the establishment of tree belts in the landscape and the interception of groundwater by those belts of trees;

3 highly transmissive aquifers respond faster to revegetation and tend to have lower salinities, which favours vegetation growth;
4 to establish recharge control, revegetation should occur first in systems with the largest rates of recharge.

Having established the basic principles concerning the link between aquifers and vegetation, we can now discuss the central question: Where in a landscape should we plant trees to maximise their impact on recharge or discharge of an aquifer? This is the final part of the chapter.

Where, and in what formation, should we plant trees to ameliorate salinity?[2]

Trees in the landscape

Trees can be visualised as wicks, linking a source of water (an aquifer, the capillary fringe above an aquifer, or soil water) with the atmosphere. By judicious placement of trees in the landscape, it may be possible to influence landscape hydrology and hence manage salinity (Coram 1998), although revegetation by trees alone is unlikely to be economically viable for many parts of Australia because of the slow rate of tree growth and low financial return to the landholder (George et al. 1999).

Option 1: tree belts on hillslopes

Salinity in the landscape, either directly at the surface or within the rooting zone below the surface, arises because of the movement of saline groundwater. Hillslopes represent points on the landscape where trees could be placed to intercept both surface and subsurface flows before they reach the valley floor. This section discusses the ecohydrology of trees on hillslopes in relation to salinity.

In deciding to plant trees on a hillslope, the following questions must be answered in the affirmative (Silberstein et al. 2002a):

1 Is there water in the upslope region that can move downslope?
2 Is the slope sufficient to cause significant lateral flow of surface and subsurface water?
3 Is the soil permeable enough to allow water to flow laterally as subsurface flow?
4 Is subsurface flow occurring within the rooting zone of the trees?

Any answer in the negative will preclude the use of limited hillslope plantings to ameliorate salinity.

In deciding whether there is likely to be water available to move downslope, we need to know annual rainfall and annual crop water use. Where rainfall significantly exceeds pasture water use, it is likely that water is available but the movement of this water will be determined by the slope and hydraulic conductivity of the soil. Neither low slopes nor low soil conductivities favour lateral flows of water. On steep slopes surface run-off can be reduced by creating capture zones using low soil banks, thereby increasing infiltration.

White et al. (2002) examined water use of belts of trees planted along contours of hills in Western Australia. They estimated that the 8 m wide belts of eucalypts (four different species) used 595 mm of water per year. Of this, 440 mm was lost as transpiration, 100 mm as interception loss and 55 mm as soil evaporation. Most importantly, rainfall was only 445 mm y^{-1} and

2 The structure and content of this section of Chapter 8 draws extensively from the excellent chapters and reports by Stirzaker (2002a, b), Stirzaker et al. (2002), Silberstein et al. (2002b), Coram (1998) and LeFroy and Stirzaker (1999).

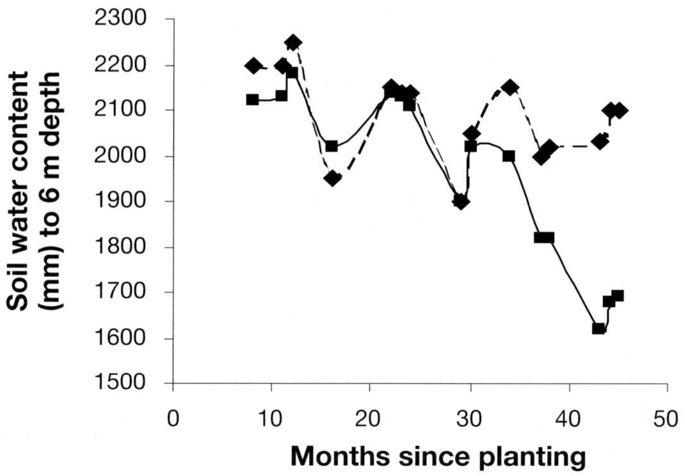

Figure 8.10 The total amount of water stored in the upper 6 m within a belt of trees (squares; solid line) was lower than that observed in adjacent cropland (diamonds; dashed line)

Source: Redrawn from Knight et al. (2002)

most of the difference between rainfall and water use (150 mm) was accounted for by uptake of groundwater flowing under the trees down the slope. They concluded that, where groundwater is available, belts of trees planted on slopes are effective in reducing leakage.

Knight et al. (2002) compared water extraction by belts of perennial trees and shrubs in south-eastern Australia with the extraction in adjacent crop fields. Figure 8.10 shows that the soil water content within the belt (measured to depths of 6 m) was less than that of the adjacent crop field about 2.5 years after planting the belt of trees and shrubs. Four years after planting, the soil beneath the belt was 400 mm drier than the profile beneath the crop. This is much larger than any single rain event can deliver and therefore it is unlikely that leakage would occur even during the heaviest rainfall at this location.

Several generalisations arise from field studies and modelling of tree plantings on slopes. These are:

- tree belts should be closely spaced when annual rainfall is much larger than pasture water use and there is the potential for a large amount of 'spare' water;
- steep slopes favour tree belts that are far apart;
- short wet or winter seasons favour closely spaced tree belts.

Option 2: trees planted over a shallow saline water table

Shallow aquifers are those that have the saturated zone about 1–4 m below ground level and they occur when recharge exceeds discharge. Such water should be available to trees since most trees can put roots down to these depths. However, root growth is poor in saturated soil because of the lack of oxygen in saturated soils (MacFarlane & Williamson 2002). Roots tend to stop growing at the junction between the capillary fringe and the saturated zone. The capillary fringe is the zone of water above the saturated water table where capillary forces between water and soil have caused some water to move upwards. It can be anything between 5 cm and 1 m in thickness and is the zone usually associated with water uptake by roots when roots are using groundwater. Above the capillary zone lies the fully aerated soil where most of the tree and almost all the crop roots are located.

Trees that are known to use groundwater, such as *Eucalyptus camaldulensis* (River Red Gum), do not usually use groundwater exclusively or use groundwater all year. Most use of groundwater is **facultative**. Facultative groundwater use means that uptake occurs when all other sources are exhausted (i.e. the soil above the capillary zone is very dry). Groundwater use declines with increasing salinity of the groundwater and therefore only groundwater that has a low salinity can be consistently used by trees. The switching between groundwater and soil water by trees is extensively discussed in Chapter 6, with an example from Western Australia.

The ability of trees to survive when located above shallow saline water tables is dependent on three principal factors. These are:

1 the salinity of the groundwater;
2 the sensitivity of the species to salinity;
3 the hydrology of the aquifer and soil above it.

Discharge of shallow saline groundwater by trees is enhanced most when the following conditions apply (Thorburn 1999):

- roots can access the capillary fringe;
- the water table is 3–5 m deep. If the water table is closer to the surface, accumulation of salt in the root zone is problematic;
- salinity is low, with a total electrical conductivity of less than 10 dS m^{-1};
- trees have some tolerance of salt and waterlogging;
- the permeability of the soil is low at field capacity so that the rate of salt movement to the root zone is low. However, the lateral extent of the impact of the trees is reduced;
- the impact of the trees on groundwater recharge is larger upslope of saline areas;
- the impact of trees on groundwater depth is likely to be larger in drier years than wetter years.

There are three major scenarios for planting trees above shallow saline groundwaters. The first is to plant on a regional scale above regional sized highly transmissive aquifers. The second is to plant on a regional scale on regional sized aquifers of low transmissivity. Finally, there is the option of planting at a local scale on local scale perched aquifers. These options are now discussed.

Option 3a: regional scale plantings on regional aquifers of high transmissivity

Regional scale planting of trees is expensive, both in terms of the cost of the plant material and the cost of labour. Lost earnings from the crops that are replaced by trees also have to be considered.

Highly transmissive aquifers can rapidly replace the water used by the trees and therefore, on average, the impact of the trees on the depth of the water table in this scenario is minimal. The time taken for trees to reach a critical salinity that causes reduced growth and yield declines as the level of salinity of the water table increases and the rate of water use by the trees increases. Figure 8.11 summarises the modelled relationship between the time taken to reach a critical threshold salinity and groundwater salinity as a function of the rate of tree water use (Stirzaker et al. 2002a). Clearly, the accumulation of salt within the root zone occurs faster when trees are transpiring a large volume of water every day, than when trees are transpiring only small volumes. Similarly, when the groundwater is highly saline the rate at which salt accumulates in the root zone is faster than when the groundwater is only moderately saline. Thus the **salt balance** of the new landscape (now containing trees) must be considered over the long term. The salt balance is determined by the rate of transpiration and the salt content of the groundwater.

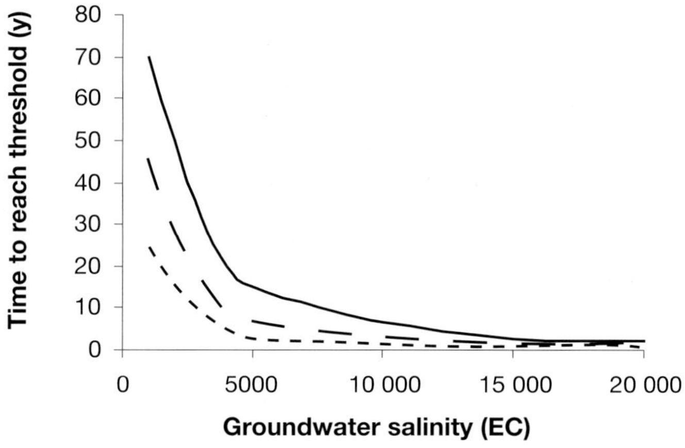

Figure 8.11 The time taken to reach the threshold salinity that causes tree death decreases as the salinity of the groundwater increases and the rate of tree water use increases. The solid line represents water use of 100 mm y^{-1}. The dashed line corresponds to a tree water use of 200 mm y^{-1} and the dotted line represents the rate of water use of 400 mm y^{-1}

Source: Stirzaker et al. (2002a)

Option 3b: regional scale planting on a regional scale aquifer of low transmissivity

Where the aquifer has a low transmissivity, the water used by a plantation can reduce groundwater depths by several metres within a relatively short time (3–10 years). In the Kyabram aquifer in Victoria (Silberstein et al. 1999), a 2.4 ha plantation caused groundwater depth to increase from 1 m to 5 m and piezometric heads revealed that the region under the plantation was a local discharge zone. However, the salinity of the aquifer was about 3 to 5000 EC, which makes this site unsustainable. **Salinity around the roots is increasing** and **tree water use is declining over time**. In the near future, salinity around the roots will be too high and tree mortality will occur. If the trees are cut down after extensive mortality has occurred, and pasture or crops are planted, the additional salt that has accumulated in the soil will severely restrict crop growth and yield. Thus, where the shallow groundwater has a salinity of 3000 EC or more it is inadvisable to plant trees to ameliorate the problem.

Option 3c: local scale planting on a local scale perched aquifer

Local plantings on a local perched aquifer can cause significant increases in the depth of the water table as the trees access the water. In addition, rainfall can leach salt laterally and therefore accumulation of salt in the root zone is avoided. In a study conducted in Western Australia, a perched water table of 2–4 m thickness and less than 8 m depth, with an EC of 8000 µS cm^{-1}, discharged at the base of a sandy hill. When five rows of trees were planted upslope of the discharge zone, groundwater depths increased within 5 years and the land was reclaimed (Stirzaker 2002a). This outcome was achieved rapidly and effectively because the aquifer was thin and highly transmissive and its initial salinity was relatively low. Similarly, a 2 ha plantation of eucalypts located immediately above a seep reduced the frequency of occurrence of a perched water table within the plantation compared to its occurrence upslope or downslope of the plantation (Greenwood et al. 1995). Chloride content of a deep aquifer was reduced by about 20% within seven years of the plantation being established (Greenwood et al. 1995).

The placement of trees in the landscape (hillslope, above small or large aquifers) for management of the water balance is clearly very important. However, it is also important to determine the spatial arrangement of trees within a paddock. Should trees be planted as scattered individuals across the entire catchment, or as narrow strips? Are other arrangements better or worse at minimising leakage of water past the root zone? The following section deals with these questions.

Sustainability of trees growing over shallow water tables

Salts will accumulate in the root zone of trees planted above shallow saline water tables because the **roots exclude the salt** and extract the water for transpiration (Thorburn 1999). Three main mechanisms exist to remove this salt. First, leaching by rain, floods or irrigation with water containing minimal salt can leach salt into deeper layers. This can, however, be a problem if the shallow water table then rises further up the profile and remains too close to the surface. Second, salt can diffuse down a concentration gradient to deeper layers if the salt concentration in the upper layers is much larger than that of deeper layers. This is a slow process. Finally, a rising water table, followed by a falling water table, can remove a fraction of the salt in the upper profile. Thus the value of planted trees in increasing discharge from shallow water tables is not always guaranteed over time. Generally, plantings above thin and mostly unconfined aquifers with shallow water tables can reduce groundwater throughflow but if the aquifer is too deep (>10 m) the impact is likely to be minimal (Thorburn 1999). Trees planted on mostly confined aquifers will have most impact if planted on recharge zones or, if planted on discharge zones, the rate of tree water use exceeds the rate of upwards movement of water (Thorburn 1999).

What is the optimum planting arrangement for salinity amelioration?

Trees can be planted in the landscape as block plantings, as wide belts of trees interspersed with pasture/crops (agroforestry) or as narrow strips (alley farming) of trees. The question is: what is the optimum planting arrangement that will minimise leakage past the root zone and maximise crop or pasture growth? Trees planted in blocks (woodlots) increase the amount of tree to tree competition and minimise the length of **interface between trees and crops/pasture**. Thus, if 25% of a 100 ha paddock is planted with trees in a single block, the interface between tree and pasture is 20 m per hectare. If the trees are in belts of 50 m width and 200 m spacing, the tree/pasture interface is 100 m per hectare. If narrow alleys of 12.5 m width and 50 m spacing are used the interface is 400 m per hectare. The larger the interface between trees and pasture, the larger the potential impact of the trees on reducing discharge from the landscape – but also the larger the potential for negative impacts (shading, competition for water and nutrients) on pasture and crop growth (Fig. 8.12). It is important to note that in a block planting, where the canopy of adjacent trees touch, total tree water use will not exceed total rainfall (unless there is significant run-on or lateral subsurface flow to the block) if groundwater availability is zero. The roots of adjacent trees are assumed to transpire all the available water and leakage is assumed to be zero. In contrast, in widely spaced trees (individuals or belts of trees) the tree roots can access water that has passed through the root zone of the pasture/crop (leaked water) and therefore these trees can transpire more than rainfall because their 'catchment' is larger than the canopy area, a situation that is not true for trees planted in close spaced blocks. Thus, if rainfall is 500 mm y^{-1} and crop/pasture water use is 450 mm y^{-1}, leakage is 50 mm y^{-1}. If trees can transpire 500 mm of rainfall plus 50 mm leaking past the crop/pasture, leakage is zero and the aim (of reducing recharge) has been achieved, often with only a fraction of the total land area being covered in trees. White et al. (2002) and Knight et al. (2002) showed how access to shallow groundwater, or roots growing out from the belt into the crop zone and competing for water, can increase the amount of water available to the trees.

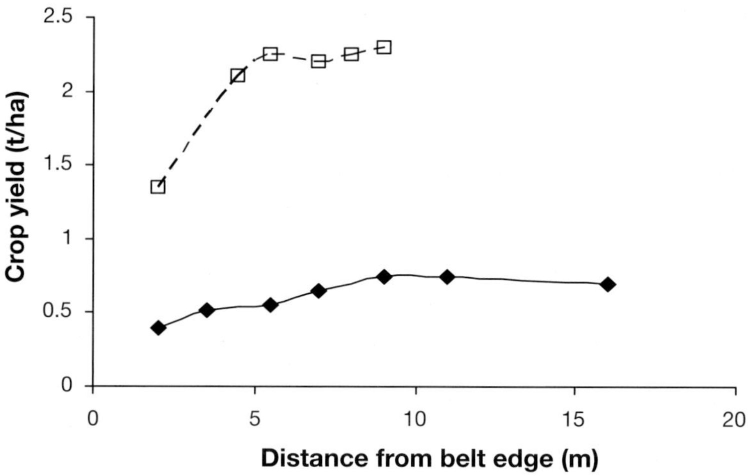

Figure 8.12 The influence of a row of trees on crop yield is usually negative for a short distance from the tree belt. The two lines show the effect of a row of trees on canola yield at two different sites in the Murray-Darling Basin

Source: Data from Knight et al. (2002)

As indicated above, trees compete with adjacent crops for water, light and nutrients. Crop growth is reduced within a certain distance of the trees (Figs 8.12, 8.13). This **zone of influence** is typically 20–40 m from the trees, although in some cases there is an enhancement of growth further from the trees which can compensate for the negative zone of influence. The reason for the enhancement of yield is because of the windbreak effect and a reduction in photo-inhibition of photosynthesis arising from the shade effect of the trees. **Root pruning** can reduce the competition for water and nutrients, reducing the distance over which a negative impact is observed (Ong et al. 2002).

Calculating the distance over which tree roots can reduce water leakage past the crop/pasture roots is difficult. To estimate the distance, it is necessary to know the LAI of native

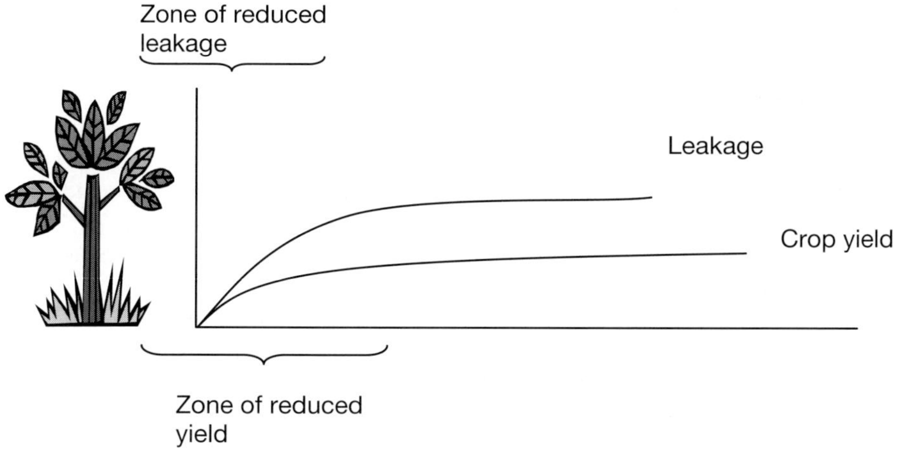

Figure 8.13 Tree belts reduce crop yield (zone of reduced yield) and leakage of water past the root zone (zone of reduced leakage)

Source: Redrawn from Stirzaker et al. (2002a)

vegetation for the site and the LAI of a belt of trees. If the LAI of native woodland close to the site is 1 and the LAI of a belt of planted healthy mature trees is 2, then to support the doubled LAI of the planted trees in a 5 m wide belt the trees must access twice the amount of water that falls on the 5 m wide belt. Therefore the zone of access of water must be at least twice the width of the tree belt, i.e. 2.5 m either side of the belt. If the LAI of the planted tree belt is less than that of a patch of native vegetation close by, we must expect significant leakage of water from the planted trees. The planted trees are not being as efficient as desired (in terms of reducing leakage and hence groundwater recharge).

Knight et al. (2002) observed that the LAI of the belt of trees planted in the Murray-Darling Basin was 12 times larger than the LAI observed in adjacent remnant native vegetation. This result was ascribed to the **mining of historic water**, i.e. water stored deep in the profile as a result of leakage from crops in previous years. It is likely that when this store of water is depleted, rapid and significant reductions in LAI would occur, with concomitant reductions in water use by the belt, unless the roots reach the groundwater at depth.

Conceptually, we need to consider the trees as having both a beneficial and negative zone of influence on the surrounding paddock. The beneficial area is the area of 'no leakage' and the negative impact is the reduced crop yield because of reduced water availability close to the trees (Fig. 8.13). This is discussed below.

Competition for water versus benefits of reduced leakage

To establish whether trees should be planted as a belt or woodlot, we need to establish whether the area of 'no leakage' (the benefit of the trees) exceeds the area of 'no yield' (the cost to crop yield arising from the trees' presence). In some sites, for example near Esperance in Western Australia, a 32% reduction in crop yield was observed in a 30 m wide strip adjacent to a tree belt. This can be used to calculate a no yield zone from the following empirical simplified descriptive equation:

No yield zone = (proportional decline in yield × width of transect) + ((tree belt width)/2)

In the Esperance example, the no yield zone = $(0.32 \times 30) + 9 = 18.6$ m.

Leakage past the root zone was reduced by 93% at 25 m, giving a no leakage distance of (0.93×25) + (half the tree belt width) = 23 m + 9 = 32 m.

The ratio (no leakage:no yield) is 32/18.6 = 1.7. Since this is much larger than 1, mixing the trees has a greater benefit on reducing leakage than the reduced crop yield and therefore tree belts are warranted (Stirzaker et al. 2002).

Knight et al. (2002) examined the width of leakage control at three sites in the Murray-Darling Basin in south-eastern Australia. They showed that crop yields were significantly reduced for 5–7 m from the belt, which corresponded to increasing soil drying in this area. Reduced leakage was generally observed within 4–5 m of the belt (and within the belts). At one site, 6 m wide belts controlled leakage over 14 m wide strips but reduced crop yield by 60% in the 4 m either side of the belt, causing an overall decline of 14% in crop yield.

Alley farming, where belts of trees are planted across the landscape interspersed with alleys of crops between the tree belts, has been proposed to reduce leakage to groundwater (Stirzaker et al. 2002b; Knight et al. 2002). Applying the idea of **ecological optimality** (whereby vegetation structure evolves in concert with soil properties and climate to maximise water use while ensuring long-term survival; Eagleson 1982), Ellis et al. (2005) developed a model to calculate the equivalent no drainage width (the width of the zone extending under and past a belt of trees where drainage past the root zone is zero). From a range of experimental studies they were able to show close agreement between theory and field data and showed that the width of the effective no drainage zone could be estimated from the leaf area per unit length of the alley, divided by the LAI of the natural vegetation close by (which is assumed to have reached optimality of structure). From this, the width of land hydrologically occupied by a belt of trees can be calculated (Ellis et al. 2005). It was shown that a belt of trees 4–8 m wide (between stems) produced an effective no drainage zone 20–60 m wide.

These are simplified approaches to considering tree belts in the landscape. A more complex and realistic approach requires that the commercial value of the trees (as wood) plus the benefits to soil erosion (reduced wind erosion) and increased biodiversity be considered.

Conclusion

In conclusion, it is clear that using trees in the landscape to ameliorate salinity requires consideration of several interacting factors, including:

- rainfall;
- hydrology;
- crop and tree growth;
- soil physics and topography;
- economics.

Only through consideration of these factors can an attempt be made to plant trees in the right place, with the optimum configuration for the maximum benefit to reduced leakage and hence groundwater recharge. However, it is clear that in a significant number of cases trees can be used to ameliorate the development of dryland salinity. There are multiple benefits from this, including the reclamation of salinised agricultural land, the prevention of more land becoming salinised, the economic value of the wood and the reduction in soil erosion through the trees acting as windbreaks. There are also increased amenity, tourist, aesthetic and wildlife preservation values to increased numbers of trees in the landscape.

In the next chapter, we move from consideration of practical and theoretical aspects of the relationships between terrestrial vegetation (mostly trees) and water (especially groundwater) in landscapes to the policy and legal frameworks guiding the management of these resources. It is the implementation of these policies and management goals that, to some degree, have driven the scientific interest in examining the relationships between trees and water.

Further reading

Abbs K & Littleboy M (1998). Recharge estimation for the Liverpool Plains. *Australian Journal of Soil Research* **36**: 335–357.

Bell RW Schofield, Loh IC & Bari MA (1990). Groundwater response to reforestation in the Darling Range of Western Australia. *Journal of Hydrology* **119**: 179–200.

Bosch JM & Hewlett JD (1982). A review of catchment experiments to determine the effect of vegetation changes on water yield and evapotranspiration. *Journal of Hydrology* **55**: 3–23.

Brown EKSNAG (2001). Plantations, farm forestry and water. Proceedings of a National Workshop, 20–21 July 2000, Melbourne. In *Water and salinity issues in agroforestry.* RIRDC/LWA/FWPRDC, Australia.

Burch GJ, Bath RK, Moore ID & O'Loughlin EM (1987). Comparative hydrological behaviour of forested and cleared catchments in south-eastern Australia. *Journal of Hydrology* **90**: 19–42.

Clarke CJ, George RJ, Bell RW & Hatton TJ (2002). Dryland salinity in south-western Australia: its origins, remedies, and future research directions. *Australian Journal of Soil Research* **40**: 93–113.

Dahlhaus PG, MacEwan RJ, Nathan EL & Morand VJ (2000). Salinity on the south-eastern Dundas Tableland, Victoria. *Australian Journal of Earth Sciences* **47**: 3–11.

Dahlman RG & MacEwan RJ. Dryland salinity in south west Victoria: challenging the myth. In *Collected case studies in engineering geology, hydrogeology and environmental* geology (eds MJ Knight & EJ Minty). Engineering Geology Specialist Group, Geological Society of Australia, Sydney, pp. 165–177.

George RJ, MacFarlane DJ & Speed RJ (1995). The consequences of a changing hydrologic environment for native vegetation in south-western Australia. *Nature Conservation* **4**: 9–22.

O'Loughlin E & Nambiar EKS (2001). Plantation, farm forestry and water. A report for the RIRDC/LWA/FWPRDC Joint Venture Agroforestry Program. RIRDC Publication 01/137. RIRDC, Canberra.

Raper GP (1998). Agroforestry water use in Mediterranean regions of Australia. *Water and salinity issues in agroforestry* **2.** RIRDC, Australia.

Timms W, Acworth RI & Berhane D (2001). Shallow groundwater dynamics in smectite dominated clay on the Liverpool Plains of New South Wales. *Australian Journal of Soil Research* **39**: 203–218.

Walker GR, Zhang L, Ellis TW, Hatton TJ & Petheram C (2002). Estimating impacts of changed land use on recharge: review of modelling and other approaches appropriate for management of dryland salinity. *Hydrogeology Journal* **10**: 68–90.

Chapter 9

Policy and guidelines for managing water in relation to ecological health in Australia

This chapter deals briefly with the policies concerning water rights and management of groundwater dependent ecosystems in Australia. Groundwater dependent ecosystems are a good example of the interface between ecology and hydrology and also illustrate the need for a more integrated and holistic approach to landscape and resource management. A significant example of a recent development in landscape and water resource management is that of the application of water rights. The meaning of water rights, the different types of water rights and a brief comparison of water rights in several states of Australia are discussed. At the end of the chapter is a summary of many of the attributes of water rights covering much of Australia.

The 20 National (Australian) Principles for the allocation of water to ecosystems are presented and there is some discussion of the need for a more explicit inclusion of groundwater and groundwater dependent ecosystems in these principles.

After reading this chapter readers should be able to discuss the following issues:

- the problems inherent in resources that remain 'commonly owned' (the 'tragedy of the commons');
- the meaning and importance of 'ecosystem services';
- the development of the concept of water rights and attributes that define a perfect set of water rights;
- the six most common types of water rights;
- the application of 'ecologically sustainable development' and the 'precautionary principle' to water management and the guiding principles that inform these philosophies;
- the main features of the 20 National Principles for allocating water to ecosystems in Australia and the manner in which they could be improved to better reflect the importance and unique attributes of groundwater dependent ecosystems;
- the principal impediments to the sustainable management of groundwater dependent ecosystems.

Introduction

Access to clean water is deemed a fundamental human right by most people. However, demand for water is increasing faster than the amount of water available from surface water stores, and many groundwater stores are over-utilised in many places. There is increasing competition for water, and regulation of water supplies is increasingly needed. The legislation covering the supply, movement, access and purity of water in most of the world is complex and rapidly changing and in Australia highly variable across the states and territories. However, basic principles and philosophies underlie the policies governing access to water. The following section describes these principles and philosophies.

Defining water rights, environmental flows, sustainable development and ecosystem services[1]

The management of water and other natural resources is the responsibility of state, territory and federal governments. The **COAG (Council of Australian Governments) Water Reforms** (1994) endorsed a strategic framework for efficient, effective and sustainable reform of the water industry. Four key premises underpinned this framework and much of the subsequent policy development of the past 15 years. These premises are: first, the concept of **water rights** was accepted, and the rights were viewed as independent of land title and land tenure. Second, the environment was, for the first time in a legal and political context, accepted as a legitimate user of water. An **allocation of water** to the maintenance of ecosystem health (**environmental flows**) was accepted as being as important as the allocation of water to human consumptive use. Third, both **sustainable and adaptive management** principles were used in developing the National Principles governing water allocations (see below). Finally, the concept of **ecosystem services** started to inform land and resource agencies and their policies. What do these terms mean?

Water rights are a legal authority to remove water from a river, lake, reservoir, water course or aquifer and to use that water for material benefit. A material benefit might be through commercial or domestic use. Water rights are granted by a regulatory authority, usually as a licence, a permit, a concession or an access entitlement. **Environmental flow**, or an allocation of water to the environment, is usually expressed as the minimum flow (volume of water) required through a specific location within a specified time to maintain the health of an ecosystem. **Ecosystem health** is perhaps one of the most difficult ecological concepts to define and is usually expressed in terms of ecosystem composition or ecosystem processes (Ludwig 1997). Morgan (2001) listed several ecosystem condition attributes that can be used to assess landscape health, including the following:

- current extent of native vegetation;
- degree of connectivity in native vegetation;
- conservation status of native vegetation;
- extent of dryland salinity;
- degree of changed hydrological conditions;
- distribution and density of weeds and feral animals;
- number and distribution of threatened species.

He also listed the following attributes that can be used to assist in quantifying landscape health:

- current rates of clearing;
- trends in the numbers, density and distribution of weeds and feral animals;
- trends in the extent of dryland salininty;
- trends in inappropriate fire regimes.

---

1 The structure and content of this chapter, including all the tables in the appendix, are extensively based on a 2003 report published by the Productivity Commission called *Water Rights Arrangements in Australia and Overseas*, Commission Research Paper, Productivity Commission, Melbourne. This book is not extensively referenced for ease of reading but the contribution of this source is acknowledged here.

As has been argued recently, the allocation of water to ecosystem health has generally been interpreted too narrowly and is often taken to mean the volume of water required to move through a river to maintain river health (Murray et al. 2003). More properly, it should mean the allocation of water to maintain the health of all water dependent ecosystems, including groundwater dependent ecosystems, not just rivers.

Sustainable management has many definitions, but the broadest is 'management that meets the needs of the present without compromising the ability of future generations to meet their own needs'. Sustainable forest management, for example, maintains the ecological integrity, and hence health, of the forest for future generations. **Adaptive management** is the process whereby management decisions, ecosystem health, and supply and demand of resources (water especially) are periodically reviewed in order that allocations, decisions and resources can be altered as conditions change.

Ecosystem services are the values assigned to ecosystems above and beyond the purely economic value of the wood, sheep, beef, apples, fish yield or other classically defined economic products of land and water. Thus, ecosystem services include the following intangible (i.e. difficult to quantify in classic economic models) benefits:

- reduction of soil erosion;
- amelioration of dryland salinity by reducing the rise of water tables;
- capture and retention of carbon in trees;
- reduction in siltation and eutrophication of lakes and rivers;
- maintenance of biodiversity (genetic, species, ecosystem and habitat diversity);
- flood mitigation;
- maintenance of clean water supplies;
- pollution uptake, capture or retention.

This is not an exhaustive list and serves merely to illustrate some of the 'services' that an ecosystem can provide of material (albeit difficult to quantify) benefit to humans. Some would argue that such an anthropocentric view is insufficiently broad and is detrimental to a more holistic philosophy of the management of natural resources. Because ecosystem services are rarely captured within normal commercial markets and their value poorly represented in dollar values, they frequently receive too little consideration in the formulation of policy (Costanza et al. 1997). A valuable way to think about ecosystem services is to think about what happens (at societal level) when these services are seriously degraded and then become absent. Costanza et al. (1997) assigned a value for global ecosystem services of $16–54 trillion per year.

Any discussion of the policies and guiding principles behind the management of water, in relation to ecosystem health, must first consider the concept of water rights as they are evolving in Australia and elsewhere. Although water has, until the recent past, been considered a **common good** (i.e. without ownership and with limited regulation or stewardship of access), it is clear that this view is no longer tenable. The analogy of the 'commons' of fisheries in oceans is useful here – it was the lack of ownership and stewardship that led to the catastrophic decline in oceanic fish resources since the mid 20th century. The development of water rights may just allow the integrated and sustainable management of both water and ecosystems in the future.

Having established the meaning of the key concepts that inform policy in water resource management, the rest of this chapter deals with water rights in Australia. It also includes a description of recent developments in managing groundwater and groundwater dependent ecosystems. The two are interlinked and codependent.

Water rights in Australia

A recent review highlighted the complexities of policy and legislation in relation to water rights in Australia (Productivity Commission 2003); this document provided an in-depth analysis of the issues of allocating water rights in Australia. Much of what follows in this section is based upon that review.

Water use can be classified as one of the following:

- consumptive;
- non-consumptive;
- in-stream.

Consumptive use means that not all (in some cases, none) of the water used is returned directly to the source of the water. Irrigation water and water used to produce beer and wine are consumptive uses of water. Non-consumptive use is when all the water used is returned to the source – for example, water used to generate electricity in hydroelectric schemes is non-consumptive. Although the quality of water used by the hydroelectric generating plant is altered (it is generally much warmer at discharge than intake), the volume lost from the source is negligible. In-stream uses of water include fishing, sailing and aquaculture.

For much of the past 200 years in Australia, water has been viewed as an open access resource, or a common resource. Thus, open and relatively unregulated access to water has been a characteristic of water management, although local, usually temporary, restrictions to access have occurred. Open access resources are deemed not to be owned by anyone and it is the absence of ownership and of a regulatory framework that results in overexploitation and resource degradation. In recognition of the need for sustainable regulation of water resources, Australia is moving rapidly to a more realistic **cost structure for the provision of water** and the application of water rights. Water rights serve to protect the resource for future use and place obligations on the users of water to protect the rights of others (including the requirement to allocate water to environmental health). Water rights place an upper limit to the volume (usually expressed as a fixed fraction of the available water; see below) that can be extracted and used by an individual or specified business. Key factors in the public's acceptance of water rights are the **transparency**, **consistency** and **predictability** (stability) in provision of water in the future. Without these attributes, an acceptance by water users of the need for water rights will remain elusive. A significant challenge to acceptance is the fact that the volume of water that falls as rainfall in any single catchment is extremely variable between seasons and between years. Therefore water rights should not attempt to define a fixed volume of water for use because the sustainable volume available is so variable. Water allocations should be expressed as a percentage of the volume of water available, which varies over time.

What makes a water right equitable and efficient in its application is now discussed.

Defining efficient and equitable water rights

Water rights, ideally, should have eight attributes that make them fair, acceptable and transparent.

1 **Universal applicability to all available water resources** (surface water, groundwater, overland flows) within the jurisdiction. This has not been achieved in Australia. Rights to surface waters (rivers, lakes) might become valueless if the overland flows that sustain some surface water bodies become diverted to private dams, for example. Hence the need for an integrated and universally applied set of water rights. Table 9.1 defines the water

Table 9.1 Water resources covered by the water rights system in Australian jurisdictions, 2003

Jurisdiction	Water sources covered
New South Wales	Rivers, lakes or estuaries or any place where water occurs naturally on or below the surface of the ground. This includes overland flows contained in a dam only where the volume collected exceeds 10% (or greater if so prescribed) of average run-off
Victoria	Water in a waterway or bore. Unlicensed collection of overland flows is limited to stock and domestic purposes only. Uses over and above this must be licensed
Queensland	Water in a water course, lake or spring, and underground water. Groundwater sources may only be included within the rights system where a Water Resource Plan has identified the need to regulate water extraction. Overland flows are only licensed where a Water Resource Plan identifies this need. Otherwise, the collection of overland flows is limited only by dam height specifications[1]
South Australia	Water in a water course, lake or well and water overflowing land that has been collected in a dam or reservoir in prescribed areas. Use of water in non-prescribed areas is subject to common law rights
ACT	Water in a water way (defined as a river, creek, stream or other channel, a lake, pond, lagoon or marsh), groundwater and overland flows

[1] Currently there is a moratorium on the development of on-farm dams

Source: Reproduced from Productivity Commission (2003).

resources covered in various Australian states and territories. Water resources are not defined identically in all jurisdictions.

2 **Predictability in the volume available** to a user. Given the highly variable volume of water available each year (floods and droughts being recurrent features of Australia's landscape), this is often hard to achieve. Using a fixed percentage of water available and specifying the timing of extraction of water (e.g. to protect seasonal low points in water flows) can be used to manage this issue.
3 **Enforced compliance** so that the rights of one person cannot be stolen by another without fear of legal retribution. In the absence of an effective enforcement of water rights, compliance is likely to be low and therefore the value of the rights conferred reduced.
4 **Legal recognition** of the water rights of the individual is an absolute requirement for water rights to have any meaning to resource managers and individual users.
5 **Specified duration** of the applicability of the water rights. Perpetual or long term rights increase their predictability and encourage investment in infrastructure and processes that maximise efficiency of water use. Short term rights do not have these attributes but do allow for a faster response in adaptive management of natural resources. When conditions change and a change in allocation or other management option is required, it is easier to effect this change when rights are assigned on short term licences. In New South Wales, most water rights are for a fixed term with the option of a renewal of the rights by the water resource agency. In Queensland, South Australia, Victoria (some water rights) and the Australian Capital Territory, most rights are ongoing but they can be revoked or modified.
6 **Exclusivity of access to water.** Water rights are exclusive if, at the margin, they limit third party costs of benefits, arising from exercising the right, to a socially acceptable level. If water rights do not have this, the rights holder could be unaware that they are causing damage to a third party. For example, overextraction of water has a negative impact on other users and on the environment.

7 **Detachment from land title and use restrictions**. The right to use the water is separate and free from any need to own land and free from any restriction on how the right can be exercised.
8 **Divisible and tradable rights allow** the water right to be subdivided and traded to others. Such trading can improve the efficiency of water use by providing an incentive to conserve water and trade the surplus generated from the savings. Trading can be permanent, whereby the right to ownership is sold to a new holder. Temporary trading also occurs seasonally or for short periods of time (1–2 years). In 2001–02, nearly 1000 GL of water were traded in the Murray-Darling Basin alone (more than 90% was traded temporarily).

Different types of water rights

Six different types of water rights can be recognised, broadly grouped as:

1 stock and domestic;
2 surface water (including in-stream use for hydroelectric generation);
3 harvest rights for overland flows;
4 groundwater;
5 indigenous;
6 environmental flow or allocation requirements.

Access to water is allowed within all classes of water rights. These rights are defined as either a volume or a fraction of the water that the right holder may take. Water rights can also specify a particular part of the year during which access is allowed and at what rate water can be accessed. The permit/licence will also define the duration of the right to access. Without these limits, water resources may be inefficiently used or overused, thereby reducing the benefit to the community from the use of the water.

Water rights also define a right holder's **priority of access** relative to other right holders. By specifying priority of access it is possible to ration water use from year to year in response to fluctuations in water availability.

The specifications of water rights are often linked to resource planning arrangements. In some parts of Australia, water rights confer on holders the right to construct the infrastructure to extract, distribute and store water. Often, separate approvals are needed to build this infrastructure.

Water rights also impose **liabilities** on their holders to protect third parties, such as other right holders or the environment. Rights come with obligations and responsibilities.

Having established some of the attributes of water rights and the different types of water rights, it is now possible to review briefly the status of water rights in different parts of Australia.

Comparing some water rights in Australia

The tables in the appendix (Tables 9.3–9.7, reproduced from the Productivity Commission 2003) provide a summary of the principal attributes of water rights enshrined in New South Wales, Victoria, Queensland, South Australia and the Australian Capital Territory. Data were not available for Western Australia and the Northern Territory. It is clear that there is a divergence in the development of water rights within Australia, a situation that must be addressed in the near future if the resource is to be managed most effectively at a national level.

Central to management of water resources in Australia is the development of resource plans.

Resource plans

Resource plans are statutory instruments that give water resource agencies the authority to allocate water between competing uses, and are commonly used in Australia (Productivity Commission 2003). These plans are written so that a range of policy objectives (e.g. satisfying the needs of non-consumptive uses as well as ensuring certainty of supply to consumptive uses), can be reasonably met. In preparing the plans, water resource agencies must consider the **triple bottom line**, that is, they must consider the **environmental, economic and social benefits and costs** of the proposed allocations.

When resource plans are prepared it is expected that water will be reallocated between consumptive and non-consumptive uses from time to time. Consequently, periodic reviews of the plans must be undertaken and contingency plans established to address water scarcity during droughts. Resource plans are often written for all surface and groundwater sources, including overland flows, and there is normally a **hierarchy of resource plans**. First, a planning framework is legislatively established. This provides for the development of strategic plans that will be used to guide operational plans.

Strategic plans are usually broad in scope and may cover riverine health and integrated resource management. Community values (social, cultural, amenity, aesthetic and economic values) are also incorporated. Strategic plans can include a framework for the issuing of water rights and establish the objectives, criteria and rules for the management of flow and use of water. Strategic plans often require an associated operational plan to guide implementation of the objectives and outcomes stated in the strategic plan (Productivities Commission 2003).

Operational plans are generally catchment specific and cover the management of diversions and flows. They may also define the rules governing the allocation of water. For example, in Queensland, objectives stated in Water Resource Plans are achieved through an associated Resource Operations Plan which contains rules governing water trading, infrastructure operations, environmental flow management and monitoring requirements. Such plans may apply to whole catchments or to specific areas within a catchment. In New South Wales, catchment scale Water Sharing Plans incorporate environmental flow requirements, the allocations of water to holders of water rights and rules governing water transfers. They also include procedures for the administration of water rights and constraints on the development of water works.

Having established a broad appreciation of the developments of water rights in Australia, it is now possible to consider the case of managing water for environmental health, with specific examples in the development of policy and guidelines for managing groundwater dependent ecosystems.

Policies and guidelines for managing groundwater to maintain the health of GDEs

In 1994 a set of **National Principles for the Provision of Water for the Environment** was agreed. These have been called the Environmental Flows guidelines and were established jointly by **ARMCANZ** (Agriculture and Resource Management Council of Australia and New Zealand; now defunct and replaced by the Natural Resources Management Ministerial Council) and **ANZECC** (Australian and New Zealand Environment and Conservation Council). The National Principles provide a policy framework to manage water with the aim of ensuring ecosystem health. Although the principles were deemed to apply to surface and subsurface water, they have been used primarily in managing surface waters (Murray et al. 2003). While there has been extensive and ongoing research to determine the environmental flow requirements of surface water ecosystems (predominantly rivers, wetlands and marshes), the determination of water requirements of groundwater dependent ecosystems is in its infancy. Indeed, groundwater dependent ecosystems and their requirement for groundwater flows have only recently been recognised (Evans & Hatton 1998; Murray et al. 2003). Consequently it is not surprising that the 1994 National Principles do not adequately reflect the unique nature of groundwater dependent ecosystems.

Underpinning the development and philosophy of the draft National Principles are the principles and philosophies of ecologically sustainable development and the precautionary principle.

Ecologically sustainable development requires that economic development should be balanced against the protection of biological diversity, **intergenerational equity** (don't spoil today what your children will need tomorrow) and the maintenance of essential ecological processes. It has seven guiding principles:

- effective decision-making processes that integrate both long and short term economic, environmental, social and equity considerations (the triple bottom line);
- a lack of complete scientific certainty should not be used to postpone measures to prevent environmental degradation (the precautionary principle, see below);
- global dimensions of environmental impacts of actions and policies should be recognised and considered;
- a strong, growing and diverse economy can enhance the capacity for environmental protection and this should be recognised;
- the need to enhance and maintain international competitiveness in an environmentally sound manner should be recognised;
- flexible and cost effective policy instruments are needed;
- broad community involvement is important and should be encouraged and supported.

The **precautionary principle** (erring on the side of conservatism), in turn, is based upon six guidelines:

- an objective risk assessment, identifying at each stage the degree of scientific uncertainty, is required at the start of any development;
- all stakeholders (including the community at large) should be involved in resource management decisions and the procedures used must be as transparent as possible;
- actions must be proportionate to the risk that will be limited or eliminated by that action;
- actions should be guided by a cost benefit assessment (advantages and disadvantages) with an eye to reducing the risk to a level acceptable to all stakeholders;
- responsibility for providing scientific evidence in risk assessments must be assigned to someone;
- actions/management decisions should be provisional, pending the results of further scientific research that provides a better understanding of the ecology and allows more objective risk assessment.

A related concept awaiting full implementation in the water and natural resource management arena is that of **integrated catchment management**. In integrated catchment management, land, water and other natural resources are managed as a coordinated and interacting system for an entire water catchment, rather than as a set of separate and disparate attributes of a catchment. Integrated catchment management requires partnerships among different levels of government and non-government agencies, while economic, social and environmental values are determined through inputs from scientists and the wider community.

The principles of ecologically sustainable development, the precautionary principle and integrated catchment management underpin the National Principles for Provisions of Water to Ecosystems. These are now discussed.

Draft National Principles for Provision of Water to Ecosystems

The 20 draft ARMCANZ/ANZECC National Principles for Provision of Water to Ecosystems are summarised in the following:

Principle 1: River regulation and consumptive use of both surface and groundwater resources can have negative impacts on ecology.

Principle 2: The best available scientific information should be used to establish the ecological water requirements of water dependent ecosystems.

Principle 3: When determining ecological water requirements, the needs of downstream or other connected ecosystems and the ecological significance of the major features of the natural water regime should be taken into account.

Principle 4: Environmental water allocations should be legally recognised.

Principle 5: In allocating water to the maintenance of ecological health, existing rights of other users should be recognised.

Principle 6: Extraction of water should be capped and action undertaken (including reallocation) when environmental water allocations are insufficient or there is scientific evidence of environmental damage.

Principle 7: Water markets should be governed such that no adverse effects on ecological values occur unless there is a net ecological benefit.

Principle 8: Additional allocations of water for any use should only occur on the basis that natural ecological processes and biodiversity are maintained.

Principle 9: All water uses should be managed in a way that recognises ecological values.

Principle 10: Water sharing rules and operation schedules should, wherever possible, recognise significant inter-annual variability in the water requirements of water dependent ecosystems.

Principle 11: Structural developments and building works should occur when they can deliver a net benefit to ecological values and should meet current standards of best practice.

Principle 12: Actions undertaken to achieve water savings should occur only where they provide a net improvement in ecological values.

Principle 13: Demand management, water pricing and water use metering strategies should be applied to assist in maintaining ecological values.

Principle 14: All water in the catchment should be considered when making water allocation and management decisions.

Principle 15: Broad scale land use changes should consider the implications for environmental water allocations, as well as other water uses.

Principle 16: Allocation of water to the environment should be responsive to monitoring outcomes and improvements in scientific understanding of ecological water requirements.

Principle 17: Benefits of environmental water allocations should be optimised by using additional management options.

Principle 18: All aspects of management of environmental water allocations should be transparent and clearly defined.

Principle 19: Research to improve our understanding of ecological water requirements is vital.

Principle 20: All stakeholders (commercial, social, environmental, community, cultural) should be involved in the decision making process for water allocations.

While these 20 principles include consideration of all water sources in a catchment, it is clear that surface waters, not groundwaters and associated groundwater dependent ecosystems, are the focus of management and research efforts. However, the National Water Initiative, an agreement arising from the Council of Australian Governments meeting in 2004, does more explicitly address groundwater and groundwater dependent ecosystems. Thus the National Water Initiative's objective is to develop a nationally compatible, market, regulatory and planning based system for managing groundwater and surface waters that optimises the social, economic and environmental outcomes. The initiative has many elements; some of the key ones are to define and improve water access entitlements across Australia, improve the specifications for environmental outcomes in managing water resources, to establish a public register of water entitlements and trades, to acknowledge links between surface and groundwaters, and continue implementation of full cost recovery pricing for water.

The **National Groundwater Committee** (NGC) has representatives from state and territory governments of Australia and other interested parties, including CSIRO and the Federal Department, Environment Australia. It reports to the federal government through the Land and Water Biodiversity Committee. That committee reports to the Natural Resource Management standing committee, which is part of the Natural Resource Management Ministerial Council (NRMMC). To enhance the management of groundwater and groundwater dependent ecosystems, the following key points must be addressed within this hierarchy of state and federal government agencies:

- groundwater, surface water and land management must become integrated so that surface and groundwater and groundwater dependent ecosystems become incorporated into state and territory legislation and policy guidelines;
- a state by State and national register of groundwater dependent ecosystems should be developed so that statutory reserves and prioritisation of groundwater dependent ecosystems can begin;
- a national comprehensive, adequate and representative and/or ecosystem services approach to promote the conservation of groundwater dependent ecosystems is required;
- management plans should explicitly acknowledge groundwater dependent ecosystems;
- an accepted framework for assessing groundwater dependent ecosystem health requires development and deployment within government resource management agencies. Possible indicators for groundwater dependent ecosystem health, and associated resource condition targets and management action targets, are given in Table 9.2. The resource condition target refers to the desired condition of the resource and is the desired outcome of the management action, described in the final column.

Table 9.2 Examples of indicators that can be used to manage groundwater in relation to groundwater dependent ecosystem health, and associated resource condition targets and management action targets

Indicators (examples only)	Resource condition target (examples only)	Management action target (examples only)
Pressure/water level	30 m above Australian height datum	Allocation limited to <50% of recharge of groundwater
Rate of change of pressure/level	0.05 m y^{-1} decline for up to 5 years	Review and change allocation annually
Flux/groundwater discharge to be maintained	1500 ML y^{-1} km^{-1}	Prohibit abstractions within 250 m of a groundwater dependent ecosystem
Water quality indicators	Increase of <10 EC units y^{-1}	Prohibit specific land uses

It is important that a single set of National Principles for allocating water to ecosystems be developed, rather than separate surface water and groundwater principles. However, there are many impediments to the development of an integrated management of water and groundwater dependent ecosystems. These are now briefly discussed.

Impediments to integrated management of water and GDEs

There are several impediments to the appropriate and successful integration of the management of water resources and environmental health, as required in the National Principles and recent legislation in most Australian states.

There are **legislative impediments**, such as water resources that cross state boundaries (e.g. the Great Artesian Basin, the Murray-Darling Basin) and require a single coherent management but are regulated by different, competing state authorities.

Political impediments exist, with state politics focused on state issues and a state electorate, with short term electoral time frames dominating thinking. In contrast, management of major trans-state resources requires a more holistic, integrated, long term vision. Regional development pressures frequently override broader environmental concerns.

Policy development is often aspirational with little development of a whole of government approach to managing landscapes, water and other resources.

Social, cultural and economic forces can be significant retardants to the acceptance of change in how resources are managed. Economic development (mining, agriculture, horticulture) is usually viewed as most important, especially in rural areas of Australia. The traditional view that owners of a property (farm) unquestionably own the rights of access to all water on that property (surface, aquifer and overland flows) is going to require considerable education to enable change in the short term.

Existing entitlements often exceed the sustainable yield for a water resource. Reducing the current entitlements of water users is considerably more difficult than not allocating more uses to the water resource – compensation is usually expected when a previously held 'right' to access is denied for the future. An interesting technical problem is that of **double counting** of water. Double counting occurs when permission is given to one user to extract a volume of water from an aquifer and that volume represents the same water that supplies the base flow of a river in the dry season, and a second user is given the right to remove a volume of water from that river. In this scenario, the same water has been allocated twice to different users. It also ignores the requirement of an allocation to the environment.

There is very poor awareness within the community and within resource management agencies of the links among surface water, groundwater, landscapes and ecosystem health. Consequently, many locations do not implement best practice approaches in resource management. Because of this, and the lack of funding for natural resource management agencies, the application of sustainable management of water and ecosystem health is patchy.

Turf wars among traditional discipline based agencies are common. Government departments and agencies with responsibility for oversight of the mining industry are usually separate from those with responsibility for agriculture or water or forestry. Total catchment management, and the integrated approach required to achieve it, remain elusive.

Conclusion

Water regulation in rural and urban Australia has been inadequate over the last 200 years, rarely consistent and bedevilled by a legacy of European heritage. However, since 1990 there have been increased community acceptance of the need for change and improvements in the science of ecosystem management. This new knowledge and readiness to embrace change are feeding into the policies and guidelines being prepared for the next decade. There is hope for a sustainable future in water management in Australia – the National Water Initiative certainly provides a framework for moving forward.

Chapters 1 to 8 have examined, with a very Australian focus, the ecophysiology and hydrology of landscapes. Chapter 9 has provided some philosophical, policy and legislative perspectives to the issues of managing water and groundwater dependent ecosystems in Australia. The next chapter condenses all those discussions in a review of the ecohydrology of South Africa. South Africa has many problems similar to Australia. Chapter 10 provides a counterpoint to Australia, with an overview of the ecology, hydrology and policy frameworks for South Africa.

Further reading

King J, Brown C & Sabet H (2003). A scenario-based holistic approach to environmental flow assessments for rivers. *River Research and Applications* **19**: 619–630.

National Land and Water Resources Audit (2001). *Australian dryland salinity assessment 2000.* Commonwealth of Australia.

Syme GJ, Nancarrow BE & McCreddin JA (1999). Defining the components of fairness in the allocation of water to environmental and human uses. *Journal of Environmental Management* **57**: 51–70.

Water and Rivers Commission (2000). *Environmental water provisions policy for Western Australia.* Water and Rivers Commission Statewide Policy No. 5.

Appendix

Table 9.3 Specification of water rights, Australian jurisdictions, 2003

Jurisdiction	Right	Measure	Reliability and priority
New South Wales	Environmental Water	Flow rules and/or volumes are specified in Water Sharing Plans	Has first priority in water use, except in severe water shortages
	Domestic and Stock Rights (surface water)	Not specified but limited to certain use	Has second priority in water use, except in severe water shortages
	Native Title Rights	Not specified but limited to certain uses	Has second priority in water use, except in severe water shortages
	Access Licences[1] (surface and groundwater)	Share and extraction rate, except local and major urban suppliers that have a specified volume. Share and extraction rate in any year is determined by the rules of Water Sharing Plan. Plans specify the share of the resource available to users once environmental and stock and domestic requirements have been met. Licensed water shares can only be taken via an approved water supply work and can only be applied to land consistent with a water use approval	The following priorities must be observed in relation to licences. Local water utility licences, major utility licences, and domestic and stock licences have priority over all other licences Regulated river high security licences have priority over both regulated river general security licences and regulated river supplementary water licences Regulated river general security licences have priority over regulated river supplementary water licences Water allocations must be diminished at a lesser rate for higher priority licences than lower priority licences
Victoria	Bulk Entitlements	Nominal volume	Varies between bulk entitlements. Entitlements are perpetual. Priorities between right holders for access to water can be qualified by the minister if a water shortage is declared
	Stock and Domestic Rights (surface and groundwater)	Not specified but limited to certain uses	Has first priority in use
	Water Rights (surface water)	Nominal volume	Varies between river systems but generally high (around 97% of water right) because shortages are met by adjusting sales water claims. Priorities are proportionate unless qualified by the minister if a water shortage is declared

Jurisdiction	Right	Measure	Reliability and priority
	Take and Use Licences[2] (surface water)	Nominal volume	Varies between river systems but generally high on regulated systems (around 97% of licence volume) because shortages are met by adjusting sales water claims. Priorities are proportionate unless qualified by the minister if a water shortage is declared
	Sales Water (surface water)	Percentage of water right or take and use licence (surface water) up to a maximum allocation	Depends upon the availability of water in the river system and thus the volume allocated to water rights or licences varies between systems and from year to year
	In-stream Use Licences (surface water)	Nominal volume	Varies between river systems but generally high on regulated systems (around 97% of licence volume) because shortages are met by adjusting sales water claims. Priorities are proportionate unless qualified by the minister if a water shortage is declared
	Take and Use Licences (groundwater)	Nominal volume	Depends upon aquifer. Priorities are proportionate
Queensland	Stock and Domestic Rights (surface and groundwater)	Not specified but limited to certain uses	Varies according to the rules of the Resource Operations Plan, but usually high priority
	Water Licences[3] (not subject to an ROP)[4]	May be area of land irrigated, flow conditions or volume	Not specified
	Water Allocations and Water Licences (subject to an ROP)	Share of the available resource once environmental and stock and domestic needs are met	Determined by rules of Resource Operations Plan and security level of allocation held. For example: High class A priority group has priority over high class B priority group High class B priority group has priority over medium priority group Medium priority group has priority over the risk priority group Rules and security levels may vary between Water Resource Plans

Jurisdiction	Right	Measure	Reliability and priority
South Australia	Stock and Domestic Rights (surface and groundwater, in prescribed areas)	Not specified but limited to certain uses[5]	First priority after environment
	Unlicensed Water Rights (surface and groundwater in non-prescribed areas)	If resource becomes overcommitted, the minister may restrict use by prescribing the resource. Otherwise, water may be freely taken provided there is no detrimental effect on other users	Not specified
	Water Licences (surface or groundwater, for taking or holding, in prescribed areas)	Nominal volume[6]	Full licence entitlements are usually available each year. Minister may restrict water use because of drought, water quality concerns or ecological reasons. Licences are restricted proportionately unless an alternative scheme has been made by the governor as recommended by the minister
	Environmental Allocations	Nominal volume or share	Some allocations (such as those for the Lower Murray Swamps) possess the same (high) reliability as Water Licences
ACT	Stock and Domestic Rights (surface and groundwater)	Not specified but limited to certain uses	In most sub-catchments, stock and domestic right holders have first priority to water
	Licences to Take Water (surface and groundwater)	Nominal volume or rate of flow that may be taken	Not defined in terms of priority
	Allocations (surface and groundwater)	Nominal volume or rate of flow that may be taken	Minimum environmental flows have priority over other uses (except stock and domestic). Commercial and irrigation uses have second priority. Recreational uses have last priority

1 Access licences are held by irrigation companies in irrigation districts. Individuals within each district hold water shares in the company
2 Licences may also be issued for in-stream use
3 Area based licences are being converted to volumetric
4 ROP = Resource Operations Plan
5 In areas where the Murray-Darling Basin Cap applies, stock and domestic rights are given a volumetric allocation. In other areas, stock and domestic use is not quantified. In prescribed areas, stock and domestic use may be licensed but this has not occurred
6 In the Eyre Peninsula, Southern Basins and Musgrave Water Allocation Plans, Water Licences (for taking water) are specified as a share of the available resource
Source: Productivity Commission (2003)

Table 9.4 Recording water right titles, Australian jurisdictions, 2003

Jurisdiction	Right	Title record
New South Wales	Environmental Water	Recorded in gazetted Water Sharing Plans
	Domestic and Stock Rights	Attached to land title record
	Native Title Rights	Attached to land title record
	Access Licences[1]	Torrens Title like register administered by Land and Property Information[2] Register records Licence Licences applied for, granted, renewed, transferred, surrendered, suspended or cancelled, and any legal or equitable interest held in a licence Details of the share and extraction components Links to approved works Expiry date of licence Relevant links to water sharing plan
Victoria	Bulk Entitlements	Register maintained by the Dept of Sustainability and Environment. Updated as amendments occur
	Stock and Domestic Rights	Attached to land title record
	Water Rights (surface)	Register kept by rural water authority. Specifies owner, land to which it attaches, and total volume of water. Annual sales water allocation for the season specified
	Take and Use Licences (surface and groundwater)	Rural water authority must keep a record of licences on issue
Queensland	Stock and Domestic Rights	Not attached to land title record except if it requires a licence
	Water Licences (not subject to an ROP)[3]	No explicit requirement for a registry. Issued and managed by Dept of Natural Resources & Mining
	Water Licences (subject to an ROP)	Registry of licences held by NR&M in the Water Entitlements Registration Database

Jurisdiction	Right	Title record
	Water Allocations (subject to an ROP)	Water allocation register that operates as a module of the Queensland Resource Registry. Records the owner's details, volume, reliability, location, purpose and the resource operations plan under which the water allocation is managed
South Australia	Stock and Domestic Rights (in prescribed areas)	Attached to land title record
	Unlicensed Water Rights (in non-prescribed areas)	Attached to land title record
	Water Licences (to take or hold, in prescribed areas)	Minister keeps a register in the form in which the Minister thinks fit. The register records provisions for transfers and sale of water allocations and any third party interests[4]
ACT	Stock and Domestic Rights	Attached to land title record
	Licences to Take Water, and Allocations	Environment Protection Authority has established a register of allocations and licences granted and transferred

1 Access licences are held by irrigation companies in irrigation districts. Individuals within each district hold water shares in the company
2 Land and Property Information is a NSW State Government Business Enterprise providing land, property and valuation information and services. These services were previously provided by the Land Titles Office, the Land Information Centre and the Office of the Valuer General
3 ROP = Resource Operations Plan
4 A Torrens Title system for water licences and online internet access to water licensing information is planned

Source: Productivity Commission (2003)

Table 9.5 Duration of rights, Australian jurisdictions, 2003

Jurisdiction	Right	Duration	Cancellation
New South Wales	Environmental Water	10 years by virtue of them being established in a 10 year Water Sharing Plan	The Minister may revoke or amend environmental water rights by revoking or modifying a Water Sharing Plan
	Stock and Domestic Rights	Perpetual	Minister may temporarily suspend a right if it is in the public interest
	Access Licences[1]	20 years for local and major utility licences, and 15 years for all other types of licences	Minister may suspend or cancel an Access Licence for the following reasons: Holder has failed to comply with the licence conditions, or has been convicted of an offence against the provisions of the *Water Management Act 2000* Any charges payable in respect of a licence have not been paid Minister may compulsorily acquire Access Licences if they think that the public interest requires it
Victoria	Bulk Entitlements	Perpetual	Minister may temporarily suspend, reduce, increase or otherwise alter any rights if the Minister has declared that a water shortage exists in the area or supply system concerned. A water shortage may be declared if the Minister thinks the volume or quality of water available in the area or system is or will be inadequate for any reason
	Stock and Domestic Rights (surface and groundwater)	Perpetual	As above
	Water Rights (surface water)	Perpetual	As above
	Take and Use Licences (surface and groundwater)	Up to 15 years	As above. Minister (or rural water authority) may revoke a licence if in the opinion of the Minister (or rural water authority) there has been a failure to comply with any condition to which the licence is subject
	In-stream Use Licences	Up to 15 years	As above
Queensland	Stock and Domestic Rights	Perpetual	Can be limited where subdivisions may lead to proliferation
	Water Licences (not subject to an ROP)[2]	Up to 10 years, sometimes longer for stock purposes	Licences may be cancelled by the Chief Executive, although holders have a right of review and appeal

Jurisdiction	Right	Duration	Cancellation
	Water Licences (subject to an ROP)	For a specified period. Water Resource Plan subject to 10 year reviews	May be varied, amended or cancelled by the Dept of Natural Resources and Mines. Licences may be forfeited if conditions are not complied with.
	Water Allocations (subject to an ROP)	Ongoing. Water Resource Plan subject to 10 year reviews	May be cancelled if the conditions of the allocation are not complied with
South Australia	Stock and Domestic Rights (in prescribed areas)	Perpetual	No provisions
	Unlicensed Water Rights (in non-prescribed areas)	Perpetual	No provisions
	Water Licences (in prescribed areas)	Perpetual	Minister may cancel a Water Licence (for taking or holding water) if the holder contravenes or fails to comply with its conditions or, in the case of taking water, takes water in excess of the specified entitlement
ACT	Stock and Domestic Rights	Perpetual	No provisions
	Licences to Take Water	Perpetual	A licence may be cancelled if an allocation on which to base the taking of water does not exist or licensee does not have lawful access to the place from where water is to be taken
	Allocations	Ongoing subject to a 10 year review of the Water Resource Management Plan	No provisions governing cancellation of an allocation but an allocation may be reduced, either wholly or in part, if it is necessary or desirable to do so: Because a reduction in the flow of the waterway makes it necessary To prevent a reduction, or further reduction, in the quality of water To prevent damage, or further damage, to an ecosystem that depends on the water in a water way

1 Access licences are held by irrigation companies in irrigation districts. Individuals within each district hold water shares in the company
2 ROP = Resource Operations Plan

Source: Productivity Commission (2003)

Table 9.6 Separate from land title and use restrictions, Australian jurisdictions

Jurisdiction	Water right	Separate from land title	Separate from use
New South Wales	Stock and Domestic Rights	No	No
	Native Title Rights	No	No
	Access Licences[1] (surface and groundwater)	Yes[2]	Yes[3]
	Works and use approvals	No	No
Victoria	Stock and Domestic Rights	No	No
	Bulk Entitlements	Yes	Yes
	Water Rights	Yes[4]	No
	Take and Use Licences (surface and groundwater)	Yes[4]	No
	Sales Water	Yes[4]	Yes
	In-stream Use Licences	Yes	No
Queensland	Stock and Domestic Rights	No	No
	Water Licences (not subject to an ROP)[5]	No	No
	Water Allocations (subject to an ROP)	Yes	No
	Water Licences (subject to an ROP)	No	Yes[6]
South Australia	Stock and Domestic Rights	No	No
	Unlicensed Water Rights (in non-prescribed areas)	No	Yes
	Water Licences (taking and holding, in prescribed areas)	Yes	Yes
ACT	Stock and Domestic Rights	No	No
	Licences to Take Water	Yes[7]	No
	Allocations	Yes	Yes

[1] Access licences are held by irrigation companies in irrigation districts. Individuals within each district hold water shares in the company

[2] Under the supply contract arrangements within each irrigation district, an individual must be a shareholder of the irrigation company to obtain a water share. To be a shareholder, individuals must own land within the irrigation district

[3] Major utility and local urban supply licences may only be used for the supply of urban water

[4] The *Water Act 1989* (Vic.) does not specify that a person must be a landholder to be granted a water right or licence to take water (surface). However, rural water authorities generally only approve applications if the applicant is an owner or occupier of land within the irrigation district to which the application applies. To hold a water right, a person must be the owner or occupier of a landholding within the irrigation district. While water rights may be transferred between owners or occupiers, rights remain attached to the landholding specified in the register and the register is updated to reflect the transfer

[7] ROP = Resource Operations Plan

[6] Water licence may include a condition about the purpose for which the water may be taken

[7] While legally separate from land title, Licences to Take Water are generally site specific

Source: Productivity Commission (2003)

Table 9.7 Divisibility and transferability, Australian jurisdictions, 2003

Jurisdiction	Right	Divisible	Transferable
New South Wales	Environment Health Water	Yes	Only within the environment
	Supplementary Environmental Water	Yes	On loan only and must be repaid
	Adaptive Environmental Water	Yes	Yes
	Stock and Domestic Rights	No	No
	Native Title Rights	No	No
	Access Licences[1] (surface and groundwater)	Yes	Licences are transferable if provided for by the relevant Water Sharing Plan. However, major utility or local water utility licences may only be traded for a maximum of one year. Water shares within an irrigation district are only transferable to others within the district
Victoria	Bulk Entitlements	Yes	Bulk entitlements may be transferred to another rural water authority, the owner or occupier of a holding in an irrigation district, the holder of a Take and Use licence or persons outside the state of Victoria
	Stock and Domestic Rights	No	No
	Water Rights (surface water)	Yes	Water rights may only be transferred to land owners or occupiers within the irrigation district. They may only be transferred out of the district if they are first converted to another type of right
	Take and Use Licences (surface and groundwater)	Yes	Yes
	Sales Water	Yes	Yes
	In-stream Use Licences	Yes	Yes
Queensland	Stock and Domestic Rights	No	No
	Water Licences (not subject to an ROP)[2]	No	Can be transferred to other land in accordance with *Water Regulation 2002*
	Water Allocations (subject to an ROP)	Yes	Transferable subject to the rules of the Resource Operations Plan
	Water Licences (subject to an ROP)	No	No

Jurisdiction	Right	Divisible	Transferable
South Australia	Stock and Domestic Rights (in prescribed areas)	No	No
	Water Licences (in prescribed areas)	Yes	Yes
	Water Rights (surface and groundwater in non-prescribed areas)	No	Transferable within an irrigation district. It is possible to trade outside the district if irrigation authority agrees
ACT	Stock and Domestic Rights	No	No
	Licences to Take Water	Yes	No
	Allocations	Yes	Allocations and licences may be transferred if the transfer complies with the Water Resources Management Plan at the receiving location. If an allocation is transferred to a new holder, a new licence application request for the new allocation holder is required

[1] Access licences are held by irrigation companies in irrigation districts. Individuals within each district hold water shares in the company
[2] ROP = Resource Operations Plan

Source: Productivity Commission (2003)

Chapter 10

Integrated water resource management in South Africa

This chapter introduces readers to South African groundwater dependent ecosystems (GDEs), their context within a developing country with scarce water resources and significant social and biophysical diversity. It integrates the ecology, hydrology, policies and management options as they relate to GDEs. As part of South Africa's development, water management is in the process of change. South Africa has legislated the principles of **integrated water resource management** and has begun the difficult task of water sector reform and implementation. The management of GDEs presents a challenge to catchment managers and scientists alike, and successful implementation will be a measure of the ability to achieve sustainable management in an integrated way.

Having read this chapter, readers should understand the following:

- the physical and integrated water resource management contexts for GDEs in South Africa;
- the main types of GDEs in South Africa;
- appropriate tools to use to identify and monitor GDEs in South African catchments;
- policies in place in South Africa to enable the allocation of groundwater to GDEs.

Why a chapter on South Africa?

Much of this text has been concerned with Australian issues and presented Australian case studies. However, Australia is not the only country faced with water shortages, and the problems of managing water and vegetation sustainably for future generations is very pertinent to South Africa. Indeed, many countries of arid and semi-arid regions, including Israel, much of the continent of Africa and much of the Arab world, are facing similar issues. Consequently, this chapter is included to show that the principles, techniques and approaches in Australia are being applied elsewhere. South Africa is particularly relevant because of its similarity in climate and vegetation structure to Australia, although this in itself is not of great import. What is more important is that South Africa has been able to re-examine its approach to landscape and resource management because of the political upheavals in the recent past, opening new opportunities for renewed efforts in sustainable management.

Background to South Africa

To understand water resource management in South Africa it is necessary to have a broad understanding of the country's geology and climate. Consequently this chapter starts with a review of these aspects.

South African geology and aquifers

South Africa shares an early geological history with Australia as both were part of the ancient supercontinent **Gondwanaland**. Both continents today benefit from significant reserves of non-renewable resources such as manganese and diamonds. However, the South African landscape, in contrast to much of Australia and the rest of southern Africa, includes mountainous areas with diverse environments. Figure 10.1 (Colour Plate 7) shows the main aquifer types in South Africa based on primary lithology. South African geology spans from the early Archean era to the modern day and includes some of the oldest rocks on earth. Gneisses of the Beit Bridge Complex near Zimbabwe are 3.3 billion years old and the Barberton granite-greenstone belt in the east close to Mozambique similarly dates from 3.2 billion years. The ancient rocks of the Witwatersrand supergroup (2.8 billion years old) contain gold bearing conglomerate layers formed before the earth's atmosphere became oxidizing. This basin around Johannesburg has yielded 50 000 t of gold since 1886 and is the world's richest gold field. Vast manganese deposits are found in the Kalahari desert in the 2.2 billion year old Hotazel Formation, named after the difficult working conditions of the area! Much younger Kimberlite dykes, 100 Ma, in the central and north-western parts of the country yield diamonds, making South Africa the third largest (by value) diamond producing country.

Mountain ranges include the Cape fold belt, the Drakensberg, the Magaliesberg and the Soutpansberg. The Cape Fold belt is comprised of the Table Mountain group quartzites (deposited 450 Ma) and occurs to the north and east of Cape Town as well as encircling the city itself. The Lesotho highlands, the Malutis and the Drakensberg of the eastern escarpment are underlain by basalts extruded 180 Ma during the early break-up of Gondwanaland. The Magaliesberg range is underlain by quartzites and the Soutpansberg is formed by the 1800 Ma 'red bed' sandstones of the Waterberg group.

Large expanses of the country have relatively low relief; the largest of these is the **Karoo**, a desert and semi-desert area in the interior. The Karoo supergroup contains tillites deposited during the last ice age that affected Gondwanaland, sandstones, shales, basaltic lavas and coal which provides about 90% of South Africa's electricity. The **Kalahari** is a transboundary desert in South Africa, Namibia and Botswana underlain by relatively young (less than 60 Ma) sands and calcretes. It contains the transfrontier Kalahari-Gemsbok park, managed by the three countries.

Recent unconsolidated sands occur on the coastal plain and are more extensive in the humid eastern part of the country, north of Durban. On these sandy coastal plains, important wetland areas, such as the St Lucia world heritage site, are fed by groundwater. However, over 90% of the area of South African geology is lithified and forms hard rock (secondary) aquifers controlled by secondary faulting and jointing. These include the Karoo, most of the mountain areas and the granite-greenstone belts. Primary unconsolidated aquifers are restricted to the coast, alluvial valley deposits and the Kalahari as shown in colour plate 7.

The geology of landscapes and the location and type of aquifers present influence the fate of rainfall on a given catchment. The amount and timing of rainfall and other aspects of climate also determine vegetation distribution. This is now discussed.

South African climate

Temperature and rainfall

The major factors influencing temperatures in South Africa are the relatively high altitudes in the interior, and the ocean currents. The eastern coastal regions are strongly affected by the warm south-west flowing Agulhas current. The West Coast is affected by **upwelling** cold water of the Benguela current. Daily mean maximum temperatures in January exceed 30°C in much

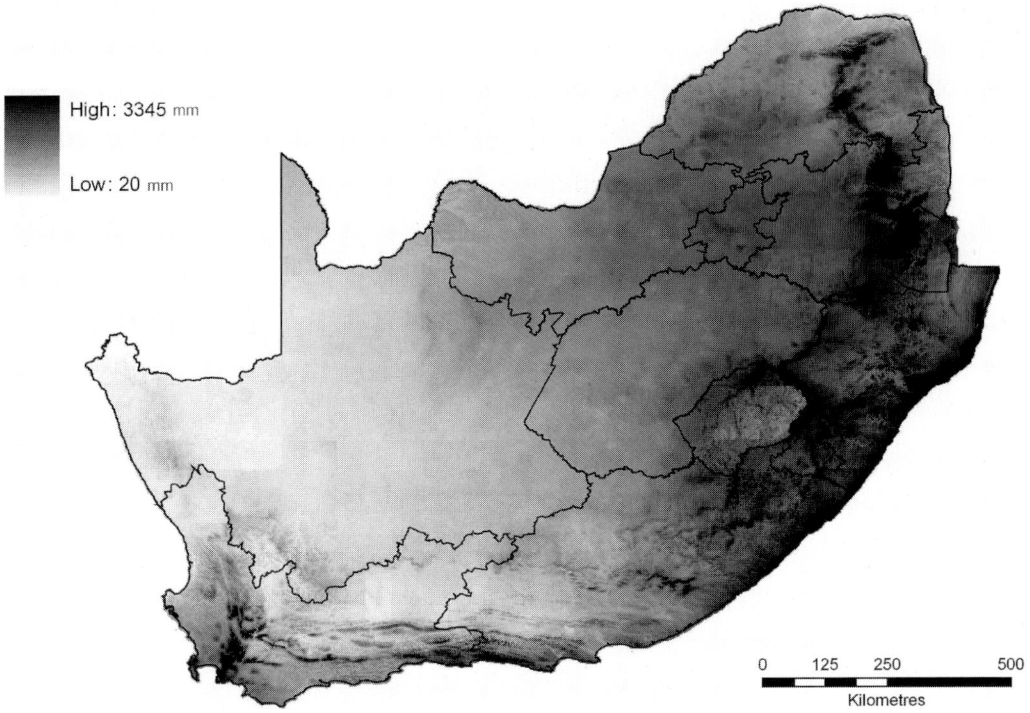

Figure 10.2 Map of South African mean annual precipitation

Source: Dent et al. (1989)

of the western and central interior but are more moderate (22–25°C) along the southern and eastern coasts and interior seaboard, and below 20°C at high altitudes on the eastern escarpment and in Lesotho. Daily mean minimum temperatures in July are 6–8°C or higher throughout the coastal regions but are ≤4°C over most of the interior. Frosts are a significant factor, with most of the interior experiencing more than 150 days of frost per year. Frosts kill the above ground parts of the grasses in the grasslands which cover much of the higher lying areas of the eastern interior. The dieback substantially reduces annual transpiration and is the main reason why catchments in these areas have quite high proportions of run-off.

South Africa has a low mean annual rainfall of about 490 mm compared with a world average of about 860 mm (Fig. 10.2, WRI 2003). The western regions and the interior, almost 60% of the country, get less than 500 mm per year and only 12%, situated along the eastern side and the southern and south-western coastal mountains, gets more than 750 mm (Fig. 10.2). Rainfall is very variable, the range typically being 40% or more of the mean in dry areas, to less than 20% in high rainfall areas (Tyson 1986). The effects of this variation are amplified by the fact that dry and wet years occur in cycles of several years. The calculated mean rainfall is heavily influenced by the occurrence of very wet years, so the rainfall in typical or normal years is less than the mean (i.e. the median is less than the mean). A narrow region in the western and southern part of the country gets its rainfall mainly in winter. This **Mediterranean climate** of cool wet winters and hot dry summers is similar to the area around Perth in Western Australia. Cold fronts formed in the circumpolar system of the southern Atlantic Ocean bring winter rainfall to the Western Cape in South Africa. The bulk of the country gets its rainfall in summer. These rain bearing systems originate mainly in the moist maritime system over the Indian Ocean. The amount of rainfall is strongly influenced by the distance from the sea and interactions between the Inter-tropical Convergence Zone (low pressure) and South Indian Ocean high pressure system. The summer rainfall systems are strongly influenced by the **El Niño** phenomenon and the upper atmosphere quasi-biennial oscillation, which result in extended dry and wet cycles. The region between the two is commonly called the all-year rainfall region but actually gets its rain from both the summer and winter rainfall systems. The north-eastern region of the country is subject to periodic cyclones which can produce torrential downpours that lead to severe flooding (e.g. February–March 2000). The evaporation of water, either from wet surfaces or through stomata of vegetation, requires radiation input and this is now briefly discussed.

Radiation and evaporation

South Africa has very high levels of solar radiation (Schulze et al. 1997). In mid-summer the maximum daily radiation (i.e. on sunny days) can reach more than 34 MJ m^{-2} d^{-1} in the arid western interior. There is a strong east–west gradient, with mean measured values ranging from 21–25 MJ m^{-2} d^{-1} in the eastern part of the country to 25–29 MJ m^{-2} d^{-1} in the western part. In winter radiation levels are lower, ranging from 15–16 MJ m^{-2} d^{-1} in the northern regions to 11–12 MJ m^{-2} d^{-1} along the southern and western coastal regions where there is a Mediterranean climate. The high levels of radiation are the main driver for potential evaporation which exceeds rainfall except in a few montane areas in the south-west and along the eastern escarpment at high altitudes (Schulze et al. 1997). Potential evaporation (based on A-pans measurements) ranges from about 1200–2000 mm y^{-1} along the southern coast and eastern side of the country to more than 2250 mm in much of the western interior and more than 3000 mm in the Nama Karoo and Kalahari regions.

Water that is not soon evaporated is frequently captured as surface run-off. This is now discussed.

Surface run-off

The effects of the low rainfall are exacerbated by the high potential evaporation, so that only 9% of the rainfall ends up in streams and rivers as surface run-off (Midgley et al. 1994). This is far lower than the world mean value of 34% (Fig. 10.3) but similar to that of Australia and Zimbabwe (WRI 2003). As rainfall increases, the amount of surface run-off and groundwater recharge increases, but the relationship is curvilinear – low rainfall results in little or no surface run-off and high rainfall results in a high proportion of run-off. Similarly, high rainfall is generally associated with high recharge and low rainfall with little or no recharge, especially below about 400 mm of rainfall (Beekman & Xu 2003). Variation in surface run-off also is much greater than the variation in rainfall; there is often no run-off at all in drought periods. In dry regions the variation from year to year may exceed 160% of the mean run-off, and even in the wettest catchments in South Africa the variation is more than 40% of the mean (Schulze et al. 2001). Internal renewable natural water resources (IRWR) (km^3 y^{-1}) are defined as average annual flow of rivers and recharge of groundwater generated from endogenous precipitation (Aquastat terminology, FAO).

The interaction of climate, especially temperature and rainfall, and vegetation, is central to the study of ecohydrology. Before we can consider the management of water resources it is important to have some description of the vegetation of South Africa.

South African vegetation

Figure 10.4 (Colour Plate 8) shows the biomes found in South Africa. Biomes are broad ecological units representing major life zones of large areas, characterised by associations of distinctive life forms and plant and animal species. In South Africa these are defined mainly by

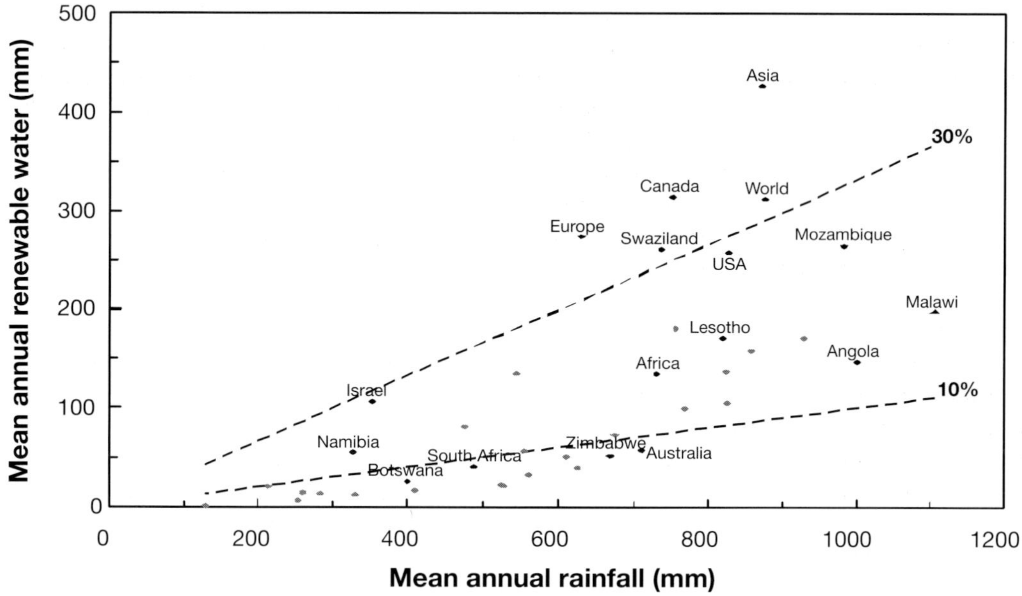

Figure 10.3 Relationship between rainfall and renewable water resources, including groundwater and river inflows, for selected countries and continents. The hollow diamonds indicate the values for the different major catchment regions of South Africa

Source: Data from Leemans & Cramer (1991), WRI (2003) and Midgley et al. (1994)

vegetation structure and climate (Low & Rebelo 1996). Biomes are equivalent to the modern concept of an ecoregion as used by the World Wildlife Fund. A recent classification recognises seven biomes:

- forest
- fynbos
- grassland
- Nama Karoo
- thicket
- savanna
- succulent Karoo.

Other classifications recognise a desert biome (northern succulent Karoo and north-western Nama Karoo) and don't distinguish the thicket biome.

The **Fynbos biome** is characterised by winter rainfall (400–3800 mm y^{-1}) with an increasing amount of summer rainfall from west to east. The vegetation is generally dominated by shrubs with an increase in grasses and a decrease in *Restionaceae* (reeds) from west to east. The flora is extremely rich in species with about 9000 species, 69% of which are confined to this biome. The **succulent Karoo** biome is also characterised by winter rainfall (20–290 mm y^{-1}) with hot summers and frequent fogs along the west coast. The vegetation is characterised by shrubs and herbaceous species. It is a global centre of diversity for leaf and stem succulent plant species. The **Nama Karoo** vegetation is characterised by summer rainfall with some winter rainfall in the west (100–520 mm y^{-1}), and very hot summer temperatures. The vegetation is dominated by low shrubs and grasses. The **grassland** is found in the summer rainfall region over a wide range of rainfall and is characterised by frequent frosts in winter. There are frequent lightning fires in spring and summer. The vegetation is dominated by grasses with a mixture of herbaceous species, and woody vegetation is confined to sheltered situations. The **savanna** biome differs from the grassland biome mainly in that frosts are less frequent or rare. Fires are also less frequent, particularly in the semi-arid savanna of the Kalahari region. The vegetation has two layers – an understorey dominated by grasses with a mixture of herbaceous species and an overstorey of woody plants (shrubs and trees) ranging from scattered plants to woodlands. The smallest biome is the **forest biome**, which covers about 0.25% of the country and occurs in a range of climates. The rainfall is always more than 525 mm y^{-1}. Forests are generally confined to sheltered areas and occur as small patches except in the southern, eastern and north-eastern regions. The forest can be divided into the afromontane type and the subtropical coastal type, which is found on the coastal lowlands in the north-east.

Having provided the context of water management in South Africa in relation to geology, climate and vegetation, we now move to a more specific discussion of GDEs in South Africa, including their identification and dependency on groundwater.

GDEs in South Africa

When hydrogeologists talk about groundwater they specifically mean the water that occurs in saturated aquifers or aquitards. Other scientists, such as ecologists or plant physiologists, may refer to all underground water as groundwater, including interflow or soil water. South African water legislation does not define the term groundwater but defines an aquifer as 'A geological formation which has structures or textures that hold water or permit appreciable water movement through it'. 'Appreciable water' is usually taken to be enough water to supply a well.

The term 'groundwater dependent ecosystem' is difficult to define and means different things in different countries. Some researchers use the term in reference to ecosystems that occur underground and within aquifers. A broad understanding of the term is only just emerging in South Africa. Typically we mean ecosystems which depend on groundwater in, or discharging from, an aquifer. If this source of water was removed or the quality of the source was changed beyond the natural range, the functioning and structure of the ecosystem would be fundamentally affected. The alternative term, aquifer dependent ecosystem, may yet emerge as the more widely used in South Africa, but in this text we will use the more widely recognised GDE. Using the term 'aquifer' removes confusion about the primary water source.

Identifying GDEs in South Africa

Historically, GDEs were identified only as they wilted and died adjacent to groundwater abstraction. Nowadays we are attempting to identify and understand GDEs in order to protect them and the aquifer attributes on which they depend. Identification and verification of ecosystem groundwater use and dependency generally requires a multidisciplinary approach as outlined in Chapters 2 and 3. Identifying GDEs requires knowledge of the catchment water balance and the soil–plant–atmosphere continuum with a more complete understanding of the role of aquifers: it can be viewed as the aquifer–soil–surface water–plant–atmosphere continuum (ASS-PAC).

Many tools are available in South Africa to identify GDEs, including tracers, plant physiological tests and remote sensing from satellites (Fig. 10.5). In addition, spatial groundwater and ecological datasets and expert hydrogeological and ecological knowledge are essential to identifying GDEs. Generally neither alone is sufficient. Groundwater data is often a limiting factor in developing countries. Expert knowledge may assist in describing conceptual aquifer models based on known geology, and inferring flow regimes and probable discharge areas. The protocol outlined below assists catchment managers in identifying terrestrial GDEs and the need for protective resource quality objectives (discussed below).

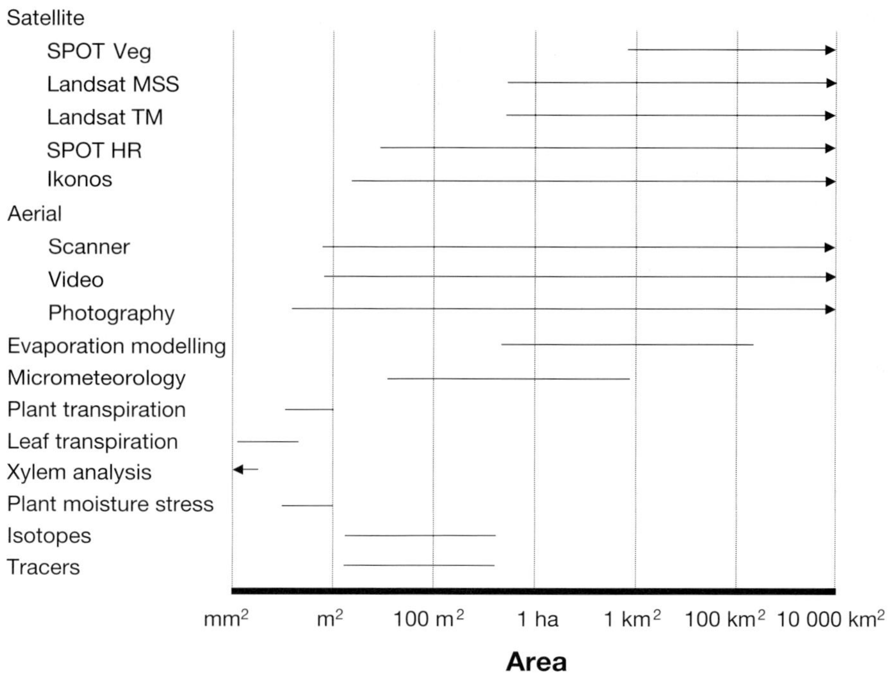

Figure 10.5 Tools to identify groundwater dependent vegetation at different scales in South Africa

Understanding ecosystem dependency on aquifers

Sustainable use of renewable resources such as groundwater is the accepted aim of integrated water resource management (IWRM). This is usually taken to mean that a resource is used today and managed so that it will still be available for use by future generations. In South African law, groundwater should be used sustainably and stakeholders in the catchment should decide the levels of impact which are an acceptable consequence of using groundwater (within sustainable limits). If we are going to pump water out of an aquifer or disrupt the flow regime or chemistry of an aquifer, we need to understand the consequences of our actions. GDEs are very important in this respect, and we need to understand which aquifers they depend on and the nature of their dependency. With greater understanding of the ecohydrologic links in a catchment, impacts of water use can be minimised.

Types of South African GDEs

Research into South African GDEs is at an early stage. There are some existing bodies of research, especially on riparian zones and environmental flows for rivers and estuaries, which form an important contribution to the developing field of GDEs. This section gives a broad outline of the types of GDEs which are known and thought to occur in South Africa.

GDEs in South Africa are currently described according to the aquifer system and the habitat type, as shown in Table 10.1. This gives an immediate description of some of the key characteristics of the GDE:

- physical aquifer characteristics that determine the quality and nature of the groundwater occurrence and discharge;
- biophysical characteristics that determine the dependency and uptake of water by the ecosystem.

Table 10.1 Preliminary classes of GDEs in South Africa and their likelihood of occurrence

Habitat types	Secondary aquifer types					Primary aquifer types		
	Karoo dykes and sills	Basement and younger granites	Extrusives	Carbonate	Fractured meta-sediments	Alluvial	Inland aeolian (Kalahari)	Coastal plain
In aquifer	Unlikely	Unlikely	Unlikely	Known	Unlikely	Probable	Unlikely	Probable
Spring	Known	Known	Known	Known	Known	Known	Unlikely	Probable
Riverine aquatic	Probable	Probable	Probable	Known	Probable	Known	Unlikely	Known
Riparian	Probable	Probable	Probable	Known	Probable	Known	Unlikely	Known
Wetland/seep	Known	Known	Known	Known	Known	Known	Unlikely	Known
Terrestrial	Probable	Probable	Probable	Probable	Probable	Probable	Known	Known
Estuarine/coastal	Unlikely	Unlikely	Unlikely	Known	Known	Known	Unlikely	Known

For example, a wetland dependent on a fractured (meta-sediment) aquifer would be expected to have different characteristics from a wetland on unconsolidated coastal sands. The fractured aquifer is more likely to have complex anisotropic flow in a large regional system with low storativity and discharge via discrete springs or spring lines. The coastal sands aquifer will typically form a shallower, smaller more homogenous system with relatively high storativity and diffuse discharge from broad supra-tidal areas.

Groundwater dependent ecosystems occur throughout the landscape at various scales. They are found in discharge areas of local and regional scale aquifer systems. They are represented across the continuum of habitat types from humid aquatic systems to arid terrestrial systems and do not, in reality, occur as discrete groupings used to classify GDEs. Riparian and hyporheic zones highlight the artificial distinction between aquatic and terrestrial, and wetland ecosystems, by definition, occur across the spectrum. However, the classification helps to order our understanding of complex interactions and group similar settings in a rational way.

In aquifer and cave ecosystems

Groundwater dependent ecosystems in aquifers and cave systems have not been well researched in South Africa compared with Australia, particularly Western Australia. However, they are known to occur in the dolomitic carbonates of South Africa, and may occur in calcretised layers of sedimentary alluvial and coastal deposits and in the **pseudo-karst** fracture and cave systems of fractured aquifers such as the Table Mountain group aquifer. These systems lack plant species but can have a range of invertebrate fauna such as shrimps, as well as a range of poorly known microbiota such as bacteria and fungi.

The dolomites in the north-western region of the country form South Africa's largest area of karst terrain (Fig. 10.1 (Colour Plate 7), Fig. 10.7). Some of the cave systems in these dolomites are well known worldwide because they include sites where remains of pre-historic hominid remains have been found (since prehistoric times humans have been a keystone groundwater dependent species). The most famous ones are at Sterkfontein and Makapansgat and now form part of the World Heritage Site known as the Cradle of Humankind. Other cave systems include the well-known limestone Cango Cave system at Oudtshoorn and the sandstone cave systems on the Cape Peninsula. There are also limestone formations in the western and southern coastal forelands in the Western Cape but no GDE investigations have been done there yet. Shrimps are visible in videos taken with a camera lowered down boreholes in faulted rock formations in the Tshipise area and a few other sites. There have been a few studies of cave ecosystems but systematic investigations of the fauna and functioning of these ecosystems are absent. Data are currently being gathered on the species composition of the in-aquifer fauna of the dolomites in Sterkfontein in Gauteng.

Spring and seep related GDEs

Spring and **seep GDEs** are often closely linked to other habitat types such as riparian and wetland GDEs. They occur in most aquifer types of South Africa, excluding the unconsolidated Kalahari sands where the water table is generally below the surface in this extensive primary aquifer. Alluvial aquifers underlain by impermeable bedrock may support springs and seeps within the Kalahari region.

Springs and seeps generally support relatively lush vegetation and in many areas are associated with biodiversity hotspots for plants, amphibians, birds and insects. In the Western Cape, groundwater fed springs at the Kammanassie reserve also support the endangered mountain zebra. Changes in spring flow are believed to have influenced zebra mortality as well as river base flow and water stress in associated riparian species. The springs are species rich and 244

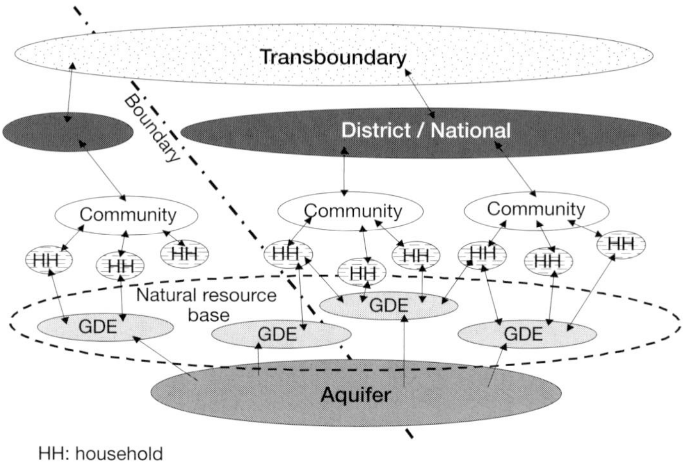

Figure 10.6 Links from aquifers to ecosystems and society

plant species from 145 genera were identified at four springs in the Reserve. In this area springs emanate from fractured meta-sediments. The springs have been grouped into three classes according to their hydrogeological setting, as shown in Figure 10.8 (Cleaver et al. 2003). This has helped researchers to assess the vulnerability of the springs and associated ecosystems to climate change, drought and abstraction.

- **Type 1:** Shallow seasonal springs and seeps emanating at perched water tables, which can be interpreted as interflow or rejected recharge. These springs are vulnerable to drought and climate change, but not abstraction from the deeper aquifer. Approximately half the springs in the area are this type.
- **Type 2:** Lithologically controlled springs, due to the presence of interbedded aquitards, located mainly at the Peninsula–Cedarberg, Goudini–Skurweberg and Nardouw–Bokkeveld contacts. These springs are vulnerable to climate change and abstraction.
- **Type 3:** Fault controlled springs. These may include hot water springs and are vulnerable to climate change and excessive abstraction.

Wetlands

Groundwater fed wetlands, like springs, occur on all the main aquifer types in South Africa, excluding the Kalahari sands. Wetlands and springs are often grouped together because there often isn't a clear distinction between them. This is particularly true in secondary aquifers such as those associated with dykes and sills, fractured rocks and the basement. A spring typically has a distinct discharge point whereas wetlands tend to have a more diffuse area of discharge. Wetlands fed by alluvial aquifers are often closely linked to the riparian zone and those on the coastal sands to estuarine and coastal brackish habitats.

Under the South African National Water Act (1998) a wetland is defined as:

> Land which is transitional between terrestrial and aquatic systems where the water table is usually at or near the surface, or the land is periodically covered with shallow water, and which land in normal circumstances supports or would support vegetation typically adapted to life in saturated soil.

This definition links wetlands to groundwater and unconfined (**phreatic**) aquifers through the use of 'water table' as one of the possible defining features. However, many wetlands occur as a

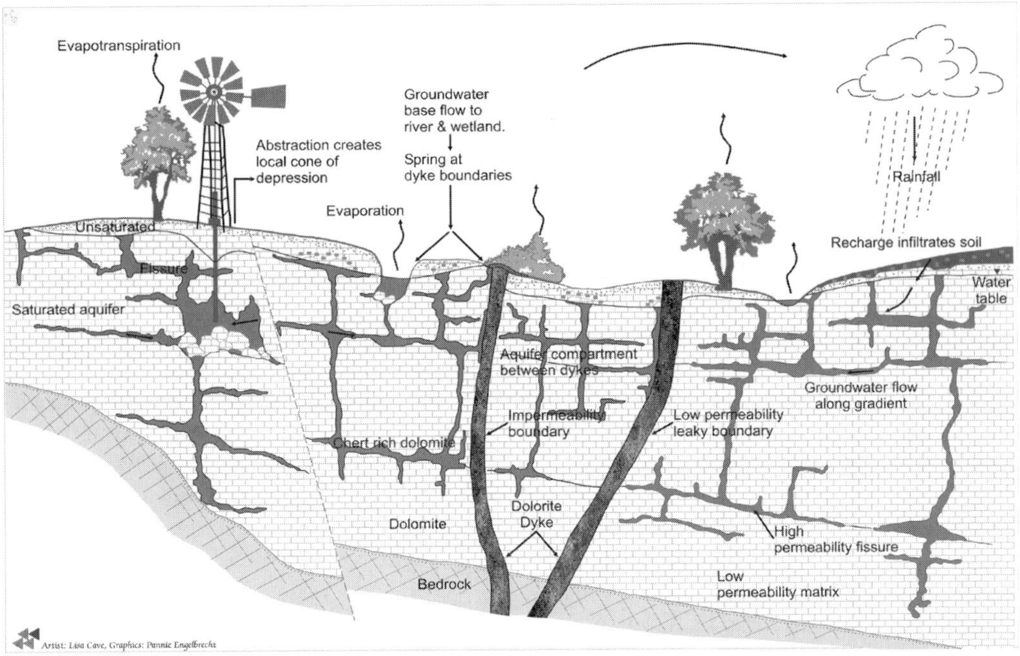

Figure 10.7 Schematic cross-section of the dolomitic aquifer of North-West Province, South Africa

Source: Colvin et al. (2004b), reproduced with the permission of the Water Research Commission

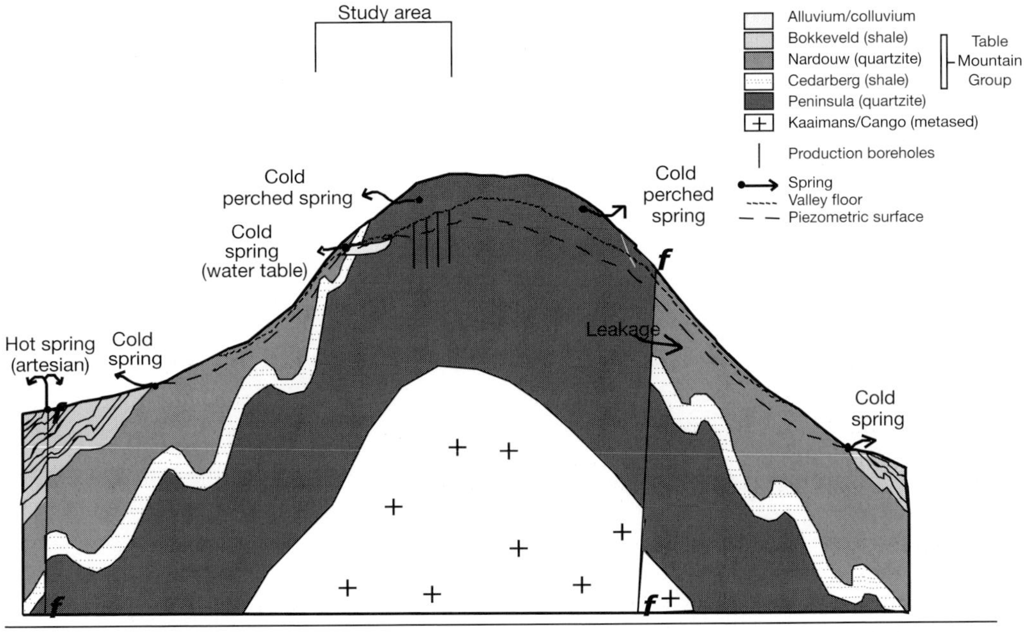

Figure 10.8 Schematic section of spring types in the Table Mountain group fractured meta-sedimentary aquifer at the Kammanassie Nature Reserve in the Western Cape, South Africa

Source: Cleaver et al. (2003), reproduced with the permission of the Water Research Commission result of surface rather than subsurface drainage and ponding in low permeability soils. It is important to understand whether wetlands are linked to aquifer discharge areas before classifying them as groundwater (or aquifer) dependent.

Wetlands associated with both **sinkholes** (points of recharge) and **springs** (points of discharge) are very important aquatic features in the dolomitic karstic area of South Africa, which lacks continuous river systems. Due to the relative geographical isolation of these habitats, many of the springs and wetlands sustain rare or endemic flora and fauna (Stephens 2003). The flow regime of the South African dolomites, and therefore the occurrence of wetlands, is largely controlled by low permeability dykes which compartmentalise the carbonates as shown in Figure 10.7.

Each **dolomitic eye** (a spring with groundwater rising to the surface) supports ecological communities evolving independently of one another, driven by particular local factors (e.g. temperature, flow). For example, the algal and diatom community in the Molopo eye are different to that in the Malmanies eye, indicating that each eye could be a unique ecosystem (Nel et al. 1995). Distinct morphological and behavioural changes have been recorded in the *Pseudocrenilabrus philander* (southern mouthbrooder) and *P. crenilabrus* in the various eyes, indicating species specification (Nel et al. 1995). Dolomite eyes are highly sensitive to invasion by exotics (e.g. black bass) and indigenous species are sensitive to the introduction of alien predatory fish, due to the water clarity (Nel et al. 1995).

Riverine aquatic ecosystems

South African rivers have been classified in terms of their geomorphology, ecology, climate and geology (Ecoregions – Kleynhans 1999), but they have not yet been comprehensively assessed in terms of their relation to aquifers. Licences for use of surface or groundwater, under new legislation, may not be issued until the quantity and quality of water required to maintain aquatic ecosystems has been determined. This water is known as the Reserve. This new legislation has stimulated research into the interconnections between rivers and aquifers. The following section on GDEs and environmental management in South Africa highlights some of the difficulties in linking sustainable management and scientific understanding of interactions in the water cycle.

An example of a riverine ecosystem where groundwater discharge sustains key ecosystems is the Sabie-Sand River system. The sources of this river system are in the mountains along the Drakensberg escarpment in Mpumalanga and the tributaries flow through plantations, irrigated farms and rural areas before passing through Kruger National Park and continuing through Mozambique to the Indian Ocean. Base flow discharge from the upper catchment, primarily groundwater (Birkhead et al. 1997), keeps this river perennial. Most studies of this ecosystem have focused on documenting the flora and fauna, the role of the hydrodynamics of different kinds of river flows in shaping the river floodplain, the proportions of rocky and sandy river bed, and tree species distributions and recruitment (Jewitt et al. 1998). The area has relatively low rainfall and the riparian forest is sustained by river discharge into the riparian aquifer. The riparian forest and associated vegetation play key roles in maintaining the biodiversity and functioning of the adjoining terrestrial ecosystems (Davies et al. 1993; Jewitt et al. 1998).

Although groundwater contribution to base flow is generally regarded as the most critical role of aquifers in surface water systems, groundwater also contributes to peak flows. Interactions between groundwater and surface water are typically dynamic, depending on the hydraulic and topographic gradients (Woessner 2000). Thus a river may be **losing water** to the aquifer at high flows and **gaining water** at low flows; the situation may vary along the length of a river. For example, the Orange River is a gaining river for the upper half of the catchment but is a losing river for the rest. So are many of the other rivers which rise in the montane regions (along the escarpment) and flow through more arid regions. Groundwater may also contribute to peak flows in rivers via displaced subsurface water during rainfall events (Sklash & Farvolden 1979; McDonnell et al. 1990; Midgley & Scott 1994; Bonell 1998). This means that a large component of storm flow passes through an associated aquifer. Research shows the pre-event component in storm flow varies between 40% and 80% in humid temperate regions (Rodhe 1987; Buttle 1994) and 25% to 35% in semi-arid regions (Turner et al. 1991). Various factors appear to influence the response of catchments to rainfall events, including climate, soil thickness, soil structure topography, vegetation and geology (McDonnell et al. 1990, Sandström 1996).

Riparian ecosystems

Riparian ecosystems are associated with groundwater dependency in a range of aquifer types, but most importantly with unconsolidated alluvial aquifers. These form important keystone ecosystems in the semi-arid and arid areas of the country, particularly where surface flows are seasonal and ephemeral.

The riparian forest ecosystems are sensitive to changes in groundwater levels. From 1979/80 to 1982/83 flows in the Kuiseb River in Namibia did not reach its delta, and the **piezometric** surface in the alluvial aquifer dropped by 3 m (Ward & Breen 1983). A number of large *Faidherbia (Acacia) albida* trees (riparian fringe woodland) died and the growth and vitality of riverine vegetation declined. Localised stands of young *F. albida* (established in 1974 and 1976) did survive, suggesting that the large dead trees had lost their ability to adjust to the lowering of the water table. *Acacia erioloba*, a non-riparian species, did not show any signs of mortality, presumably because it is better able to track changing water tables. Ward and Breen (1983) suggested that the young *F. albida* trees were able to keep their roots in contact with the falling water table, unlike the mature trees. Subsequent work (Prof D Ward, University of Stellenbosch, pers comm 2001) suggested that the older trees were dying naturally due to old age, not their inability to adapt their root systems. The trigger for the deaths was a natural drought but its effects were exacerbated by dams in the upper reaches which reduced flows and altered flood frequencies.

Figure 10.9 (Colour Plate 8) shows a normalised difference vegetation indices (NDVI) remotely sensed (satellite) image of the Limpopo valley between South Africa and Zimbabwe, and its tributary the Shashe, showing relatively lush areas in red. The lush vegetation is either irrigation-fed (groundwater source), distinguished by the circular shape of the pivot irrigated areas, or reliant on groundwater discharge from the alluvial aquifers of the Limpopo and its tributaries.

It is estimated that in excess of 10 million m^3 are abstracted annually from the alluvial aquifer shown in Figure 10.10, for both irrigation and mining. Since 1991 a mining company, one of the larger abstractors on this stretch of river, has monitored seasonal water stress in selected riparian trees (*Ficus sycamorus* and *Croton megalobotrys*) as part of its licensing conditions for groundwater abstraction. The company has redesigned its alluvial well field to reduce interference effects between pumping wells and thereby reduce drawdown impacts on dependent tree species.

Groundwater dependent ecosystem health is particularly important in this trans-boundary area where it forms an important keystone ecosystem. It is an important political issue as Botswana, Zimbabwe and South Africa intend to strengthen their eco-tourism concerns in this area with a trans-frontier conservation area.

Terrestrial GDEs

Terrestrial GDEs are those where there is no direct groundwater discharge at the soil surface but the water table is within reach of ecosystem components. This form of interaction is also called cryptic discharge (Hatton & Evans 1998). In most cases the connection is via the root

Figure 10.10 Drilling water supply boreholes on the banks of the seasonal Limpopo River. The view looks north from the South African bank to the riparian vegetation on the Zimbabwean bank in this trans-boundary GDE

system of the plants, particularly the deep rooted woody plants such as shrubs and trees. If the water table is too shallow for the typically deeper rooted woody species, they will be replaced by herbaceous growth forms including grasses, sedges and reeds or bulrushes more typical of wetlands. Cryptic discharge in terrestrial habitats has been noted in South Africa on Kalahari and coastal sands and fractured aquifers. A protocol to identify terrestrial GDEs is presented in Figure 10.12.

Large trees in the arid zone

One of the characteristic features of the arid and semi-arid Karoo and Kalahari regions of southern Africa is the occurrence of large trees. The trees occur in a variety of settings but always where there is deep weathering or a deep profile which can be penetrated by the roots of these species, and groundwater at depths of 10–30 m. The best developed tree stands are the **gallery forests** along ephemeral rivers where the floodplain alluvium is recharged during periodic floods. This combination makes them GDEs which are both terrestrial and riparian. The roots of *Acacia erioloba* have been recorded at depths of 25 m (Henkel 1931), 46 m (Timberlake 1980) and 60 m in a borehole (Jennings 1974). In many cases, though, the trees may not have access to groundwater but rely on water stored in the vadose zone and recharged periodically by heavy rains and during high rainfall periods. Typical tree species in these environments include *Acacia albida, Acacia erioloba, Acacia haematoxylon, Acacia karoo, Rhus lancea, Tamarix usneoides, Euclea pseudebenus, Salvadora persica, Ficus sycomorus* and *Ficus cordata*. These trees, singly, in stands and as gallery forests, are **keystone ecosystems** (Milton 1990; Dean et al. 1999). The trees provide an environment which supports a rich insect fauna (>700 species in one section of the Kuiseb River in Namibia, Prinsloo 1990). The gallery forests and associated vegetation sustain a rich vertebrate fauna, including reptiles, birds and large

Figure 10.11 South African, Mozambiquan and Australian scientists researching groundwater discharge to mangrove swamps in Maputo Bay (photo C. Colvin)

mammals (Milton & Dean 1999), and provide key resources for the local human populations (Jacobson et al. 1995) as well as for farming and mining. The large ungulates complete the cycle by dispersing the seeds and possibly scarifying the seed coats to enable them to germinate more readily. These species are thought to act as nutrient pumps but it is equally likely that they are providing water to shallower rooted plants via **hydraulic lift**.

Sandplain fynbos

Sandplain fynbos is a form of the species rich fynbos (fine-bush) which occurs where there are deep, unconsolidated, acid sand deposits on the western and southern coastal lowlands in the Western Cape (Low & Rebelo 1996). It is very similar to the kwongan vegetation on similar substrates in Western Australia (Chapter 5). This kind of fynbos is characterised by an overstorey (3–5 m tall) of deep rooted *Proteaceae* and an understorey and ground layer with a rich flora of woody and herbaceous species, many of which are endemic to these communities. These areas are also associated with shallow (<10–20 m water table) groundwater and are probably groundwater dependent like the analogous systems in western and south-eastern Australia. Studies have shown that higher leaf area indices and lower levels of moisture stress are evident in sandplain or strandveld areas where the water table depth is around 5 m (as opposed to 14 m, Colvin et al. 2002).

Estuarine and coastal ecosystems

Estuarine GDEs (Fig. 10.11) are known to be important in the unconsolidated coastal sand aquifers of South Africa, but are also linked to other aquifer types where they discharge at the coast. For instance, the alluvial aquifers of the north-west coast feed brackish wetlands in arid areas with ephemeral surface water flow, such as the Buffels River close to the Namibian border and the desolate skeleton coast.

On the humid east coast, groundwater in the extensive coastal sands feeds lakes and wetlands in northern KwaZulu Natal. Lake Sibiya, Lake Mzingazi and the St Lucia wetlands are examples where detailed monitoring of relative groundwater heads, lake levels and the hydrochemistry have indicated the importance of groundwater inflows (Kelbe et al. 2001). Relatively fresh groundwater discharge to these systems often maintains brackish refugia habitats during high salinity periods. Groundwater discharge into mangrove areas further north in Mozambique has been found to influence the distribution of mangroves in Maputo Bay, and this may be the case for the more limited mangrove areas in eastern South Africa.

Once the location, extent and dependency of GDEs has been established, it is possible to incorporate that information into the management of water resources, discussed below. In particular we examine the principles underpinning water management and the policies implementing those principles.

Principles of water use and management in South Africa

South Africa is a water scarce country and current levels of consumptive use may be close to sustainable limits. Statistics on water use, in particular groundwater use, are poor. It is estimated that groundwater supplies 15% of bulk water supply, around 10% of agriculture and over 50% of rural communities' supplies. Prior to the new water laws, groundwater had a variable status which differed from surface water and it could be deemed a private resource. Under the new National Water Act (Act 36 of 1998), groundwater has the same status as surface water and may not be privately owned. The Act explicitly recognises the following **principles of integrated water resource management (IWRM)**:

- water is a *scarce* and *unevenly distributed* national resource which occurs in many different forms, all of which are interdependent and part of the water cycle;
- water is a natural resource that *belongs to all people*, however, the discriminatory laws and practices of the past have prevented equal access to water and use of water resources;
- the national government should ensure the *equitable allocation* of water *for beneficial* use, including the redistribution of water;
- the government should take account of *international water* matters;
- the *protection of the quality* of water resources is necessary to ensure *sustainability* of the nation's water resources in the interests of all water users;
- the management of water resources should *integrate* all aspects of water resources and, where appropriate, should be *delegated* to a regional or catchment level so as to enable everyone to *participate*;
- every person has the right of access to basic water supply and basic sanitation necessary to ensure *sufficient water* and an *environment not harmful to health* or well-being;
- the national government is the *custodian* of the nation's water resources.

The recognition of the connection of groundwater to surface water and the requirement for sustainable use are of particular importance for the future management of GDEs in South Africa.

In future, water resource management will be carried out by 19 **Catchment Management Agencies** in South Africa. The Catchment Management Agencies and the National Department of Water Affairs, in consultation with stakeholders, need to develop a vision for water resource development in an area. This will describe what the available water should be used for, how much impact is acceptable, how the ecosystems linked to rivers and aquifers should be managed and what aspects of the resource should be protected and monitored. All significant water resources in an area need to be identified, quantified and classified according to how much impact is acceptable. **Resource quality objectives** describe aspects of the resource to be protected and the **Reserve** describes how much water needs to be set aside for basic human needs and aquatic ecosystems. Water for the Reserve is given the highest priority for allocation. International obligations to neighbouring countries are the second priority; this includes water for environmental flows as well as downstream consumptive use.

Previously unallocated groundwater resources are seen as important in redressing previous imbalances and enabling equitable access to water. Access to all of the country's natural resources was preferentially given to white people during the apartheid era, and resource redistribution is an important issue that will be high on the social and political agenda of South Africa for many years to come.

Having established the principles governing management of water resources, it is now possible to examine the policies in place to manage water in the context of IWRM and GDEs. The following section also tackles the difficult question of whether we know enough to manage such systems effectively.

Environmental management, IWRM and GDEs in South Africa

In recent years concern about conservation and sustainable use of biodiversity has extended beyond being a preoccupation of environmentalists to every sector of society. This shift in perception has been driven by the realisation that the ecological services provided by biodiversity sustain all levels of human society, whether social, political or economic.

Policies relevant to GDEs in South Africa

The South African constitution includes a basic human right which requires the environment to be protected from degradation, sustain human health well-being and secure sustainable development. This right has been reflected in the provisions of water for ecosystem and human needs in the **Water Principles**, outlined at the beginning of this chapter, the **Environmental Management Policy** and the White Paper on the **Conservation and Sustainable Use of South Africa's Biodiversity**. South Africa has also ratified a number of international agreements which could be relevant to the protection of GDEs, including the Convention on Biological Diversity, the Convention on Migratory Species and the Convention on Wetlands especially as Waterfowl Habitats (Ramsar).

Management of all environmental aspects of water resources therefore falls within the domain of policies on both water resources and environmental management. These policies have been given effect in the form of the National Water Act and the **National Environmental Management Act**. The National Environmental Management Act recognises that environmental matters, including the responsibility for dealing with Environmental Impact Assessments, are shared between the national and provincial governments. The basic requirement for such assessments is the preparation of a preliminary scoping document to determine whether there are ecologically sensitive features associated with the water resource, including GDEs. There are initiatives to develop guidelines for biodiversity assessment in Environmental Impact Assessments, which will set standards for impact assessment and will influence management by identifying aspects that need to be monitored.

A proposal to develop a water resource is therefore subject to three tests: whether it impacts the requirements of the ecological Reserve, whether it impacts the resource quality objectives, and whether environmental standards and goals under the National Environmental

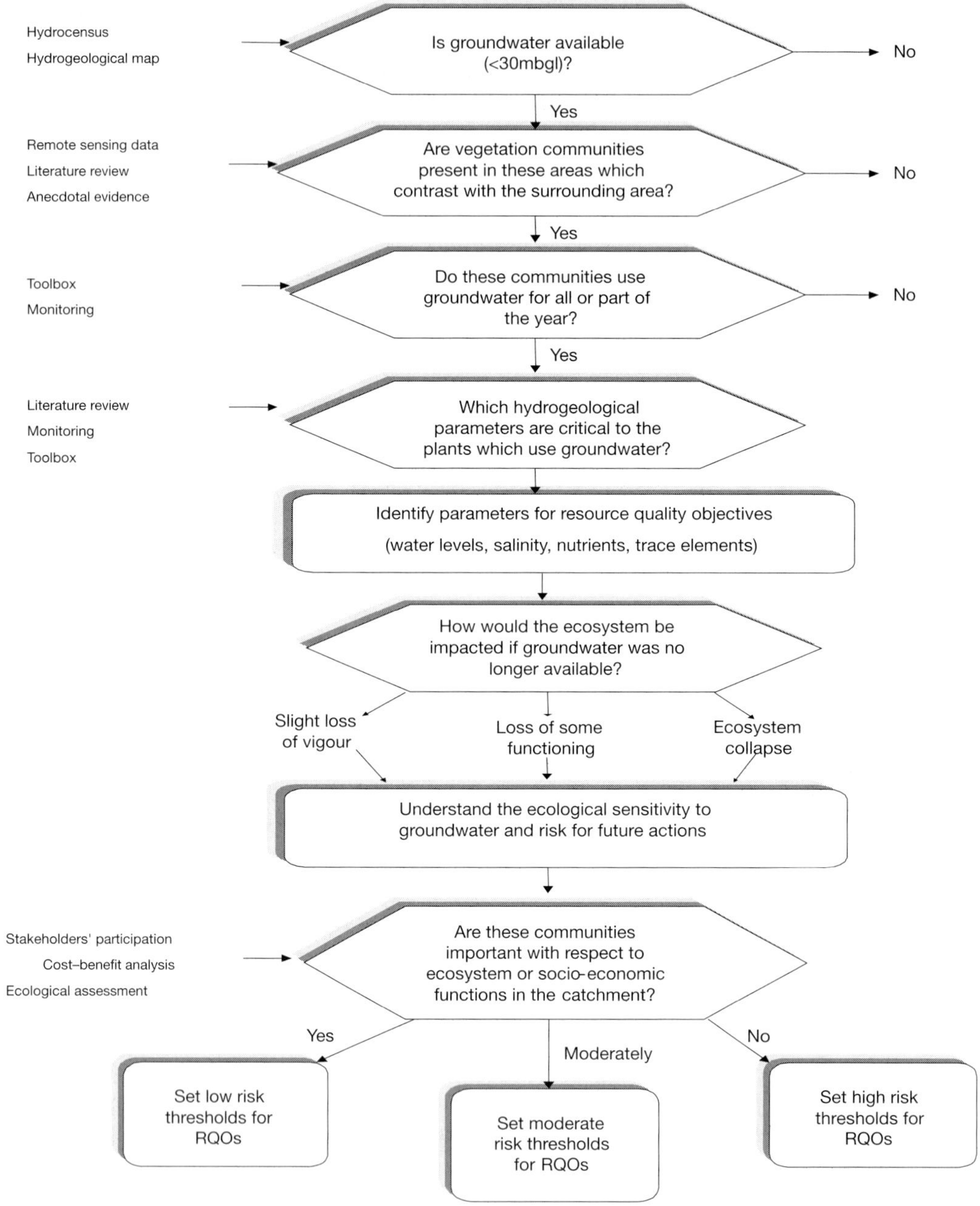

Figure 10.12 A protocol to identify terrestrial GDEs in South Africa

Source: Colvin et al. (2004)

Management Act are affected. These measures should safeguard the conservation and ecological requirements of GDEs if fully implemented. The resource quality objectives are a particularly powerful legal instrument to protect GDEs. The Reserve was conceived by aquatic ecologists to protect and maintain the resource base, including biotic components, to ensure the sustainability of surface water resources. GDE wetlands, rivers and springs may be afforded the high priority protection of the Reserve. Resource quality objectives are not restricted to aquatic ecosystems, as described below, and may be used to protect terrestrial GDEs as shown in Figure 10.12. For both resource quality objectives and the Reserve, the allowed level of impact can vary depending on how the aquifer is classified by stakeholders. It is not expected that pristine (reference) conditions will be chosen for many aquifers. The National Water Act states that:

> The purpose of the resource quality objectives is to establish clear goals relating to the quality of the relevant water resources. In determining resource quality objectives a balance must be sought between the need **to protect** and sustain water resources on the one hand, and the need **to develop and use** them on the other.

Resource quality objectives may relate to:

- the Reserve;
- instream flow;
- water levels;
- the presence and concentrations of particular substances in the water;
- the characteristics and quality of the water resource and the instream habitat and riparian habitat;
- the characteristics and distribution of aquatic biota;
- the regulation or prohibition of instream or land based activities which may affect the quantity of water in or the quality of the resource;
- any other characteristic of the water resource in question.

This fits with the definition of **resource quality** in the National Water Act:

> 'resource quality' means the quality of all the aspects of a water resource including:
> *(a)* the quantity, pattern, timing, water level and assurance of instream flow;
> *(b)* the water quality, including the physical, chemical and biological characteristics of the water;
> *(c)* the character and condition of the instream and riparian habitat;
> *(d)* the characteristics, condition and distribution of the aquatic biota.

Resource quality objectives may be seen as goals to aim for, or thresholds or safety nets which represent the limit of acceptable impact. They may be numeric or descriptive. Following the generic description of resource quality objectives in the National Water Act, the objectives for groundwater could include (Colvin et al. 2004b):

- water levels, groundwater gradients, storage volumes, a proportion of the sustainable yield of an aquifer and quality parameters required to sustain groundwater component of Reserves for basic human needs and base flow to springs, wetlands, rivers and estuaries;
- groundwater gradients and levels required to maintain the integrity of the aquifer and the aquifer's broader functions;
- the water table or piezometric levels;
- the presence (or not) of dissolved and suspended substances (naturally occurring hydrogeochemicals and contaminants);
- aquifer parameters such as permeability, storativity and recharge; landscape features such as springs, sinkholes and caverns characteristic of the aquifer type; subsurface and surface ecosystems in which groundwater fulfils any vital function; bank storage and storage with alluvial aquifers which support riparian vegetation;
- aquatic biota in features dependent on groundwater base flow, such as rivers, wetlands and caves, or biota living in the aquifer itself or the hyporheic zone;

- terrestrial plants and ecosystems dependent on groundwater;
- land use and water use which impact recharge quantity or quality. Subterranean activities, such as mining or waste disposal, which impact the aquifer directly. This would exclude activities which impact or occur solely in aquitards. The control of land based activities by aquifer protection zoning of land use;
- any other groundwater characteristic.

It is clear that resource quality objectives can include any requirements or conditions needed to ensure that the water resource is maintained in a desired and sustainable state or condition (MacKay 1998). Resource quality objectives are particularly important for groundwater resources as there is no restriction to aquatic systems. The full ecological role of groundwater may be recognised and protected where necessary. Similarly, the critical role of groundwater in sustaining rural livelihoods can be fully recognised and protected.

Do we know enough to manage and protect GDEs?

Groundwater–surface water and groundwater–ecosystem interactions occur at various scales and may be seasonally variable. The hydrological complexity of this relationship presents a problem for the integration of ecohydrology and water resources management.

Often catchment managers need to know how much water is needed for a river and how much is available for allocation. Approximate water balance figures are required, with an understanding of assurance of supply and cost of abstraction. However, the impacts of abstraction on GDEs are often difficult to predict. Inflows of water to a catchment, in terms of rainfall, are always variable and unpredictable so available surface and groundwater cannot be known with a high degree of certainty. This makes management planning difficult. Further difficulties in protecting GDEs result from our incomplete understanding of aquifer systems and how they will respond to abstraction. GDEs are often dependent on the 'overspill' or highest groundwater levels, and essentially rely on the aquifer being full in certain places and at certain times of the year.

Setting these water levels as targets for protection (or resource quality objectives) and issuing licences to abstract groundwater contingent on the targets being met is one way of circumventing our incomplete understanding and knowledge. It requires **adaptive resource management** by water managers and users (Rogers 2002). Adaptive resource management relies on monitoring, and the capacity and flexibility to respond to changing resource conditions. In undercapacitated developing countries, such as South Africa, this presents a challenge. However, the adaptive approach of 'learning by doing' also provides a means to use groundwater with an insufficient knowledge base. A key requirement for adaptive resource management is a set of well defined and robust indicators that can be measured and used to establish whether the management plan is achieving the desired outcomes. This concept is described in the following section.

Environmental sustainability indicators

Environmental sustainability indicators perform a variety of functions, either proactive or reactive in nature. First, they act as an early warning system in that they point to areas of weakness and allow problems to be solved before they worsen. Second, they provide a useful mechanism through which the effectiveness of a policy can be monitored and measured over time. Hence it is very important that they are relevant to the concerns and responsibilities of the decision makers as well as serving as information to assess the effectiveness of policy making. Third, they can serve as tools to facilitate community action. One of the key princi-

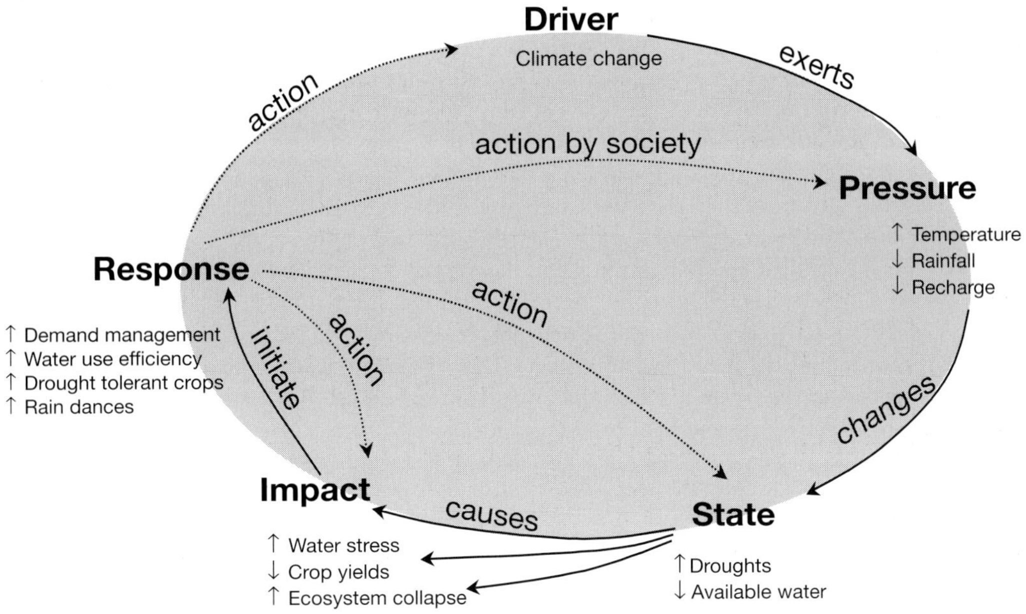

Figure 10.13 The DPSIR framework, showing climate change and drought in southern Africa

ples of sustainability indicators is that they should be developed, understood and accepted by the community.

A system adopted worldwide and within South Africa describes sustainability indicators in terms of the driver–pressure–state–impact–response (DPSIR) framework illustrated in Figure 10.13. DPSIR is a general framework for organising information about the state of the environment (Muller & Pretorius 2002). The idea originated from social studies, then was widely applied internationally, particularly for organising systems of indicators in the context of the environment and, later, sustainable development. The framework assumes cause–effect relationships between interacting components of social, economic and environmental systems:

- **d**riving forces of environmental change (e.g. industrial production);
- **p**ressures on the environment (e.g. discharges of wastewater);
- **s**tate of the environment (e.g. water quality in rivers and lakes);
- **i**mpacts on population, economy and ecosystems (e.g. water unsuitable for drinking);
- **r**esponse of the society (e.g. watershed protection).

IWRM relies on monitoring changing states and impacts of water resources, and responding by influencing anthropogenic drivers and pressures. It is always important to bear in mind that IWRM is about managing people and their impact on water resources, not about managing water resources per se. Modern cultures have given up on rain-dancing as a means of IWRM!

Conclusion

South African aquifers occur in a wide range of stratigraphic and topographic settings. Many ecosystems in the diverse biomes of South Africa are thought to be dependent on aquifer discharge. Over 80% of aquifers are secondary and therefore discrete springs and linear discharge features are common. The remaining primary aquifers include important alluvial and coastal GDEs. South Africa is moving towards IWRM, enabled by new legislation and a transforming water sector. GDEs are beginning to be considered in the context of IWRM. GDEs representing the main habitat types are known to occur in South Africa but few studies have explored the full range of GDE settings, the nature of groundwater (or aquifer) dependency and environmental water requirements. Examples of known South African GDEs include in-aquifer ecosystems in the dolomites (North-West Province), springs and seeps in the Table Mountain group sandstone (Western Cape), terrestrial keystone species such as *Acacia erioloba* in the Kalahari, lakes and punctuated estuaries on the shallow sand aquifers of the east coast in KwaZulu Natal, and riparian zones in the seasonal alluvial systems of the Limpopo River.

The identification of GDEs is often difficult but the first focus should be at a catchment scale, which is most relevant for water management and allocation. Legislation exists in the form of the National Environmental Management Act and the National Water Act, which include resource quality objectives, to protect GDEs. However, it is not known how GDEs will be valued and protected by stakeholders and whether the new catchment management agencies will be able to effectively implement protection measures.

Further reading

Allanson BR, Hart RC, O'Keeffe JH & Robarts RD (1990). *Inland waters of southern Africa: an ecological perspective.* Kluwer Academic, Dordrecht.

Dyson LL & van Heerden J (2001). The heavy rainfall and floods over the north-eastern interior of South Africa during February 2000. *South African Journal of Science* **97**: 80–86.

Schulze R, Meigh J & Horan M (2001). Present and future vulnerability of eastern and southern Africa's hydrology and water resources. *South African Journal of Science* **97**: 150–160.

Glossary

μS cm^{-1}: units for the measure of electrical conductivity of a salt solution. Salinity in ground water can be determined by measuring the electrical conductivity of a soil 'saturation extract' (a solution obtained by saturating a soil sample with water). The electrical conductivity increases with salinity because of the increasing presence of ions. Rain water, for example, has a conductivity of 20–50 μS cm^{-1}, while sea water has a conductivity of 50 000–60 000 μS cm^{-1}. Ground water becomes saline at about 6000–8000 μS cm^{-1} (or 6-8 dS m^{-1}).

adaptive management: the process of continually reviewing the allocation of water between consumptive and non-consumptive purposes as conditions change over time, such as the understanding of environmental needs evolves and/or the community values of environmental protection change.

advection: the transfer of energy by mass transfer, warm wind moving into a cold area.

albedo: the proportion of incoming solar radiation that is reflected by a surface.

allocation: the act of providing a water right to a water user or a use, or the act of modifying the volumetric entitlement of a water right. Allocations can be undertaken administratively (by planning body) or through the purchase in a market for water rights.

alluvial: material such as sediments deposited by flowing water.

alluvial aquifer: an aquifer consisting of unconsolidated sediments deposited by flowing water often occurring beneath or alongside a current channel of a river.

anabranch: a river channel apart from the main stream, usually associated with a braided river system that has channels shifting location from time to time.

aquifer (unconfined, confined and perched): an aquifer is a geological formation or structure that stores and/or transmits water, such as to wells and springs. The term usually refers to those water-bearing formations capable of yielding water in sufficient quantity for human uses.

A confined aquifer is a layer of soil or rock below the land surface that is saturated with water. There are layers of impermeable material both above and below it and the water is under pressure so that when the aquifer is penetrated by a bore, the water will rise to above the top of the aquifer sediments.

An unconfined aquifer is an aquifer whose upper water surface (water table) is at atmospheric pressure, and thus is able to rise and fall in response to rainfall recharge.

A perched aquifer occurs where a layer of groundwater exists above a semipermeable layer at a level above the watertable.

aquifuge: a body of rock that contains no interconnected openings and therefore neither absorbs nor transmits water.

aquicludes: a body of rock of low permeability that can absorb water, but cannot transmit it at a rate sufficient for economic extraction by wells.

aquitard: geological formation that may contain groundwater but is not capable of transmitting significant quantities of water under normal hydraulic gradients. May function as a confining bed.

arid zone: arbitrarily defined as those areas in Australia which receive less than 250 mm of rainfall per year in the south or 350 mm of rainfall each year in the north.

aridity index (see Eamus et al. 1999): the ratio of a region's potential annual evaporation, as determined by its receipt of solar radiation, to its average annual precipitation.

artesian: of, being, or concerning an aquifer in which water rises to the surface due to pressure from overlying water. An artesian aquifer is a confined aquifer in which water is under sufficient hydrostatic pressure to be discharged to the surface without pumping.

An artesian well is a water well drilled into a confined aquifer where enough hydraulic pressure exists for the water to flow to the surface without pumping.

Artesian water is ground water that is under pressure when tapped by a well and is able to rise above the ground level. The pressure in such an aquifer is commonly called artesian pressure, and the formation containing artesian water is an artesian aquifer.

auger: an instrument for drilling vertical holes, used for taking soil samples.

available water: the water held by the soil below field capacity and above permanent wilting point.

bank storage: water that moves from a river in flood into the adjacent geological material, some of which may move back into the river when the flood subsides.

base flow: streamflow derived from groundwater seepage into a stream. The level to which water in the stream returns when at its minimum or base level of flow between rain events, in climates with seasonal rainfall it is often treated as the dry season flow. Base flow may also include drainage from deep soil and weathered material.

Bowen ratio (micro-meterological term): for any moist surface, the Bowen ratio is the ratio of heat energy used for sensible heating (conduction and convection) to the heat energy used for latent heating (evaporation of water or sublimation of snow). The Bowen ratio ranges from about 0.1 for the ocean surface to more than 2.0 for deserts; negative values are also possible. In plant physiology, a Bowen ratio system is used to calculate canopy conductance and evapotranspiration. It consists of two sensor arms, each of which suspend a temperature sensor and air intake for the humidity sensor over the canopy. The lower arm is above the vegetation canopy (typically 1 m) The upper arm is further above the canopy (typically 2.1 m); thus temperature (T) and dew point (Td) gradients are defined over a height difference of 1.1 m from which evapotranspiration and canopy conductance can be calculated.

canopy conductance: canopy transpiration rate divided by the potential rate of evaporation from bare saturated soil under atmospheric conditions. See **Bowen ratio**.

capillary fringe or zone: the saturated zone of water immediately above the water table, containing water in direct contact with the water table but held at pressures less than atmospheric. The position of the capillary fringe varies with changes in the water table and the amount of recharge. The capillary fringe is more accurately called the tension-saturated zone because the water is held in the soil pores against gravity by capillary tension. This zone is frequently accessed by plant roots.

capillary rise: water in contact with, but rising above, the water table, caused by capillary migration due to tensional forces in the pore spaces of soil, sediment and rock material. In fine sediment, capillary rise can be 2–3 m, but maybe only centimetres in coarser grained material.

catchment: the land surface on which rain falls. When referring to a particular river, it is the land surface from which water flows into the river, sometimes through tributaries. A catchment is bounded by hills or ridges directing water into the water course.

char height: the height above ground level below which vegetation is reduced to charcoal by a fire.

chenopod shrubland: shrubland dominated by shrubs of the Chenopodiaceae family (commonly known as saltbushes and bluebushes).

chlorofluorocarbons: a family of inert non-toxic and easily liquified chemicals used in refrigeration, air conditioning, packaging, and insulation or as solvents or aerosol propellants. They drift into the upper atmosphere where their chlorine components destroy ozone. Chlorofluorocarbons are used to calculate the age of groundwater because the concentration of the CFCs in the atmosphere and the water has increased in a known way over the past 50 years.

conduction: the transfer of energy by molecule to molecule contact, in response to a temperature gradient. Hot leaves can conduct heat to surrounding air through conduction.

cone of depression: a cone-shaped lowering of the water table which forms as water is pumped from a well.

convection: the transfer of energy from one body to another through mixing. Two types are recognised: sensible heat flux (in response to a temperature gradient) and latent heat flux (through change of state among solid to liquid to gas phases).

confined aquifer: see **aquifer**.

Council of Australian Governments (COAG): the peak inter-governmental forum in Australia, comprising the Prime Minister, State Premiers, Territory Chief Ministers and the President of the Australian Local Government Association (ALGA). COAG initiates, develops and monitors the implementation of policy reforms which are of national significance and which require cooperative action by Australian governments, such as water reform.

coupling coefficient: a measure of the degree to which a canopy responds to changes in atmospheric VPD. Thus a short grass sward is said to be uncoupled from atmospheric VPD because the canopy is short and closed. In contrast, a tall open woodland which is very responsive to changes in VPD is highly coupled.

cured: the description given to grass stems, leaves and other plant material after it has lost most of its water. Cured, or dried material, is more flammable than fresh (not cured) plant material.

DBH (diameter at breast height): diameter of a tree trunk measured approximately 1.5 m from the ground.

deep seepage: the loss of water from a catchment through underlying soil or geological formations. Much of this water recharges aquifers and some remains in the deep unsaturated zone.

discharge: the volume of water that is lost from an aquifer within a given period of time. Usually measured in cubic metres per second. A discharge zone is an area where groundwater appears on the land surface due to a restriction on downslope water movement. Discharge is maximised where the watertable is very close to the soil surface, where there is little vegetation, or where soil properties tend to encourage water to move towards the surface.

disconnected stream: a stream above but not in hydrological contact with local groundwater. This is not a formal term.

diversion: 1. Extraction, or abstraction from surface water. 2. Changing the natural flow or part of the natural flow of a water source.

drawdown: the difference between the water table level during or immediately following removal of water and the water level when no abstraction is taking place.

dryland salinity: caused when percolating surface water raises the water table which brings natural salts in the soil to the surface. The replacement of deep rooted native plants, perennial trees, shrubs and grasses with annual crops and pastures (that do not use as much water) is a major cause for the increase in the volume of percolating surface water

eddy covariance: a technique used to measure whole-canopy fluxes of water vapour and CO_2. An eddy covariance system measures sensible and latent heat fluxes. Sensible heat flux is the product of the volumetric heat capacity of air and the covariance between vertical wind speed and air temperature. Latent heat flux is calculated as the product of the latent heat of vaporization and the covariance between vertical wind speed and humidity.

effluent stream: a stream which is fed directly by the surrounding groundwater (the piezometric level is above the stream elevation).

El Niño: a disruption of the ocean-atmosphere system in the tropical Pacific that causes changes in rainfall patterns.

embolism: also known as cavitation, embolism is the formation of a blockage in xylem tissue which occurs when an air bubble forms in a tracheary element, causing a break in the transpiration stream. A plant will wilt and may eventually die if too many vessels are blocked for too long.

endorheic: a term used to describe a closed or blind drainage system, one which has no outlet; many of the highveld pans are endorheic pans (Allan et al. 1995).

environmental allocations: water allocated for the specific and exclusive use of the environment. They may be defined in volumetric terms or as a share of the available resource. Allocations may possess their own legal title and be transferable.

environmental water requirements: description of the groundwater regime needed to maintain the ecological values of a dependent ecosystem. These descriptions are developed through the application of scientific methods and techniques and/or local knowledge and long-term observation.

ephemeral: lasting for a brief time, transitory. An ephemeral water body is short-lived and is only filled with water during or immediately after periods of precipitation in the catchment. An ephemeral water body usually has no baseflow since it is fed by surface flow alone.

epicormic bud: dormant organ of vegetative growth on the stem of a eucalypt species. Many eucalypts sprout from epicormic buds after a fire.

estuary: a partially or fully enclosed body of water at the mouth of a river, which is permanently or periodically open to the sea. Sea water within an estuary can be diluted, with fresh water from the river.

eutrophic lake: a lake that has an excessive supply of nutrients, mostly nitrates and phosphates. This often leads to large algal blooms (population explosion of algae).

evaporation: the total loss of moisture as water vapour from all sources, including open water, plant surfaces (interception), and from soil surfaces, and excluding transpiration. It involves the transition of water from the liquid phase to the vapour phase, and during this process energy (known as latent heat) is absorbed so evaporation can be expressed in terms of energy exchange. It is usually expressed in mm for comparison with rainfall.

evapotranspiration: the movement of water from the landscape to the atmosphere, calculated as the sum of evaporation (from bare soil, water surfaces) and transpiration (from vegetation through stomata).

facultative groundwater organisms: organisms which use groundwater but are capable of functioning without it, in contrast to obligate users of groundwater.

field capacity: the water content of the soil when the suction gradient acting upwards balances the force of gravity acting downwards and there is no drainage.

flow recession: the rate at which streamflow returns to baseflow levels after the flow has peaked due to rain.

fractured aquifer: an aquifer in which water flows through and is stored in fractures in the rock caused by folding and faulting.

fluvial: relating to or found in rivers.

fuel load: the amount of fuel (dried and cured branches, leaves and other vegetation material) in the canopy and on the ground. The higher the fuel load, the greater the possibility of a more intense fire. Fuel load is reduced with hazard reduction fires, where the material is burnt in a controlled manner on windless days. The type of fuel present will govern how a fire behaves. Fine fuel (twigs, leaves and grasses) burns to produce a short-lived flame, while heavy fuel (larger branches and fallen logs) continues to burn once the fire front has passes. Fuel load can be quantified as the oven dry weight of fuel per unit area, commonly expressed as tonnes per hectare.

gaining stream: a stream that gains water from the adjacent environment through the discharge of interflow or groundwater or both. Known as an effluent stream in American groundwater terminology (see effluent stream).

glycophyte: plant restricted to non-saline soils.

gravimetric water content: the mass of water contained in a known mass of soil. Usually determined by weighing soil before and after drying the soil in an oven to remove the water.

gravitational water: the water drained by gravity from the soil as the water content decreases from saturation to field capacity.

Great Artesian Basin: one of the largest artesian basins in the world, covering approximately one-fifth of the land area of Australia, beneath arid and semi-arid regions of Queensland, New South Wales, South Australia and the Northern Territory. The basin consists of alternating layers of waterbearing (permeable) sandstone aquifers and non-waterbearing (impermeable) siltstones and mudstones. Groundwater in the basin flows generally westward and recharge by infiltration of rainfall into the outcropping sandstone aquifers occurs mainly along the eastern margins of the basin, along the western slopes of the Great Dividing Range. Natural discharge occurs from mound springs and as direct evaporation in areas with shallow water tables. Water from the basin is often the only available supply for towns and properties.

groundwater: all water in the saturated sub-surface; water that flows or seeps downward and saturates soil or rock, supplying springs and wells, water stored underground in rock crevices and in the pores of geologic materials.

groundwater dependent ecosystems: a groundwater dependent ecosystem requires the input of groundwater to maintain its current position, function and composition. Types of groundwater dependant ecosystems include, terrestrial vegetation, river base flow systems, aquifer and cave ecosystems, wetlands, terrestrial fauna and estuarine and some near-shore marine ecosystems.

groundwater recharge: see **recharge**.

groundwater quality: the chemical, physical, and biological characteristics of groundwater, usually in the context of its suitability for a specified purpose.

Gnangara Mound (of Western Australia): one of the largest groundwater aquifers in Western Australia, located north of Perth, provides a significant proportion of Perth's irrigation and domestic water needs. As the groundwater is often close to the surface, the aquifer supports a variety of significant environmental features such as wetlands, shallow cave streams, springs, seepages and native vegetation dependent on groundwater.

halophyte: a plant tolerant to saline conditions.

hectare: an area of 10 000 m^2, approximately 2.5 acres.

hydraulic conductivity (of soil; of stems): generally, the rate at which a material allows water to move through it. Hydraulic conductivity of soil is influenced by factors such as soil particle size and shape. Hydraulic conductivity of stems is influenced by xylem vessel diameter and age of xylem tissues. The larger the hydraulic conductivity of a soil or stem, the smaller the resistance to water movement and the greater the ease with which water flows in response to pressure gradients.

hydraulic gradient: the slope of the water table or piezometric surface. Hydraulic gradient is expressed as the ratio of the change of hydraulic head divided by the horizontal distance between the two points of measurement.

hydraulic head: the height to which water in a confined or unconfined aquifer will rise if a bore is installed, due to the pressure in an aquifer.

hydrograph: a graphical representation of the amount of water flowing past a point as a function of time. Also used to refer to a graph of hydraulic head versus time.

hydrologic equilibrium: also known as catchment water balance, the balance of inflow to, outflow from, and storage in, a hydrologic unit, such as a drainage basin, aquifer, soil zone, lake, or reservoir.

hydrological cycle: also known as the water cycle, the continuous circulation of water between oceans, the atmosphere, land and vegetation. Water evaporates from the oceans, land and vegetation into the atmosphere, where clouds are formed, the water then falls as rain and returns to the land, vegetation and oceans.

hydrological year: a continuous 12-month period selected to present data so that the peak flows are in the centre of the graph.

hydrology: the study of the occurrence, properties, circulation and distribution of water on the earth and in the atmosphere.

hydroperiod: the frequency and duration of inundation or saturation of an ecosystem. In the context of wetlands, a hydroperiod describes the length of time during the year that the substrate is either saturated or covered with water.

hypogean life: organisms occurring or originating from beneath the surface of the earth.

hyporhoeic zone: the saturated and biologically active zone in and alongside a river bed. The hyporhoeic zone acts as a nutrient storage system and provides habitat for aquatic organisms during periods of low water availability or inundation, thus promoting rapid recovery of aquatic ecosystems after floods or droughts.

Kuczera curve: describes the relationship between water yield of a catchment and condition or age of forest in that catchment over time. Water yield declines sharply after a disturbance (such as clearing or a fire), reaching a minimum as young regrowth returns. Water yield will then increase gradually as the forest matures or recovers. Young regrowth forests yield less water than their old-growth counterparts.

infiltration (of water into soil): passage of water into soil by forces of gravity and capillarity. A soil's infiltration capacity depends on the properties of the soil and the moisture content. Once the water has entered the soil, further movement is known as percolation. Infiltrating water replenishes soil moisture deficiencies but infiltration should not be confused with groundwater recharge.

influent stream: also known as a losing stream, a stream that is perched above the surrounding groundwater to which it is connected and is feeding. An influent stream directly recharges groundwater. See also **gaining stream**.

integrated (or total) catchment management: the management of land, water and other natural resources as a coordinated system for an entire water catchment. It involves incorporating community determined economic, social and environmental values.

interception: precipitation held on foliage surfaces which does not reach the ground or the catchment because it is lost in evaporation.

interflow: the (rapid) lateral movement of subsurface water from rainfall through the soil layers above the water table to a stream or other point where it reaches the surface.

lacustrine: wetlands such as dams and lakes situated in topographic depressions, including permanent and seasonal forms.

irrigation salinity: caused when irrigation water percolates through the soil and raises the water table, thereby bringing natural salts either to the surface of the soil or into neighbouring surface water bodies.

latent heat of vaporization: the heat that must be absorbed to evaporate a mole of a liquid.

lateral flow: sideways movement of water within either the saturated or unsaturated zones.

leaf-area-index (LAI): the most widely used definition is the ratio of the total one-sided area of the leaves on a plant divided by the area of the canopy when projected vertically onto the ground (as though the sun was directly above). For instance a leaf-area index of 2.0 means there are 2 m^2 of projected leaf area per m^2 of ground. LAI is strongly correlated with both the productivity (growth rate) and the transpiration and rainfall interception of a plant or plant community.

leaf-to-air vapour pressure difference (LAVPD): there is a gradient of water vapour pressure or concentration from the water-saturated air inside leaves to the drier air outside. LAVPD is the difference between these extremes. The larger the vapour pressure deficit of the bulk air, the steeper the gradient and the larger the LAVPD. Large LAVPDs favour a large rate of transpiration but stomata tend to close as LAVPD increases.

leakage: movement of water down out of the root zone to deeper profiles.

lignotuber: enlarged root of some tree species, particularly eucalypt species which have specialised buds which sprout after damage to the aboveground parts of the tree, especially after fire. The lignotuber acts as a long-term storage organ, allowing regrowth after major disturbance.

losing stream: also known as an influent stream, a stream with a water level which is higher than the water table in the surrounding groundwater which it is feeding. See also: Gaining stream.

lysimeter: used to study various aspects of the hydrological cycle. A lysimeter consists of a container of soil placed below the ground surface to intercept and collect water moving downward through the soil in which vegetation can be planted, and which is isolated hydrologically from the soil around it. Lysimeters can be used to measure transpiration, evaporation and infiltration of minimally disturbed soil and associated vegetation. They are used mostly in crops and pastures.

macropore flow: the movement of water through the unsaturated and saturated zones in large gaps or channels in the soil which are too large to be affected by capillary forces. These include desiccation cracks, fissures and root channels. Flow rates are significantly more rapid than matrix flow.

matrix flow: movement of water through interconnected pores under the influence of hydraulic gradients and gravity. Matrix flow rates are influenced by factors such as permeability, capillary forces and moisture content in the unsaturated zone.

Mesophytes: plants living in environments that are neither very wet nor very dry. Most crop plants are mesophytes.

moisture stress: a measure of the moisture shortage in plant tissues and thus of the tension in the water conducting tissues (xylem) in the plant. It is usually measured in units of pressure (MPa) using a pressure bomb/chamber.

mound springs: natural outlets of artesian aquifers from which groundwater flows to the surface. Mound springs are an important resource for native flora and fauna, particularly in arid areas.

net primary productivity (NPP): the sum of the accumulation of plant biomass through photosynthesis and nutrient uptake plus any net loss of organic carbon from the plant to other compartments of the ecosystem for instance, through respiration, leaf loss or herbivory.

net rainfall: or net precipitation, is usually taken as total rainfall less the net loss due to interception.

neutron probes: used to measure soil moisture. The radioactive source in the instrument emits neutrons. When the fast neutrons encounter hydrogen in the soil, they lose energy and are slowed down. Most of the hydrogen in the soil is associated with soil water. The neutrons, having no electric charge, cannot be directly detected so a gas is used. The slow neutrons enter the nucleus of the gas, the nucleus is raised to a higher energy state and photons are emitted from the nucleus. The resulting photons can be easily detected with an electronic device and the proportional soil water content can be determined once a standard curve has been constructed for that particular soil type.

obligate groundwater dependent organisms: organisms which are only able to exist or survive by using groundwater resources, in contrast to facultative users of groundwater.

oligotrophic lake: a lake with a low nutrient concentrations in its waters.

osmotic effect: explains the effect of a salt solution on a plant cell. Dissolved salts cause plant dehydration by decreasing the osmotic potential of soil water. Water flows from high osmotic potential (low ion concentration) to low osmotic potential (high ion concentration). When a soil solution has a lower osmotic potential than a plant's cells, the plant cannot extract water from the soil, hence, the effect on a plant is similar to drought stress.

overland flow: rainwater flowing over the land surface, usually either an impervious surface such as rock, or saturated by preceding rainfall. Overland flow can also refer to intense rainfall exceeding the soil infiltration rate; the water does not enter the subsurface at any point. Also called Hortonian (infiltration excess) flow after the hydrologist Horton who first described the phenomenon.

palustrine: freshwater wetland environments other than those along rivers and lakes; includes seasonal and permanent water bodies. Palustrine can be emergent where the vegetation consists of low shrubs, reeds, mosses and lichens or forested, where vegetation is dominated by shrubs and trees, such as a paperback (*Melaleuca* spp.) swamp.

perched aquifer: see **aquifer**.

perched groundwater: unconfined groundwater separated from an underlying main body of groundwater by an impermeable layer and unsaturated zone.

perched spring: spring fed by groundwater discharge from a perched aquifer.

perched water table: the water table of a perched aquifer or perched groundwater separated from an underlying main body of groundwater by an unsaturated zone and perched on an impermeable layer.

percolation: the downward movement of water through the soil due to gravity and hydraulic forces, especially the downward flow of water in a saturated or nearly saturated soil.

Permanent wilting point: the water content of the soil at which the plants cannot extract any water so their leaves wilt. Plants will die if the soil moisture content stays at permanent wilting point for an extended period. Traditionally assigned the value of –1.5 MPa but this has little relevance to trees in Australian landscapes.

permeability: a material's ability to allow a substance to pass through it. In hydrology it refers to the ability of soil and rocks to conduct water under the influence of gravity and hydraulic forces.

phreatic surface: another word for water table, the upper limit of the zone of saturation, where the water pressure is equivalent to atmospheric pressure. See also: Water table

phreatic zone: another word for zone of saturation, a sub-surface zone in which all the interstices are filled with water under pressure greater than that of the atmosphere. The Phreatic zone is separated from the unsaturated zone in unconfined aquifers, by the water table.

phreatic water: another word for groundwater. Water that occupies pores, cavities, cracks and other spaces in crustal rocks. It includes water precipitated from the atmosphere which has percolated through the soil and fossil water retained in sedimentary rocks since their formation.

phreatophytes: Plants whose roots extend downward to the water table to obtain water from the groundwater or the capillary fringe. Phreatophytes are common in riparian habitats near rivers and streams. The term literally means water-loving plants. Obligate phreatophytes are completely dependant on access to groundwater; facultative phreatophytes have the ability to develop deep root systems, enabling them to use groundwater resources to maintain high transpiration rates.

phyllode: flattened petiole (leaf stalk) that functions as a leaf, as observed in *Acacias*.

piezometer: a well used to measure ground water fluctuations. A piezometer is used for monitoring groundwater depth or pressure at a particular depth in an aquifer or within a specific geological layer. A piezometer only measures a small portion of the aquifer. A tube is inserted into the ground to a specific depth and protrudes vertically above the level of the soil. Groundwater enters the tube through a slotted section and the pressure forces water up the tube. The height of the water is measured against a reference point or a pressure gauge can be fitted to the tube. In a nested piezometer, two or more tubes are installed in the same borehole and the slotted section of each tube is set at a different depth so changes in water pressure in different layers of the aquifer can be measured.

piezometric level: the elevation to which groundwater levels rise in boreholes in confined or semi-confined aquifers.

piezometric surface: an imaginary surface representing the piezometric level or hydraulic head throughout all or part of a confined or unconfined aquifer.

potential evaporation: the maximum rate at which water will evaporate from a saturated simple surface under defined atmospheric conditions. Generally it is expressed as the amount of water that could be evaporated with the energy available under a given set of circumstances and assuming that water is freely available.

precautionary principle: the principle of taking action to minimise potentially serious risks rather than waiting until further information becomes available.

precipitation: deposition of water on the earth's surface or movement of water from the atmosphere (where it generally occurs in the vapour phase) to the landscape (where it occurs as liquid or solid) as rain, snow, hail, sleet, dew, mist and frost. Precipitation is measured in units of depth (mm). Precipitation results in the release of energy (known as latent heat).

preferential flow: see **macropore flow**.

primary aquifer: an aquifer in which water moves through the primary openings of the geological formation.

primary openings: interstices that were formed contemporaneously with the formation of the sedimentary deposit or rock that contains them, e.g. interstitial pores. Synonymous with primary porosity. Forms primary permeability.

primary salinity: the natural salinity of a soil caused by salt deposits. Some salt is released from weathering rocks (particularly those formed from marine sediments), and some is carried by rain from surrounding oceans and deposited on the soil.

quickflow: extra stream flow occurring after a storm.

radiation: transfer of energy by electromagnetic waves, the only form of energy transfer that doesn't require a medium, therefore the only way energy can be transferred through outer space.

Ramsar: Ramsar is a city in Iran, on the shores of the Caspian Sea, where the Convention on Wetlands was signed in 1971, thus the Convention's informal nickname, 'Ramsar Convention on Wetlands'. The Ramsar list contains wetlands of international significance selected by member states according to a number of criteria such as high biodiversity, presence of rare and threatened species and the types of fish and waterbirds supported.

recharge: water added to an aquifer. For instance, rainfall that infiltrates and then percolates into the ground. Elevated slopes and areas with shallow soils are common recharge areas. Conditions which aid recharge include highly permeable soils, soils overlying fractured rocks, shallow-rooted or absent vegetation and rainfall exceeding evapotranspiration.

regulated river or stream: river or stream with flow controlled through the use of weirs, locks and dams. Also known as supplemented river or stream.

resource quality: the quality of all aspects of a water resource including the quality, pattern, timing, water level and assurance of stream flow, the water quality (including the physical, chemical and biological characteristics of water), the characteristics and condition of the stream and riparian habitat and the characteristics, condition and distribution of aquatic biota.

resource quality objectives: provide goals for water resource management which are set by the Minister during the process of classification of significant water resources. Resource quality objectives relate to all aspects of water resource quality as listed above.

rest water level: the level of water in a bore not affected by pumping.

riparian: of, on or relating to the banks of a natural course of water. A riparian habitat includes the physical structure and associated vegetation associated with a watercourse which is commonly characterised by alluvial soils, and which might be intermittently inundated or flooded.

riparian vegetation: plants situated on the banks of a river or other water body. Riparian vegetation is significantly different to that of the surrounding area.

riverine wetlands: wetlands associated with rivers and streams, including inland deltas. Riverine wetlands can be permanent or seasonal.

roots: the descending axis of a plant which is usually below ground. Root functions include anchorage, absorption and conduction of water and minerals, and sometimes food storage.

run-off: run-off is usually synonymous with streamflow. It is the sum of surface runoff (excess rainfall minus rainfall retained by a surface) and groundwater flow that reaches streams. The term is usually used to refer to the volume of surface water that leaves a catchment in a period of time.

salination: the process whereby soluble salts accumulate within the soil.

sapflow and sap velocity: sapflow is movement of water through the xylem of a plant, usually measured as litres of water per day. Measuring sapflow provides an estimation of transpiration. Sap velocity is the rate of sapflow, from which sap flow and sapflux density can be calculated. Sapflow, sap velocity and sapflux density are influenced by a number of environmental conditions such as soil moisture content, temperature, solar radiation and vapour pressure deficit.

sapwood area: the area of the outer part of the wood of a trunk, in which the sap flows. It is composed of xylem vessels.

saturated zone: an area of the profile below the water table that has reached maximum water holding capacity. Saturated zone is more accurately defined as the level at which the pressure in the water is greater than atmospheric pressure, therefore, the saturated zone does not include the capillary fringe.

schlerophyllous vegetation: vegetation adapted to low soil nutrients and low water availability. Many Australian plants are schlerophyllous, such as eucalyptus and acacias. Schlerophyllous vegetation is typified by small, hard, rigid leaves so schleromorphy was originally thought to be an adaptation only to low water availability.

scorch height: the height above ground to which foliage has been browned by a fire. This is roughly 10 times the height of the flames in the fire.

seasonal river: a water course which only flows reliably during certain periods of the year as determined by the seasonal distribution of rainfall. A seasonal river generally flows between 20% and 80% of the time; these rivers have a limited baseflow component with little or no groundwater input.

secondary aquifer: an aquifer in which water moves through the secondary openings of the geological formation.

secondary openings: interstices that were formed by processes that affected the rocks after they were formed, e.g. faults and fissures. Synonymous with secondary porosity. Forms secondary permeability.

secondary salinity: additional salinity in the soil and rivers brought about by human activities, also known as anthropogenic salinity or irrigation salinity. Primary salinity in Australia has been intensified by changes in land use since European settlement.

seep: slow escape of groundwater, usually used to describe a small, diffuse wetland area where interflow and groundwater emerges at a slow rate, to become surface flow.

semi-arid zone: receiving more rainfall than the arid zone, lands where rainfall is too low and unreliable for crops to be grown with certainty.

semi-confined aquifer: an aquifer that is confined by layers of semi-permeable material through which recharge and discharge may occur, also known as a leaky aquifer.

sensible heat flux: the process where excess heat energy is transferred into the atmosphere from a surface. The sensible heat flux (H) is a function of the specific heat of air, air density, eddy diffusivity for heat and the air temperature integrated over a given height or depth. See also **Bowen ratio**.

soil: the upper surface layer of the earth consisting of fragmented rock or unconsolidated sediments, living organisms, organic matter, water and gases. Soil properties arise from the interaction of its parent material, time, climate, fauna and flora.

soil hydrophobicity: the tendency for a soil particle or soil mass to resist water influx. Soil hydrophobicity can be enhanced when organic matter in litter and upper mineral soil layers is volatised during a fire. Some of the material moves into the soil profile and condenses to form a water-repellent layer that impedes infiltration. Soil hydrophobicity may also occur naturally in the absence of fire in some soils. The condition can have the benefit of reducing evaporation from soil as water is trapped below the surface.

soil porosity: the proportion of pore space to total soil volume.

soil saturation: all soil pores are filled with water.

soil texture: a measure of the size distribution of particles in a soil. The coarsest fraction is sand and the finest fraction is clay. Gravel, stones, pebbles and rock are not considered as part of the soil in defining soil texture. A soil with finer particles has a greater bulk density and smaller soil pores. Texture influences water infiltration, movement and storage. Small pores in fine soils such as clays hold water very strongly, while large pores in coarse sand hold water very weakly and thus sands drain faster than clay.

soil water: water held in soil pores and in the soil itself, in both liquid and vapour phases, including saturated and unsaturated states. Soil water is measured as a percentage of the soil dry weight (% by weight) but sometimes as the volume of water as a percentage of the soil volume (% by volume) or as the depth of water per metre depth of soil (m/m).

source area: the saturated zone along an effluent stream which contributes water to streamflow.

specific yield (of an aquifer): the ratio of the volume of water a rock or soil will yield by gravity drainage to the volume of the rock or soil, measured when the water table drops. Gravity drainage may take many months to occur because some water is too strongly adsorbed to the earth material to drain quickly.

spring: a place, usually a distinct point or small area, where groundwater reaches the surface. A spring generally occurs as a result of topographical, lithological or structural controls on groundwater movements.

stable isotope: an isotope that does not undergo radioactive decay. A radioactive isotope spontaneously emits charged particles from the nucleus to produce a more stable nucleus.

stemflow: rainfall captured by the above-ground parts of a tree that flows down the trunk (stem) to the ground. Stemflow does not include the water retained in the bark.

stomatal conductance (G_s): a measure of the ease with which water vapour escapes from plant leaves through stomatal pores. Influenced by the stomatal pore size and their number. A high conductance occurs when stomata are wide-open.

stormflow: increased runoff and water flow which is associated directly with a particular rainfall even. Stormflow is the same as the quickflow or direct runoff.

storm response ratio: the volume of runoff relative to the volume of rainfall that generated it.

streamflow: water flowing in a stream or river. Also known as run-off.

stygofauna: fauna (usually invertebrates) associated with groundwater ecosystems.

sub-surface water: all water which occurs beneath the surface of the earth, including soil water, water in the unsaturated zone and groundwater.

subterranean: situated, occurring or operating beneath the surface of the earth.

surface flow: movement of water above the Earth's surface, as runoff or streams. Often derived from precipitation but also from springs.

surface water: water that occurs or flows on the surface, including streams, rivers, estuaries, lakes, and overland flooding.

temperate zone: areas experiencing mild temperatures. The weather is more changeable in the temperate zone than in the tropics; and there are four seasons. Rain falls throughout the year but doesn't usually last very long. In Australia, the temperate zone occurs in the south western corner, and the south eastern corner, including Tasmania. The temperate zone can be further divided according to the summer temperatures and rainfall and winter rainfall.

tension saturated zone: another term for the capillary fringe.

throughflow: the proportion of the gross precipitation that reaches the ground and is not intercepted by, or retained in, the canopy.

thermocouple psychrometer: a device for measuring the humidity in soil air, using a wet bulb and a dry bulb thermometer. Indicates soil water potential. See water potential.

time-domain-reflectometry (TDR) probes: sensors used to measure soil moisture content. The principle of TDR probes involves converting the travel time of a high frequency, electromagnetic pulse into volumetric water content. The instrument generates a fast-rise pulse and sends it at the speed of light down a transmission line consisting of two parallel waveguides that are inserted or buried in the soil. The wave is reflected from the open ends of the waveguides and returns along the original path. By microprocessor, the travel time of the wave is used to directly calculate the dielectric constant of the soil. The transit time and the dielectric constant are dependant on the moisture content of the soil.

transmissivity: the product of hydraulic conductivity and saturated thickness. The measure of the amount of water that can be transmitted horizontally by the full saturated thickness of the aquifer under a hydraulic gradient of 1.

transpiration: the process of water loss from leaves to the atmosphere. Water absorbed by plant roots is pulled through the conductive tissue of the plant as the transpiration stream, eventually reaching the leaves. Liquid water is evaporated in the sub-stomatal cavity and diffuses through open stomata.

tropical zone: occurs close to the equator and hence, seasons are less pronounced than in the temperate zone. Tropical ecosystems in Australia include rainforests and savannas in northern Western Australia, The Northern Territory and Queensland. Tropical areas generally have a dry and wet season which vary in length.

unconfined aquifer: see **aquifer**.

unsaturated zone: a layer which has not yet reached maximum water-holding capacity. The soil and underlying material where the soil pores are only partially filled with water. The layer may also include bodies of fractured rock and bedrock. Water in the unsaturated zone is under pressure less than that of the atmosphere, some of which is held by capillary tension.

vadose zone: another name for the unsaturated zone.

vapour pressure deficit: a measure of the water content of the air based on the difference between the partial pressure of water vapour if the air were saturated and the ambient water vapour pressure. Air absorbs water vapour and the water vapour, being a gas, exerts a pressure in addition to that of the air. The higher the air temperature the more water vapour can be present in the atmosphere. When the air contains as much water vapour as it can, it is saturated. A large vapour pressure deficit means the air is dry.

vegetation structure: refers to the types of plant form in an ecosystem, for instance shrub, grass or tree. Vegetation structure can have a number of layers such as the over-storey, under-storey and ground-cover which may have different levels of cover density (closed or open).

water course: any collection of water. A watercourse can be a natural channel in which water flows regularly or intermittently, a river, a spring a wetland, lake or dam into which, or from which water flows.

water cycle: the continuous sequence of water evaporating into the earth's atmosphere where it condenses and returns to the earth as rainfall.

water table: the upper boundary of a free (unconfined) groundwater body at atmospheric pressure or elevation at which the soil is saturated from below. The water table occurs between the zone of saturation and the zone of aeration and is also known as the free water surface or ground water level. At the level of the water table the hydrostatic pressure is equal to atmospheric pressure (the water table therefore excludes the capillary fringe).

water potential (of soil, leaf or pre-dawn): the free-energy status of water. A concept applied to water in soils, plants and the atmosphere. Soil water potential is a measure of the difference between the free energy state of soil water and that of pure water. Technically it is defined as the amount of work that must be done per unit quantity of water to transport that water, at a specified elevation and at atmospheric pressure to the soil with reference to water in a defined state.

Leaf water potential is the total potential for water in a leaf, consisting of the balance between osmotic (or solute) potential (due to the presence of solutes), turgor pressure (the hydrostatic pressure in a cell) and the matric potential (arising from forces of colloids and capillaries in the cell wall). The determination of water deficit in leaves can be measured with a 'pressure bomb' or by psychrometric methods. When measured before dawn (pre-dawn leaf water potential) the leaf water potential equates with soil water potential as water potentials in the soil and leaf come into (partial or full) equilibrium overnight.

water yield (of a catchment): the streamflow in a given interval of time derived from a unit area of catchment. It is calculated by dividing the observed water flow at a given location by the drainage area above that location and is usually expressed in cubic metres per second per square kilometre.

wetland: an area that is regularly flooded and has a water table that stands at or above the land surface for at least part of the year, such as a bog, pond, estuary or marsh. A wetland has soil that is flooded for sufficiently long periods for waterlogging to become the dominant factor determining its diagnostic characteristics. Wetland vegetation is typically adapted to life in saturated soil.

wetting front: the boundary between soil wet by water from rainfall and dry soil as the water moves downwards in the unsaturated zone.

xerophytes: plants adapted to low water availability, found in areas with low rainfall.

References

Akeroyd MD, Tyerman SD, Walker GR & Jolly AD (1998). Impact of flooding on the water use of semi-arid riparian eucalypts. *Journal of Hydrology* **206**: 104–117.

Alder NN, Pockman WT, Sperry JS & Nuismer S (1997). Use of centrifugal force in the study of xylem cavitation. *Journal of Experimental Botany* **48**: 665–674.

Allison GB & Hughes MW (1983). The use of natural tracers as indicators of soil-water movement in a temperate semi-arid region. *Journal of Hydrology* **60**: 157–173.

Allison GB, Cook PG, Barnett SR, Walker GR, Jolly ID & Hughes MW (1990). Land clearance and river salinisation in the western Murray Basin, Australia. *Journal of Hydrology* **119**: 1–20.

Amoozegar A (1989). Comparisons of the Glover solution with the simultaneous-equations approach for measuring hydraulic conductivity. *Soil Science Society of America Journal* **53**: 1362–1367.

Anderson MP & Woessner WW (1992). *Applied groundwater modeling.* Academic Press, San Diego.

Armbruster M, Seegert J & Feger K (2004). Effects of changes in tree species composition on water flow dynamics: model applications and their limitations. *Plant and Soil* **264**: 13–24.

Armstrong D & Narayan K (1998). Using groundwater responses to infer recharge. Part 5. In *The basics of recharge and discharge* (eds L Zhang & G Walker). CSIRO Publishing, Melbourne.

Arthington AH, Rall J, Kennard MJ & Pusey BJ (2003). Environmental flow requirements of fish in Lesotho rivers using the drift methodology. *River Research and Applications* **19**: 641–666.

Asseng S, Fillery IRP, Dunin FX, Keating BA & Meinke H (2001). Potential deep drainage under wheat crops in a Mediterranean climate. I. Temporal and spatial variability. *Australian Journal of Agricultural Research* **52**: 45–56.

Attiwell P & Wilson B (2003). *Ecology: an Australian perspective.* Oxford University Press, Melbourne.

Atwell B, Kriedemann P, Turnbull C, Eamus D & Bieleski RL (1999). *Plants in action: adaption in nature, performance in cultivation.* Macmillan Education Australia, Melbourne.

Australian Dryland Salinity Assessment (2000). *National Land and Water Resources Audit.* Land and Water Australia, Canberra.

Australian Native Vegetation Assessment (2001). *National Land and Water Resources Audit.* Land and Water Australia, Canberra.

Australian Water Resources Assessment (2000). *Surface Water and Groundwater: Availability and Quality.* Commonwealth of Australia, Canberra.

Baddeley JA & Watson CA (2004). Seasonal patterns of fine root production and mortality in *Prunus avium* in Scotland. *Canadian Journal of Forest Research* **34**: 1534–1537.

Bari MA & Schofield NJ (1992). Lowering of a shallow saline water table by extensive eucalypt reforestation. *Journal of Hydrology* **133**: 273–291.

Barton CVM & Montagu K (2004). Detection of tree roots and determination of root diameters by ground penetrating radar under optimal conditions. *Tree Physiology* **24**: 1323–1331.

Batelaan O, De Smedt F & Triest L (2003). Regional groundwater discharge: phreatophyte mapping, groundwater modelling and impact analysis of land-use change. *Journal of Hydrology* **275**: 86–108.

Becker P, Tyree MT & Tsuda M (1999). Hydraulic conductance of angiosperms versus conifers: similar transport efficiency at the whole plant level. *Tree Physiology* **19**: 445–452.

Becker P, Meinzer FC & Wullschleger SD (2000). Hydraulic limitation of tree height: a critique. *Functional Ecology* **14**: 4–11.

Beekman H & Xu Y (2003). *Review of groundwater recharge estimation in arid and semi-arid southern Africa*. UNESCO, Paris.

Bell DT (1999). Australian trees for the rehabilitation of waterlogged and salinity-damaged landscapes. *Australian Journal of Botany* **47**: 697–716.

Berry KA & Armstrong D (1995). *Hydrogeological investigation and numerical modeling, Lake Eyre Region, Great Artesian Basin*. Western Mining Corp Ltd, Exploration Division–Minerals, HYD T004.

Birkhead AL, Olbrich BW, James CS & Rogers KH (1997). *Developing an integrated approach to predicting the water use of riparian vegetation*. Report no. 474/1/97, Water Research Commission, Pretoria.

Blackburn G & McLeod S (1983). Salinity of atmospheric precipitation in the Murray-Darling drainage division, Australia. *Australian Journal of Soil Research* **21**: 411–434.

Bolhar-Nordenkampf HR & Oquist G (1993). Chlorophyll fluorescence as a tool in photosynthesis research. In *Photosynthesis and production in a changing environment: a field and laboratory manual* (eds DO Hall, JMO Scurlock, HR Bolhar-Nordenkampf, RC Leegood & SP Long). Chapman & Hall, London.

Bonell M (1998). Selected challenges in runoff generation research in forests from the hillslope to headwater drainage basin scale. *Journal of American Water Resources Association* **34**: 765–785.

Bosch JM & Hewlett JD (1982). A review of catchment experiments to determine the effect of vegetation changes on water yield and evapotranspiration. *Journal of Hydrology* **55**: 3–23.

Brodribb TJ & Field TS (2000). Stem hydraulic supply is linked to leaf photosynthetic capacity: evidence from New Caledonian and Tasmanian rainforests. *Plant, Cell and Environment* **23**: 1381–1388.

Brodribb TJ & Holbrook NM (2003). Stomatal closure during leaf dehydration: correlation with other leaf physiological traits. *Plant Physiology* **132**: 166–217.

Brodribb TJ & Holbrook NM (2004). Diurnal depression of leaf hydraulic conductance in a tropical tree species. *Plant, Cell and Environment* **27**: 820–827.

Brooks NK, Folliott FP, Gregersen MH & DeBano FL (1997). *Hydrology and the management of watersheds*. Iowa State Press, Iowa.

Burgess SSO, Adams MA, Turner NC & Ong CK (1998). The redistribution of soil water by tree root systems. *Oecologia* **115**: 306–311.

Bussière F, Solmon F & Fouéré A (2002). Implementation and evaluation of DROP: a model for the simulation of rainfall distribution below plants described in 3D. *Agronomie* **22**: 93–103.

Butnor JR, Doolittle JA, Kress L, Cohen S & Johnsen KH (2001). Use of ground penetrating radar to study tree roots in the south-eastern United States. *Tree Physiology* **21**: 1269–1281.

Buttle JM (1994). Isotope hydrograph separation and rapid delivery of pre-event water from drainage basins. *Progress in Physical Geography* **18**: 16–41.

Caldwell MM, Dawson TE & Richards JH (1998). Hydraulic lift: consequences of water efflux from the roots of plants. *Oecologia* **113(2)**: 151–161.

Campbell GS (1985). *Soil physics with basic: transport models for soil-plant systems.* Elsevier, New York.

Canadell J, Jackson RB, Ehleringer JR, Mooney HA, Sala OE & Schulze ED (1996). Maximum rooting depth of vegetation types at the global scale. *Oecologia* **108**: 583–595.

Cech TV (2005). *Principles of water resources: history, development, management and policy.* John Wiley & Sons, New York.

Chen XY, Eamus D & Hutley LB (2004). Seasonal patterns of fine-root productivity and turnover in a tropical savanna of northern Australia. *Journal of Tropical Ecology* **20**: 221–224.

Cleaver G, Brown LR, Bredenkamp GJ, Smart MC & Rautenbach CJD (2003). *Assessment of environmental impacts of groundwater abstraction from Table Mountain Group (TMG) aquifers on ecosystems in the Kammanassie Nature Reserve and environs.* Report no. 1115/1/03, Water Research Commission, Pretoria.

Clifton CER (2001). *A framework to assess the environmental water requirements of groundwater dependent ecosystems.* 3rd Australian Stream Management Conference, Brisbane.

Colvin C (2004a). Understanding groundwater: a stakeholders' guide to North West Dolomite water management area. WRC project K5/1324. CSIR report number 2003–128.

Colvin C, Cavé L & Saayman I (2004b). *A functional approach to setting resource quality objectives for groundwater.* WRC Project no. 1235/1/04, Water Research Commission, Pretoria.

Colvin C, Le Maitre D & Hughes S (2002). Assessing terrestrial groundwater dependent ecosystems in South Africa. Report number 1090-2/2/03. Water Research Commission, Pretoria.

Cook PG & Walker GR (1990). *Effect of soil type on groundwater recharge in the Mallee region.* Report no. 28, Centre for Groundwater Studies.

Cook PG, Hatton TH, Pidsley D, Herczeg AL, Held AA, O'Grady A & Eamus D (1998). Water balance of a tropical woodland ecosystem, northern Australia: a combination of micrometeorological, soil physical and groundwater chemical approaches. *Journal of Hydrology* **210**: 161–177.

Cook PG, Herczeg AL & McEwan KL (2001). *Groundwater recharge and stream baseflow, Atherton Tablelands, Queensland.* Technical Report no. 08/01. CSIRO Land and Water.

Cook PG, Favreau G, Dighton JC & Tickell S (2003). Determining natural groundwater influx to a tropical river using radon, chlorofluorocarbons and ionic environmental tracers. *Journal of Hydrology* **277**: 74–88.

Coram J (ed.) (1998). National classification of catchments for land and river salinity control. In *Water and Salinity Issues in Agroforestry* **3**. RIRDC 98/78, Canberra.

Cornish PM & Vertessy RA (2001). Forest age induced changes in evapotranspiration and water yield in a eucalypt forest. *Journal of Hydrology* **242**: 43–63.

Costanza R, d'Arge R, de Groot R, Farber S, Grasso M, Hannon B, Limburg K, Naeem S, O'Neill RV, Paruelo J, Raskin RG, Sutton P & van den Blet M (1997). The value of the world's ecosystem services and natural capital. *Nature* **387**: 253–260.

Crockford RH & Richardson DP (2000). Partitioning rainfall into throughfall, stemflow and interception: effect of forest type, ground cover and climate. *Hydrological Processes* **14**: 2903–2920.

Davies BR, O'Keeffe JH & Snaddon CD (1993). *A synthesis of the ecological functioning, conservation and management of South African river ecosystems.* Report no. TT 62/93, Water Research Commission, Pretoria.

Davis JA & Froend R (1999). Loss and degradation of wetlands in south-western Australia: underlying causes, consequences and solutions. *Wetlands Ecology and Management* **7**: 13–23.

Davis SD, Ewers FE, Sperry JS, Portwood KA, Crocker MC & Adams GC (2002). Shoot dieback during prolonged drought. *American Journal of Botany* **89**: 820–828.

Dawes WR, Gilfedder M, Walker GR & Evans WR (2004). Biophysical modelling of catchment-scale surface water and groundwater responses to land-use change. *Mathematics and Computers in Simulation* **64**: 3–12.

Dent MC, Lynch SD & Schulze RE (1989). *Mapping mean annual rainfall and other rainfall statistics over southern Africa*. Report no. 109/1/89, Water Research Commission, Pretoria.

De Vries JJ & Simmers I (2002). Groundwater recharge: an overview of processes and challenges. *Hydrogeology Journal* **10**: 5–17.

Dewar RC (1997). A simple model of light and water use evaluated for *Pinus radiata*. *Tree Physiology* **17**: 259–265.

Diersch HJ (2004). FEFLOW: Interactive, graphics-based finite-element simulation system for modelling groundwater flow, contaminant mass and heat transport processes. Getting started; user's manual; reference manual, version 5.1. WASY, Institute for Water Resources Planning and System Research Ltd, Berlin.

Dodd J & Bell DT (1993). Water relations of the canopy species in a banksia woodland, Swan Coastal Plain, Western Australia. *Australian Journal of Ecology* **18(3)**: 281–293.

Dogramaci S, Mauger J & George R (2003). *Engineering options as tools for salinity management in the Spencer Gully catchment*. Land Use Impact Series 7. Water and Rivers Commission, Perth.

Donovan LA, Linton MJ & Richards JH (2001). Pre-dawn water potential does not necessarily equilibrate with soil water potential under well watered conditions. *Oecologia* **129**: 328–335.

Douglas MM, Townsend AS & Lake PS (2003). *Streams*. Springer-Verlag, New York.

Drake PL & Franks PJ (2003). Water resource partitioning, stem xylem hydraulic properties and plant water use strategies in a seasonally dry riparian tropical rainforest. *Oecologia* **137**: 321–329.

Duff GA, Myers BA, Williams RJ, Eamus D, O'Grady A & Fordyce IR (1997). Seasonal patterns in soil moisture, vapour pressure deficit, tree canopy cover and pre-dawn water potential in a northern Australian savanna. *Australian Journal of Botany* **45**: 211–224.

Dunin FX (1992). Integrating agroforestry and perennial pastures to mitigate water logging and secondary salinity. *Agricultural Water Management* **53**: 259–270.

Dunin FX, Williams J, Verburg K & Keating BA (1999). Can agricultural management emulate natural ecosystems in recharge control in south eastern Australia? *Agroforestry Systems* **45**: 343–364.

Dyson LL & van Heerden J (2001). The heavy rainfall and floods over the northeastern interior of South Africa during February 2000. *South African Journal of Science* **97**: 80–86.

Eagleson PS (1982). Ecological optimality in water limited natural soil-vegetation systems. I. Theory and hypothesis. *Water Resources Research* **18**: 325–340.

Eagleson PS (2002). *Ecohydrology: Darwinian expression of vegetation form and function*. Cambridge University Press.

Eamus D (2003). How does ecosystem water balance affect net primary productivity of woody ecosystems? *Functional Plant Biology* **30**: 187–205.

Eamus D & Cole SC (1997). Diurnal and seasonal comparisons of assimilation, phyllode conductance and water potential of three Acacia and one Eucalypt species in the wet-dry tropics of Australia. *Australian Journal of Botany* **45**: 275–290.

Eamus D & Prior L (2001). Ecophysiology of trees of seasonally dry tropics: comparisons among phenologies. In *Advances in ecological science* (ed H Caswell). Academic Press, California, pp. 113–197.

Eamus D, Berryman CA & Duff GA (1995). The impact of CO_2 enrichment on water relations of *Maranthes corymbosa* and *Eucalyptus tetrodonta*. *Australian Journal of Botany* **43**: 273–282.

Eamus D, O'Grady AP & Hutley LB (2000). Dry season conditions determine wet season water use in the wet-dry tropical savannas of northern Australia. *Tree Physiology* **20**: 1219–1226.

Eamus D, Froend R, Hose G, Loomes R & Murray B (2006). A functional methodology for determining the GW regime needed to maintain health of groundwater dependent vegetation. *Australian Journal of Botany* **54**: 97–114.

Eberbach PL (2003).The eco-hydrology of partly cleared, native ecosystems in southern Australia: a review. *Plant and Soil* **257**: 357–369.

Edmunds W & Tyler SW (2002). Unsaturated zones as archives of past climates: towards a new proxy for continental regions. *Hydrogeology Journal* **10**: 216–228.

Egan JL & Williams RJ (1996). Life-form distributions of woodland plant species along a moisture availability gradient in Australia's monsoonal tropics. *Australian Systematic Botany* **9**: 205–217.

Eldridge DJ & Freudenberger D (2005). Ecosystem wicks: woodland trees enhance water infiltration in a fragmented agricultural landscape in eastern Australia. *Austral Ecology* **30**: 336–347.

Ellis T, Hatton TJ & Nuberg I (1999). A simple method for estimating recharge from low rainfall agroforestry systems. In *Proceedings of Environwater 99*, 2nd Inter-Regional Conference on Water-Environment Emerging Technologies for Sustainable Land Use and Water Management. 1–3 September 1999, Lausanne, Switzerland. pp. 4–10.

Ellis T, Hatton TJ & Nuberg I (2005). An ecological optimality approach for predicting deep drainage from tree belts of alley farms in water limited environments. *Agricultural Water Management* **75**: 92–116.

Farrington P, Salama RB, Watson GD & Bartle GA (1992). Water-use of agricultural and native plants in a Western Australian wheat-belt catchment. *Agricultural Water Management* **22**: 357–367.

Feddes RA, Kabat P, van Bakel PJT, Bronswijk JJB & Halbertsma J (1988). Modelling soil water dynamics in the unsaturated zone: state of the art. *Journal of Hydrology* **100**: 69–111.

Federer CA (2002). BROOK90: a simulation model for evaporation, soil water, and streamflow. Documentation for versions 4 and 3.2/3/4. http://users.rcn.com/compassbrook/b90doc.html.

Fensham RJ, Fairfax D, Pocknee J & Kelley G (2004). Vegetation patterns in permanent spring wetlands in arid Australia. *Australian Journal of Botany* **52**: 719–728.

Finch JW (2001). Estimating change in direct groundwater recharge using a spatially distributed soil water balance model. *Quarterly Journal of Engineering Geology and Hydrogeology* **34**(1): 71–83.

Flannery TY (1994). *The future eaters*. Reed Books, Sydney.

Foran B & Poldy F (2002). *Future dilemmas: options to 2050 for Australia's population, technology, resources and environment*. CSIRO, Australia.

Fordyce IR, Duff GA & Eamus D (1997). The water relations *of Allosyncarpia ternata* at contrasting sites in the monsoonal tropics of northern Australia. *Australian Journal of Botany* **45**: 259–274.

Freeze RA (1978). Mathematical models of hillslope hydrology. In *Hillslope hydrology* (ed MJ Kirkby). Wiley, New York, pp. 177–225.

Froend R, Loomes R, Horwitz P, Bertuch M, Storey A & Bamford M (2004). *Study of ecological water requirements on the Gnangara and Jandakot Mounds under section 46 of the Environmental Protection Act. Task 2: determination of ecological water requirements*. Report to the Water and Rivers Commission. Centre for Ecosystem Management, ECU, Joondalup.

Gardner KM (1999). The importance of surface water/groundwater interactions. Issue paper EPA–910-R-99–013. Environmental Protection Agency, Washington DC.

Gash JHC (1979). An analytical model of rainfall interception by forests. *Quarterly Journal of the Royal Meteorological Society* **105**: 43–55.

Gash JHC, Lloyd CR & Lachaud G (1995). Estimating sparse forest rainfall interception with an analytical model. *Journal of Hydrology* **170**: 79–86.

George RJ & Dogramaci S (2000). *Toolibin: a life and death struggle for the last freshwater wheatbelt lake.* Hydro 2000, 3rd International Hydrology and Water Resources Symposium of the Institution of Engineers, Australia, vol. 2, pp. 733–738.

George RJ, Nulsen RA, Ferdowsian R & Raper GP (1999). Interactions between trees and groundwaters in recharge and discharge areas: a survey of Western Australian sites. *Agricultural Water Management* **39**: 91–113.

Grabarnik P, Pages L & Bengough AG (1998). Geometrical properties of simulated maize root systems. *Plant and Soil* **200**: 157–167.

Granier A, Huc R & Barigah ST (1996). Transpiration of natural rainforest and its dependence on climatic factors. *Agricultural and Forest Meteorology* **78**: 19–29.

Greacen EL, Walker GR & Cook PG (1989). *Procedure for filter paper method of measuring soil water suction*. Technical Report no. 108. CSIRO Division of Soils, Adelaide.

Greenwood EAN, Klein JD, Beresford JD & Watson GD (1985). Differences in annual evaporation between grazed pasture and eucalyptus species in plantations on a saline farm catchment. *Journal of Hydrology* **78**: 261–278.

Greenwood EAN, Biddiscombe EF, Rogers AL, Beresford JD & Watson GD (1995). Growth of species in a tree plantation and its influence on salinity and groundwater in the 400 mm region of south western Australia. *Agricultural Water Management* **28**: 231–243.

Groom PK (2003). Groundwater-dependency and water relations of four *Myrtaceae* shrub species during a prolonged summer drought. *Journal of the Royal Society of Western Australia* **86**: 31–40.

Groom PK (2004). Rooting depth and plant water relations explain species distribution patterns within a sandplain landscape. *Functional Plant Biology* **31(5)**: 423–428.

Groom PK, Froend RH & Mattiske EM (2000a). Impact of groundwater abstraction on a Banksia woodland, Swan Coastal Plain, Western Australia. *Ecological Management and Restoration* **1**: 1–12.

Groom PK, Froend RH, Mattiske EM & Koch B (2000b). Myrtaceous shrub species respond to long-term decreasing groundwater levels on the Gnangara Groundwater Mound, northern Swan Coastal Plain. *Journal of the Royal Society of Western Australia* **83**: 75–82.

Groom PK, Froend RH, Mattiske EM & Gurner RP (2001). Long-term changes in vigour and distribution of *Banksia* and *Melaleuca* overstorey species on the Swan Coastal Plain. *Journal of the Royal Society of Western Australia* **84**: 63–69.

Habermehl MA (1988). The Great Artesian Basin. *Australian Geology and Geophysics* **5**: 9–38.

Harding GB & Bate GC (1991). The occurrence of invasive Prosopis species in the north-western Cape, South Africa. *South African Journal of Science* **87**: 188–192.

Harris CR (1992). Mound springs: South Australian conservation initiatives. *Rangeland Journal* **14**: 157–173.

Harris PP, Huntingford C, Cox PM, Gash JHC & Malhi Y (2004). Effect of soil moisture on canopy conductance of Amazonian rainforest. *Agricultural and Forest Meteorology* **122**: 215–227.

Hatton T & Evans R (1998). *Dependence of ecosystems on groundwater and its significance to Australia.* Occasional Paper no. 12/98, Land and Water Resources Research and Development Corporation, CSIRO Australia.

Hatton T & George R (2001). The role of afforestation in managing dryland salinity. In *Plantations, farm forestry and water* (eds EK Sadanandan Nambiar & AG Brown). Water and salinity issues in agroforestry no 7. RIRDC Report 01/20, RIRDC, Canberra.

Hatton TJ & Nulsen RA (1999). Towards achieving functional ecosystems mimicry with respect to water cycling in southern Australian agriculture. *Agroforestry Systems* **45**: 203–214.

Hatton T, Reggiani P & Hodgson G (2002). *The role of trees in the water and salt balances of catchments.* RIRDC, Canberra.

Hatton TJ, Ruprecht J & George R (2004). Pre-clearing hydrology of the Western Australia wheatbelt: target for the future? *Plant and Soil* **257**: 341–356.

Hayashi M & Rosenberry DO (2002). Effects of groundwater exchange on the hydrology and ecology of surface waters: *Ground Water* **40**: 309–316.

Haydon SR, Benyon RG & Lewis R (1996). Variation in sapwood area and throughfall with forest age in Mountain Ash. *Journal of Hydrology* **187**: 351–366.

Hayhoe H (1980). Analysis of a diffusion model for plant root growth and an application to plant-soil water uptake. *Soil Science* **131**: 334–343.

Healy RW & Cook PG (2002). Using groundwater levels to estimate recharge. *Hydrogeology Journal* **10**: 91–109.

Hendrick RL & Pregitzer KS (1996). Application of minirhizotrons to understand root function in forests and other natural ecosystems. *Plant and Soil* **185**: 293–304.

Henkel JS (1931). The camelthorn (*Acacia giraffae* Burch). *Rhodesia Agricultural Journal* **28**: 71–73.

Hillel D (1977). *Computer simulation of soil-water dynamics: a compendium of recent work.* IDRC, Ottawa.

Hingston FJ & Gailitis V (1976). The geographic variation of salt precipitation over western Australia. *Australian Journal of Soil Research* **14**: 319–335.

Hodgson L & Finlayson B (1990). Single well dilution: its use in groundwater investigations for salinity management. *Australian Journal of Soil and Water Conservation* **3**: 37–43.

Howard KWF & Lloyd JW (1979). The sensitivity of parameters in the Penman evaporation equations and direct recharge balance. *Journal of Hydrology* **41**: 329–344.

Hubbard R, Bond B & Ryan M (1999). Evidence that hydraulic conductance limits photosynthesis in old *Pinus ponderosa* trees. *Tree Physiology* **19**: 165–172.

Hudson N (1981). *Soil conservation.* Cornell University Press, New York.

Hutley LB, O'Grady AP & Eamus D (2000). Evapotranspiration from Eucalypt open-forest savanna of northern Australia. *Functional Ecology* **14**: 183–194.

Hutley LB, O'Grady AP & Eamus D (2001). Monsoonal influences on evapotranspiration of savanna vegetation of northern Australia. *Oecologia* **126**: 434–443.

Jacobson PJ, Jacobson KM & Seely MK (1995). *Ephemeral rivers and their catchments: sustaining people and development in western Namibia.* Desert Research Foundation of Namibia, Windhoek.

Jewitt GPW, Heritage GL, Weeks DC, Mackenzie JA, van Niekerk A, Görgens AHM, O'Keeffe JH, Rogers K & Horn M (1998). *Modelling abiotic-biotic links in the Sabie River.* Report no. 777/1/98, Water Research Commission, Pretoria.

Jolly I & Cook P (2002). Time lags in salinity control. *Natural Resource Management* **5**: 16–21.

Jolly I & Walker G (1995). A sketch of salt and water movement in the Chowilla floodplain. CSIRO Division of Water Resources, Adelaide.

Jonckheere I, Fleck S, Nackaerts K, Muys B, Coppin P, Weiss M & Baret F (2004). Review of methods for in situ leaf area index determination: Part I. Theories, sensors and hemispherical photography. *Agriculture and Forest Meteorology* **121**: 19–35.

Jothityangkoon C, Sivapalan M & Farmer DL (2001). Process controls of water balance variability in a large semi-arid catchment: downward approach to hydrological model development. *Journal of Hydrology* **254**:174–198.

Kaleris V (1998). Quantifying the exchange rate between groundwater and small streams. *Journal of Hydraulic Research* **36**: 913–932.

Keane D (1997). The sustainability of use of groundwater from the Great Artesian Basin with particular reference to the south western edge of the basin and impact on the Mound Springs. Environmental Engineering thesis, RMIT, Melbourne.

Keig G & McAlpine JR (1974). *WATBAL: a computer system for the estimation and analysis of soil moisture regimes from simple climatic data.* CSIRO Division of Land Use Research, Technical Memorandum 74/4.

Kelbe B, Germishuyse T, Snyman N & Fourie I (2001). *Geohydrological studies of the primary coastal aquifer in Zululand.* Report no. 720/1/01, Water Research Commission.

Kennett-Smith A, Cook PG & Walker GR (1994). Factors affecting groundwater recharge following clearing in the south western Murray Basin. *Journal of Hydrology* **154**: 85–105.

Keywood MD, Chivas AR, Fifield LK, Cresswell RG & Ayers GP (1997). The accession of chloride to the western half of the Australian continent. *Australian Journal of Soil Research* **35**: 1177–1189.

King RS, Nunnery KT & Richardson CJ (2000). Macroinvertebrate assemblage response to highway crossings in forested wetlands: implications for biological assessment. *Wetlands Ecology and Management* **8**: 243–256.

King J, Brown C & Sabet H (2003). A scenario-based holistic approach to environmental flow assessments for rivers. *River Research and Applications* **19**: 619–630.

Kingsford RT (2000). Ecological impacts of dams, water diversions and river management on floodplain wetlands in Australia. *Austral Ecology* **25**: 109–127.

Kinhill (1997). *Olympic Dam expansion project environmental impact statement.* Prepared for WMC (Olympic Dam Corporation) Pty Ltd by Kinhill Engineers Pty Ltd.

Kirkby M (1988). Hillslope runoff processes and models. *Journal of Hydrology* **100**: 315–339.

Kleynhans CJ (1999). Appendix R1: ecoregional typing. In *Resource directed measures for protection of water resources: vol. 3. River ecosystems. Version 1.0.* Department of Water Affairs and Forestry, Pretoria.

Knight A, Blott K, Portelli M & Hignett C (2002). Use of tree and shrub belts to control leakage in three dryland cropping environments. *Australian Journal of Agricultural Research* **53**: 571–586.

Kodric-Brown A & Brown JH (1993). Highly structured fish communities in Australian desert springs. *Ecology* **74**: 1847–1855.

Koide RT, Robichaux RH, Morse SR & Smith CM (1991). Plant water status, hydraulic resistance and capacitance. In *Plant physiological ecology* (eds RW Pearcy, J Ehleringer, HA Mooney & PW Rundel). Chapman & Hall, New York, pp. 161–183.

Konikow LF & Mercer JW (1988). Groundwater flow and transport modeling. *Journal of Hydrology* **100**: 379–409.

Kramer PJ (1983). *Water relations of plants.* Academic Press, New York.

Lagergren F & Lindroth A (2002). Transpiration responses to soil moisture in pine and spruce trees in Sweden. *Agricultural and Forest Meteorology* **112**: 67–85.

Leemans R & Cramer W (1991). *The IIASA database for mean monthly values of temperature, precipitation and cloudiness on a global terrestrial grid.* Research report RR–91-18. November 1991. International Institute of Applied Systems Analyses, Laxenburg, Austria.

LeFroy EC & Stirzaker RJ (1999). Agroforestry for water management in the cropping zone of southern Australia. *Agroforestry Systems* **45**: 277–302.

Le Maitre DC, Scott DF & Colvin C (1999). A review of information on interactions between vegetation and groundwater. *Water SA* **25**: 137–149.

Lincoln RJ, Boxshall GA & Clark PF (1982). A dictionary of ecology, evolution and systematics. Cambridge University Press, Cambridge.

Liu S, Leslie LM, Speer M, Bunker R & Mo X (2004). The effects of bushfires on hydrological processes using paired-catchment analysis. *Meteorology and Atmospheric Physics* **86**: 31–44.

Lothian JA (1988). Wetland conservation in South Australia. In *The conservation of Australian wetlands* (eds AJ McComb & PS Lake). Surrey Beatty, Sydney, pp. 147–166.

Low AB & Rebelo AG (1996). *Vegetation of South Africa, Lesotho and Swaziland.* Department of Environmental Affairs and Tourism, Pretoria.

Lu P, Yunusa IAM, Walker RR & Muller WJ (2003). Regulation of canopy conductance and transpiration and their modelling in irrigated grapevines. *Functional Plant Biology* **30**: 689–698.

MacArthur WM (1993). History of landscape development. In *Reintegrating fragmented landscapes* (eds RJ Hobbs & DA Saunders). Springer Verlag, New York, pp. 10–22.

MacFarlane DJ & Williamson DR (2002). An overview of water logging and salinity in south-western Australia as related to the Ucarro experimental catchment. *Agricultural Water Management* **53**: 5–29.

Macinnis-Ng C, McClenahan K & Eamus D (2004). Convergence in hydraulic architecture, water relations and primary productivity across seasons in Sydney. *Functional Plant Biology* **31**: 1–11.

MacKay HM (1998). Towards a classification system for water resources in South Africa. http://iwqs.pwv.gov.za/wg/waterlaw/index.html.

Majdi H (1996). Root sampling methods: applications and limitations of minirhizitron technique. *Plant and Soil* **185**: 255–258.

Majdi H & Andersson P (2005). Fine root production and turnover in a Norway spruce stand in northern Sweden: effects of nitrogen and water manipulation. *Ecosystems* **8**: 191–199.

McAlpine JR (1970). *Estimating pasture growth periods and droughts from simple water balance models.* Proceedings of 11th International Grassland Congress, Surfers Paradise, Queensland. UQP, Brisbane.

McDonald MG & Harbaugh AW (1988). *A modular three-dimensional finite-difference groundwater flow model.* US Geological Survey, Techniques of Water-Resources Investigations Book 6, Chapter A1.

McDonnell JJ, Bonell M, Stewart MK & Pearce AJ (1990). Deuterium variations in storm rainfall: implications for stream hydrograph separation. *Water Resources Research* **26**: 455–458.

McJannet DL, Vertessy RA & Clifton CA (2000). Observations of evapotranspiration in a break of slope plantation susceptible to periodic drought stress. *Tree Physiology* **20**: 169–177.

McMahon TA, Finlayson BL, Haines AT & Srikanthan R (1992). *Global runoff: continental comparisons of annual flows and peal discharges.* Catgena Verlaf, Cremlingen-Destedt, West Germany.

McPherson EG & Peper PJ (1998). Comparison of five methods for estimating leaf area index of open grown deciduous trees. *Journal of Arboriculture* **24**: 98–111.

Medhurst JL, Battaglia M & Beadle CL (2002). Measured and predicted changes in tree and stand water use following high-intensity thinning of an 8-year-old *Eucalyptus nitens* plantation. *Tree Physiology* **22**: 775–784.

Melcher PJ, Zwieniecki MA & Holbrook ML (2003). Vulnerability of xylem vessels to cavitation in Sugar Maple. Scaling from individual vessels to whole branches. *Plant Physiology* **131**: 1775–1780.

Mencuccini M & Grace J (1995). Climate influences the leaf-area/sapwood-area ratio in Scots pine. *Tree Physiology* **15**: 1–10.

Mensforth LJ, Thorburn PJ, Tyerman SD & Walker GR (1994). Sources of water used by riparian *Eucalyptus camaldulensis* overlying highly saline groundwater. *Oecologia* **100**: 21–28.
Midgley JJ & Scott DF (1994). The use of stable isotopes of water (D and ^{18}O). in hydrological studies in the Jonkershoek Valley. *Water SA* **20**: 151–154.
Midgley DC, Pitman WV & Middleton BJ (1994). *The surface water resources of South Africa 1990*, vols 1–6. Pretoria.
Milton SJ (1990). Life styles of plants in four habitats in an arid Karoo shrubland. *South African Journal of Ecology* **1**: 63–72.
Milton S & Dean R (1999). Big old camelthorn trees: key resource for Kalahari wildlife. *African Wildlife* **53**: 10–12.
Molz FJ (1981). Models of water transport in the soil-plant system: a review. *Water Resources Research* **17**: 245–260.
Monteith JL (1965). Evaporation and environment. *Symposium of the Society of Experimental Biologists* **19**: 205–234.
Monteith JL (1986). How do crops manipulate water supply and demand? *Philosophical Transactions of the Royal Society London, Series A* **316**: 245–259.
Monteith JL, Huda AKS & Midya D (1989). RESCAP: a resource capture model for sorghum and pearl millet. In *Modelling the growth and development of sorghum and pearl millet* (eds SM Virmani, HLS Tandon & G Alagarswamy). ICRISAT Research Bulletin 12, Patancheru, India, pp. 30–34.
Morris JD & Collopy JJ (1999). Water use and salt accumulation by *Eucalyptus camaldulensis* and *Casuarina cunninghamiana* on a site with shallow saline groundwater. *Agricultural Water Management* **39**: 205–227.
Morton S & Landsberg J (2003). The ecology of Australian deserts. Oxford University Press. In *Ecology: an Australian perspective* (eds P Attiwell & B Wilson). Oxford University Press, Melbourne.
Morton SW, Doherty MD & Barker RD (1996). *Natural heritage values of the Lake Eyre Basin in South Australia: World Heritage assessment.* Canberra, World Heritage Unit, Department of Environment, Sports and Territories/CSIRO Division of Wildlife and Ecology.
Mudd GM (1998). The long-term sustainability of mound springs in South Australia: implications for Olympic Dam. In *Proceedings of Uranium Mining and Hydrogeology II* (eds B Merkel & C Helling). Freiberg, Germany, pp. 575–584.
Mudd GM (2000). Mound springs of the Great Artesian Basin in South Australia: a case study from Olympic Dam. *Environmental Geology* **39**: 463–476.
Muller EA & Pretorius R (2002). Environmental indicators for National State of the Environment reporting: South Africa. CSIR Report ENV-P-R 2002–030 prepared for the Department of Environment Affairs and Tourism, Pretoria.
Mummery J & Hardy N (1994). Australia's biodiversity: an overview of selected significant components. Department of Environment, Sports and Territories, Canberra.
Murray BR, Zeppel MJB, Hose GC & Eamus D (2003). Groundwater dependent ecosystems in Australia: it's more than just water for rivers. *Ecological Management and Restoration* **4**: 110–113.
Murray B, Hose GC, Eamus D & Licari D (2006). Valuation of groundwater dependent ecosystems: a functional methodology incorporating ecosystem services. *Australian Journal of Botany* **54**: 221–229.
Myers BA, Duff GA, Eamus D, Fordyce IR, O'Grady A & Williams RJ (1997). Seasonal variation in water relations of trees of differing leaf phenology in a wet-dry tropical savanna near Darwin, northern Australia. *Australian Journal of Botany* **45**: 225–240.

National Land and Water Resources Audit (2001). *Australian dryland salinity assessment 2000.* Commonwealth of Australia, Canberra.

Nel P, van Heerden P, Momberg M, Newberry C, Newbury R & de Wet T (1995). *Report on the state of the environment in the North West Province.* Preliminary Survey, August 1995. Environmental Conservation, North West Province, South Africa.

Newman EI (1966). A method of estimating the total length of root in a sample. *Journal of Applied Ecology* **3**: 139–145.

Nielsen DM (2005). *Practical handbook of groundwater monitoring.* Taylor & Franics, Florida.

Niklas KJ (2005). Modelling below- and above-ground biomass for non-woody and woody plants. *Annals of Botany* **95**: 315–321.

Niknam SR & McComb J (2000). Salt tolerance screening of selected Australian woody species: a review. *Forest Ecology and Management* **139**: 1–19.

O'Grady A (2000). Patterns of tree and stand water use in the Eucalypt forests of northern Australia. PhD thesis, Charles Darwin University.

O'Grady AP, Eamus D & Hutley LB (1999). Transpiration increases during the dry season: patterns of tree water use in eucalypt open-forests of northern Australia. *Tree Physiology* **19**: 591–597.

O'Grady A, Eamus D, Cook P, Lamontagne S, Kelley G & Hutley L (2002). *Tree water use and sources of transpired water in riparian vegetation along the Daly River, Northern Territory.* Report to Environment Australia as part of the Healthy Rivers Program.

O'Grady AP, Cook PG, Howe P & Werren G (2006). An assessment of groundwater use by dominant tree species in remnant vegetation communities, Pioneer Valley, Mackay, Queensland. *Australian Journal of Botany* **54**: 155–171.

Ong CK, Wilson J, Deans JD, Mulayta J, Raussen T & Wajja-Musukwe N (2002). Tree-crop interactions:manipulation of water use and root function. *Agricultural Water Management* **53**: 171–186.

Overton IC & Jolly ID (2004). *Integrated studies of floodplain vegetation health, saline groundwater and flooding on the Chowilla Floodplain, South Australia.* Technical Report no. 20/05, CSIRO Land and Water.

Overton IC, Rutherford JC, Austin J & Jolly ID (2005). *Flood extent, groundwater recharge and vegetation response to a new weir in Chowilla Creek, South Australia.* Technical Report 2005, CSIRO Land and Water.

Pagès L, Asseng S, Pellerin S & Diggle A (2000). Modelling root system growth and architecture. In *Root methods: a handbook* (eds AL Smit, AG Bengough, C Engels, M van Noordwijk, S Pellerin & SC van de Geijn). Springer Verlag, Berlin, pp. 113–146.

Petheram C, Walker G, Grayson R, Thierfelder T & Zhang L (2002). Towards a framework for predicting impacts of land-use on recharge: 1. A review of recharge studies in Australia. *Australian Journal of Soil Research* **40**: 397–417.

Poole ML, Turner NC & Young JM (2002). Sustainable cropping systems for high rainfall areas of southwestern Australia. *Agricultural Water Management* **53**: 201–211.

Ponder WF (1986). Mound springs of the Great Artesian Basin. In *Limnology in Australia* (eds P De Deckker & WD Williams). CSIRO Publishing, Melbourne, pp. 403–420.

Prior LD & Eamus D (2000). Seasonal changes in hydraulic architecture and xylem embolism in *Eucalyptus tetrodonta* and *E. miniata*. *Plant, Cell and Environment* **23**: 955–965.

Prior LD, Eamus D & Duff GA (1997a). Seasonal trends in carbon assimilation, stomatal conductance, pre-dawn leaf water potential and growth in *Terminalia ferdinandiana*, a deciduous tree of northern Australian savannas. *Australian Journal of Botany* **45**: 53–69.

Prior LD, Eamus D & Duff GA (1997b). Seasonal and diurnal patterns of carbon assimilation, stomatal conductance and leaf water potential in *E. tetrodonta* saplings in a wet-dry savanna in northern Australia. *Australian Journal of Botany* **45**: 241–258.

Prior LD, Eamus D & Bowman DMJS (2004). Tree growth rates in north Australian savanna habitats: seasonal patterns and correlations with leaf attributes. *Australian Journal of Botany* **52**: 303–314.

Raisin G, Bartley J & Croome R (1999). Groundwater influence on the water balance and nutrient budget of a small natural wetland in north-eastern Victoria, Australia. *Ecological Engineering* **12**: 133–147.

Rich PM (1989). A manual for analysis of hemispherical canopy photography. Los Alamos National Laboratory Report LA–11733-M.

Rich PM (1990). Characterizing plant canopies with hemispherical photographs. *Remote Sensing of Environment* **5**: 13–29.

Richards JH & Caldwell MM (1987). Hydraulic lift: substantial nocturnal water transport between soil layers by *Artemisia tridentata* roots. *Oecologia* **73**: 486–489.

Roberts S, Vertessy R & Grayson R (2001). Transpiration from *Eucalyptus sieberi* forests of different age. *Forest Ecology and Management* **143**: 153–161.

Rodhe A (1987). *The origin of streamwater traced by oxygen–18*. Report Series A41, Division of Hydrology, Department of Physical Geography, Uppsala University.

Rogers KH (2002). *Operationalizing multi-party strategic adaptive management (SAM) of the Sabie River*. WRC Report no. 1097/1/02, Water Research Commission, Pretoria.

Rogers KH & O'Keeffe J (2003). River heterogeneity: ecosystem structure, function and management. In *The Kruger experience. Ecology and management of savanna heterogeneity* (eds JT du Toit, KH Rogers & HC Biggs). Island Press, Washington, pp. 189–218.

Roshier DA, Whetton PH, Allan RJ & Robertson AI (2001). Distribution and persistence of temporary wetland habitats in arid Australia in relation to climate. *Austral Ecology* **26**: 371–384.

Russell-Smith J, Start T & Woinarski J (2001). Effects of fire in the landscape. In *Savanna burning* (eds R Dyer, P Jacklyn, I Partridge, J Russell-Smith & D Williams). Tropical Savannas CRC Darwin, pp. 29–49.

Rutter AJ, Kershaw KA, Robins PC & Morton AJ (1971). A predictive model of rainfall interception in forests. I. Derivation of the model from observations in a plantation of Corsican Pine. *Agricultural Meteorology* **9**: 367–384.

Rutter AJ, Morton AJ & Robins PC (1975). A predictive model of rainfall interception in forests. II. Generalization of the model and comparison with observations in some coniferous and hardwood stands. *Journal of Applied Ecology* **12**: 367–380.

Ryan M & Yoder BJ (1997). Hydraulic limits to tree height and tree growth. *BioScience* **47**: 235–242.

Salama RB (1998). Physical and chemical techniques for discharge studies. Part 1. In *The basics of recharge and discharge* (eds L Zhang & G Walker). CSIRO Publishing, Melbourne.

Sandnes A, Eldhuset TD & Wollebaek G (2005). Organic acids in root exudates and soil solution of Norway spruce and silver birch. *Soil Biology and Biochemistry* **37**: 259–269.

Sandström K (1996). Hydrochemical deciphering of streamflow generation in semi-arid East Africa. *Hydrological Processes* **10**: 703–720.

Scanlon BR, Healy RW & Cook PG (2002). Choosing appropriate techniques for quantifying groundwater recharge. *Hydrogeology Journal* **10**: 18–39.

Schlegel P, Huwe B & Teixeira WG (2004). Modelling species and spacing effects on root zone water dynamics using Hydrus-2D in an Amazonian agroforestry system. *Agroforestry Systems* **60**(3): 277–289.

Schulze BR & McGee OS (1978). Climatic indices and classification in relation to the biogeography of southern Africa. In *Biogeography and ecology of southern Africa* (eds MJA Werger & AC van Bruggen). The Hague, pp. 19–52.

Schulze ED, Caldwell MM, Canadell J, Mooney HA, Jackson RB, Parson D, Scholes R, Sala OE & Trimborn P (1998). Downward flux of water through roots (ie inverse hydraulic lift) in dry Kalahari sands. *Oecologia* **115**: 460–462.

Schulze RE (1997). Climate. In *Vegetation of southern Africa* (eds RM Cowling, DM Richardson & SM Pierce). Cambridge University Press, Cambridge.

Schulze R, Maharaj M, Lynch SD, Howe BJ & Melvil-Thompson B (1997). *South African atlas of agrohydrology and climatology.* Water Research Commission, Pretoria.

Schulze RE, Meigh J & Horan M (2001). Present and future vulnerability of eastern and southern Africa's hydrology and water resources. *South African Journal of Science* **97**: 150–160.

Scott DF (1993). The hydrological effects of fire in South African mountain catchments. *Journal of Hydrology* **150**: 409–432.

Scott DF (2004). The contrasting effect of wildfire and clear-felling on the hydrology of a small catchment. *Hydrological Processes* **11**: 543–555.

Scott DF & Le Maitre DC (1998). *The interaction between vegetation and groundwater: research priorities for South Africa.* Report no. 730/1/98, Water Research Commission, Pretoria.

Scott ML, Shafroth PB & Auble GT (1999). Responses of riparian cottonwoods to alluvial water table declines. *Environmental Management* **23**: 347–358.

Shafroth P, Stromberg J & Patten D (2002). Riparian vegetation response to altered disturbance and stress regimes. *Ecological Applications* **12**: 107–123.

Shiklomanov IAJ & Rodda JR (2003). *World water resources at the beginning of the 21st century.* Cambridge University Press, Cambridge.

Shuttleworth WJ & Wallace JS (1985). Evaporation from sparse crops: an energy combination theory. *Quarterly Journal of the Royal Meteorological Society* **111**: 839–855.

Silberstein R, Vertessy R, Morris J & Feikema P (1999). Modelling the effects of tree soil moisture and solute conditions on long-term tree growth and water use: a case study from the Shepperton irrigation area, Australia. *Agricultural Water Management* **39**: 283–315.

Silberstein R, Vertessy R, McJannet D & Hatton T (2002a). Tree belts on hillslopes. In *Trees, water and salt* (eds R Stirzaker, R Vertessy & A Sarre). Report no. 01/086, RIRDC, Canberra, pp. 58–76.

Silberstein R, Vertessy R & Stirzaker R (2002b). *The basics of catchment hydrology.* RIRDC.

Siso S, Camarero JJ & Gil-Pelegrin E (2001). Relationship between hydraulic resistance and leaf morphology in broadleaf *Quercus* species: a new interpretation of leaf lobation. *Trees* **15**: 341–345.

Sklash MG & Farvolden RN (1979). The role of groundwater in storm runoff. *Journal of Hydrology* **43**: 45–65.

Smettem KRJ, Kirkby C & Chittleborough DJ (1994). Hydrologic response of undisturbed soil cores to simulated rainfall. *Australian Journal of Soil Research* **32**: 1175–1187.

Smith DM & Allen SJ (1996). Measurement of sap flow in plant stems. *Journal of Experimental Botany* **47**: 1833–1844.

Sophocleous M (2000). From safe yield to sustainable development of water resources: the Kansas experience. *Journal of Hydrology* **235**: 27–43.

Specht RL & Specht A (1999). *Australian plant communities: dynamics of structure, growth and biodiversity.* Oxford University Press, Melbourne.

Stanford JA & Ward JV (1993). An ecosystem perspective of alluvial rivers: connectivity and the hyporhoeic zone. *Journal of the North American Benthological Society* **12**: 48–60.

Stephens A (2003). *Institutional arrangements for groundwater management in dolomitic terrains. Phase 1: situation analysis.* draft synthesis report prepared for Water Research Commission, Pretoria.

Stirzaker R, Cook FJ & Knight JH (1999). Where to plant trees on cropping land for control of dryland salinity: some approximate solutions. *Agricultural Water Management* **39**: 115–133.

Stirzaker R, Ellis T & Lefroy T (2002a). Mixing tree belts with agriculture. In *Trees, water and salt* (eds R Stirzaker, R Vertessy & A Sarre). Report no. 01/086, RIRDC, Canberra, pp. 78–92.

Stirzaker R, Vertessy R & Sarre A (eds) (2002b). *Trees, water and salt: an Australian guide to using trees for healthy catchments and productive farms.* Report no. 01/086, RIRDC, Canberra.

Sun S & Dickinson GR (1995). Salinity effects on tree growth, root distribution and transpiration of *C. cunninghamiana* and *E. camaldulensis* planted on a saline site in tropical north Australia. *Forest Ecology and Management* **77**: 127–138.

Thomas DS & Eamus D (1999). The influence of predawn leaf water potential on stomatal responses to atmospheric water content at constant C_i and on stem hydraulic conductance and foliar ABA concentrations. *Journal of Experimental Botany* **50**: 243–251.

Thomson VP & Leishman MR (2005). Post fire vegetation dynamics in nutrient-enriched and non-enriched sclerophyll woodland. *Austral Ecology* **30**: 250–260.

Thorburn P (1998). *Groundwater uptake and salt accumulation by trees over shallow water tables.* Short Report, RIRDC, Canberra.

Thorburn P (ed) (1999). Agroforestry over shallow water tables: the impact of salinity on sustainability. *Water and salinity issues in agroforestry* no. 4. RIRDC publication 99/36, Canberra.

Thorburn P & Walker GR (1994). Variations in stream water uptake by *E. camaldulensis* with differing access to stream water. *Oecelogia* **100**: 293–301.

Thorburn P, Hatton T & Walker GR (1993). Combining measurements of transpiration and stable isotopes to determine groundwater discharge from forests. *Journal of Hydrology* **150**: 563–587.

Thorburn, P, Walker GR & Woods PH (1992). Comparison of diffuse discharge from shallow water tables in soils and salt flats. *Journal of Hydrology* **136**: 253–274.

Tiktak & A van Grinsven HJM (1995). Review of sixteen forest-soil-atmosphere models. *Ecological Modelling* **83**: 35–53.

Timberlake J (1980). *Handbook of Botswana acacias.* Division of Land Utilisation, Ministry of Agriculture, Botswana.

Townsend SA & Douglas MM (2000). The effect of three fire regimes on stream water quality, water yield and export coefficients in a tropical savanna (northern Australia). *Journal of Hydrology* **229**: 118–137.

Townsend SA & Douglas MM (2004). The effect of a wildfire on stream water quality and catchment water yield in a tropical savanna excluded from fire for 10 years. *Water Research* **38**: 3051–3058.

Tsuda M & Tyree MT (2000). Plant hydraulic conductance measured by the high pressure flow meter in crop plants. *Journal of Experimental Botany* **51**: 823–828.

Turner JV, Bradd JM & Waite TD (1991). The conjunctive use of isotopic techniques to elucidate solute concentration and flow processes in dryland salinized catchments. In *Proceedings of International Symposium on use of isotope techniques in water resources development.* IAEA, Vienna, pp. 33–59.

Tyree MT & Sperry JS (1989). Vulnerability of xylem to cavitation and embolism. *Annual Review of Plant Physiology and Molecular Biology* **40**: 19–38.

Tyree MT, Patino S, Bennink L & Alexander J (1995). Dynamic measurements of root hydraulic conductance using a high pressure flow meter in the laboratory and field. *Journal of Experimental Botany* **46**: 83–94.

Tyson PD (1986). *Climate change and variability in southern Africa.* Oxford University Press, Cape Town.

Tyson PD & Gatebe CK (2001). The atmosphere, aerosols, trace gases and biogeochemical change in southern Africa: regional integration. *South African Journal of Science* **97**: 108–118.

Tyson PD & Preston-Whtye RA (2000). *The weather and climate of southern Africa.* Oxford University Press, Cape Town.

Underwood AJ (1991). Beyond BACI: experimental designs for detecting human environmental impacts on temporal variations in natural populations. *Australian Journal of Marine and Freshwater Research* **42**: 569–587.

van den Honert TH (1948). Water transport as a catenary process. *Faraday Society Discussions* **3**: 146–153.

van Gardingen PR, Jackson GE, Hernandez-Daumas S, Russell G & Sharp L (1999). Leaf index estimates obtained for clumped canopies using hemispherical photography. *Agricultural and Forest Meteorology* **84**: 243–257.

van Ittersum MK, Leffelaar P, van Keulen H, Kropff MJ, Bastiaans L & Goudriaan J (2003). On approaches and applications of the Wageningen crop models. *European Journal of Agronomy* **18**: 201–234.

Verburg K, Ross P & Bristow KL (1996). *SWIMv2.1 user manual.* Divisional Report no. 130, CSIRO Division of Soils.

Vertessy RA, Hatton TJ, O'Shaughnessy PJ & Jayasuriya MDA (1993). Predicting water yield from a Mountain Ash forest catchment using a terrain analysis based catchment model. *Journal of Hydrology* **150(2–4)**: 665–700.

Vertessy RA, Benyon RG, O'Sullivan SK & Gribben PR (1995). Relationships between stem diameter, sapwood area leaf area and transpiration in young Mountain Ash forests. *Tree Physiology* **15**: 559–567.

Vertessy RA, Hatton TJ, Reece P, O'Sullivan SK & Benyon RG (1997). Estimating stand water use of large Mountain Ash trees and validation of the sap flow measurement technique. *Tree Physiology* **17**: 747–756.

Vertessy RA, Watson F, O'Sullivan S, Davis S, Campbell R, Benyon R & Haydon S (1998). *Predicting water yield from a Mountain Ash forest catchment.* Report no. 98/4, CRC for Catchment Hydrology, Victoria.

Vertessy RA, Watson FGR & O'Sullivan SK (2001). Factors determining relations between stand age and catchment water balance in Mountain Ash forests. *Forest Ecology and Management* **143**: 13–26.

Wallace JS & Biscoe PV (1983). Water relations of winter wheat. 4. Hydraulic resistance and capacitance in the soil-plant system. *Journal of Agricultural Science Cambridge* **100**: 591–600.

Wang E & Smith CJ (2004). Modelling the growth and water uptake function of plant root systems: a review. *Australian Journal of Agricultural Research* **55**: 501–523.

Ward JD & Breen CM (1983). Drought stress and the demise of *Faidherbia albida* along the lower Kuiseb River, central Namib Desert: preliminary findings. *South African Journal of Science* **79**: 444–447.

Warrick AW (1988). Additional solutions for steady state evaporation from a shallow water table. *Soil Science* **146**: 63–66.

Warrick AW, Wierenga PJ, Young MH & Musil SA (1998). Diurnal fluctuations of tensiometric readings due to surface temperature changes. *Water Resources Research* **34**: 2863–2869.

Watson F, Vertessy R, McMahon T, Rhodes B & Watson I (2001). Improved methods to assess water yield changes from paired-catchment studies: application to the Maroondah catchments. *Forest Ecology and Management* **143**: 189–204.

Wheeler BD (1999). Water and plants in freshwater wetlands. In *Eco-hydrology: plants and water in terrestrial and aquatic environments* (eds AJ Baird & RL Wilby). Routledge, London, pp. 127–180.

White DA, Dunin FX, Turner NC, Ward BH & Galbraith JH (2002). Water use by contour-planted belts of trees comprised of four eucalyptus species. *Agricultural Water Management* **53**: 133–152.

Williams D & Cook G (2001). Savanna fire regimes. In *Savanna burning* (eds R Dyer, P Jacklyn, I Partridge, J Russell-Smith & D Williams). Tropical Savannas CRC Darwin, pp. 15–28.

Williams RJ, Duff GA, Bowman DMJS & Cook GD (1996). Variation in the composition and structure of tropical savannas as a function of rainfall and soil texture along a large-scale climatic gradient in the NT, Australia. *Journal of Biogeography* **23**: 747–756.

Williams M, Malhi Y, Nobre AD, Rastetter EB, Grace J & Pereira MGP (1998). Seasonal variation in net carbon exchange and evapotranspiration in a Brazilian rainforest: a modelling analysis. *Plant, Cell and Environment* **21**: 953–968.

Williams PR, Congdon RA, Grice AC & Clark PJ (2003). Effect of fire regime on plant abundance in a tropical eucalypt savanna of north-eastern Australia. *Austral Ecology* **28**: 327–338.

Woessner WW (2000). Stream and fluvial plain ground water interactions: rescaling hydrogeologic thought. *Ground Water* **38**: 423–429.

WRI (2003). *Earth trends: the environmental portal.* Data tables on water resources and freshwater ecosystems, World Resources Institute.

Zencich S, Froend RH & Turner JVG (2002). Influence of groundwater depth on the seasonal sources of water accessed by Banksia tree species on a shallow, sandy coastal aquifer. *Oecologia* **131**: 8–19.

Zhang L & Dawes W (1998). *WAVES: an integrated energy and water balance model.* Technical Report no. 31/98, CSIRO Land and Water.

Zhang L, Dawes WR & Walker GR (1999). *Predicting the effect of vegetation changes on catchment average water balance.* Technical Report no. 99/12, CRC for Catchment Hydrology, Canberra.

Zwieniecki MA, Melcher PJ & Holbrook NM (2001). Hydrogel control of xylem hydraulic resistance in plants. *Science* **291**: 1059–1062.

Index